ROUTLEDGE LIBRARY EDITIONS:
NUCLEAR SECURITY

Volume 2

TACTICAL NUCLEAR WEAPONS

TACTICAL NUCLEAR WEAPONS
European Perspectives

STOCKHOLM INTERNATIONAL PEACE
RESEARCH INSTITUTE

Routledge
Taylor & Francis Group

LONDON AND NEW YORK

First published in 1978 by Taylor & Francis Ltd

This edition first published in 2021
by Routledge
2 Park Square, Milton Park, Abingdon, Oxon OX14 4RN

and by Routledge
52 Vanderbilt Avenue, New York, NY 10017

Routledge is an imprint of the Taylor & Francis Group, an informa business

British Library Cataloguing in Publication Data
A catalogue record for this book is available from the British Library

ISBN: 978-0-367-50682-7 (Set)
ISBN: 978-1-00-309763-1 (Set) (ebk)
ISBN: 978-0-367-51319-1 (Volume 2) (hbk)
ISBN: 978-1-00-305336-1 (Volume 2) (ebk)

Publisher's Note
The publisher has gone to great lengths to ensure the quality of this reprint but
points out that some imperfections in the original copies may be apparent.

Disclaimer
The publisher has made every effort to trace copyright holders and would welcome
correspondence from those they have been unable to trace.

Tactical Nuclear Weapons: European Perspectives

sipri

Stockholm International Peace Research Institute

SIPRI is an independent institute for research into problems of peace and conflict, with particular attention to the problems of disarmament and arms regulation. It was established in 1966 to commemorate Sweden's 150 years of unbroken peace.

The financing is provided by the Swedish Parliament. The staff, the Governing Board and the Scientific Council are international. As a consultative body, the Scientific Council is not responsible for the views expressed in the publications of the Institute.

Governing Board

Governor Rolf Edberg, Chairman (Sweden)
Professor Robert Neild, Vice Chairman
   (United Kingdom)
Mr Tim Greve (Norway)
Academician Ivan Málek (Czechoslovakia)
Professor Leo Mates (Yugoslavia)
Professor Gunnar Myrdal (Sweden)
Professor Bert Röling (Netherlands)
The Director

Director

Dr Frank Barnaby (United Kingdom)

sipri

Stockholm International Peace Research Institute
Sveavägen 166, S-113 46 Stockholm, Sweden
Cable: Peaceresearch, Stockholm
Telephone: 08-15 09 40

Tactical Nuclear Weapons: European Perspectives

sipri

Stockholm International Peace Research Institute

Taylor & Francis Ltd
London
1978

Crane, Russak & Company, Inc.
New York

First published 1978 by Taylor & Francis Ltd., London
and Crane, Russak & Company, Inc., New York

Copyright © 1978 by SIPRI
Sveavägen 166, S-113 46 Stockholm

ISBN 0 8448 1303 6

Library of Congress Catalog Card Number 77-25753

Printed and bound in the United Kingdom by
Taylor & Francis (Printers) Ltd, Rankine Road,
Basingstoke, Hampshire RG24 0PR

Preface

Tactical nuclear warfare is a topical issue. The nuclear-weapon laboratories have developed a new generation of nuclear-weapons and strong pressures are growing for their deployment. And new nuclear doctrines are being elaborated and propagandized in an attempt to rationalize this deployment. But the introduction of new types of tactical nuclear weapons in Europe could have serious and potentially disastrous consequences.

One type of tactical nuclear weapon is the enhanced-radiation weapon or 'neutron bomb', designed to produce a burst of neutrons and gamma rays capable of delivering a high enough radiation dose at distances of several hundred metres to incapacitate exposed persons in a matter of hours and kill most of them in a few days. The new weapons would have the same radiation killing capacity as existing types of tactical nuclear weapons for only about one-fifth of the explosive power. Damage to objects and property away from the battlefield would be significantly reduced. Another new type of tactical nuclear weapon is designed to suppress radiation and enhance blast for use against small hardened military targets. A third type produces large quantities of short-lived radioisotopes. The aim is to deny the enemy access to the area over which the radioactivity is spread for several days while avoiding long-term contamination.

On first sight, smaller and more accurate nuclear weapons may seem more humane and militarily preferable to the relatively high-yield tactical nuclear weapons currently deployed. But some of these new types of weapons would blur the distinction between nuclear and conventional weapons and their use make escalation to strategic nuclear war extremely likely. Indeed, the argument for these new weapons is that their use in wartime is more credible (and therefore 'acceptable') than current types of tactical nuclear weapons. This perception could easily lead to the exceedingly dangerous idea that some types of tactical nuclear war were 'winnable'.

The fact has to be faced that *any* use of nuclear weapons is almost certain to escalate until all available nuclear weapons are used. To believe otherwise is to believe that one side will surrender before it has used all the weapons in its arsenals. History shows that this is most unlikely to happen.

This book contains an extensive description of tactical nuclear weapons in Europe and the nuclear doctrines of NATO and the Warsaw Treaty Organization. A number of relevant arms control proposals are analysed and their potential effects on European security discussed. The consequences of the modernization of nuclear weapons are analysed, including the consequences for neutral countries. Finally, two specific military options for Europe, more or less representing extremes of the current range of thinking, are given. One takes into account the coming availability of relatively accurate and cheap defensive conventional weapons (precision-guided muntions), and recommends the extensive use of these in Europe by small mobile groups instead of reliance on nuclear weapons. The other describes an entirely new nuclear force in Europe armed with specially tailored nuclear weapons.

If any conclusion can be drawn from this book it is that reason demands the elimination of nuclear weapons not only from Europe, but from the world's arsenals.

Because of the current importance of the topic, SIPRI organized a meeting, which took place in Stockholm from 4 to 8 October 1976, on 'Tactical Nuclear Warfare'. This book is the result of the meeting. It is intended as a contribution to the continuing debate on nuclear policies.

Various approaches to the problem of tactical nuclear warfare in Europe were discussed at the meeting. It was generally agreed that present military doctrines in Europe are untenable, and a range of solutions was put forward and analysed. In the light of the discussion the participants modified their papers into the form in which they appear here.

The participants are listed on pages xiii and xiv and the book was edited by Frank Barnaby, Director of SIPRI.

Contents

x

Tables and Figures

Chapter 4. Arms control and tactical nuclear forces and European security

Chapter 7. 'Mini-nukes' and enhanced radiation weapons

Chapter 9. Tactical nuclear weapons and European security

Chapter 10. The new nuclear force

List of participants

Dr H. Afheldt
Max-Planck-Institut
Riemerschmidstrasse 7
813 Starnberg
FR Germany

Dr F. Barnaby
Stockholm International Peace Research Institute
Sveavägen 166
S-113 46 Stockholm
Sweden

Professor F. Calogero
Università degli studi
Instituto di Fisica 'Guglielmo Marconi'
Piazzale delle Scienze 5
Rome
Italy

Professor J. Coffey
Center for Arms Control and International Security Studies
160 Mervis Hall
University of Pittsburgh
Pittsburgh Penn. 15260
USA

Professor M. Hattori
Institute for Atomic Energy
Rikkyo University
Nagasaka Yokosuka
Japan

Dr M. Leitenberg
Center for International Studies
170 Uris Hall
Cornell University
Ithaca
New York 14853
USA

Professor J. Miettinen
Department of Radio Chemistry
University of Helsinki
Unioninkatu 35
Helsinki 17
Finland

Professor M. Milshtein
Academy of Sciences of the USSR
Leninsky Prospekt 14
117901 GSP
Moscow V-71
USSR

Dr W. Multan
Polish Institute of International Affairs
1A, Warecka Street
00950 Warsaw
Poland

Dr J. Prawitz
Försvarsdepartementet
Regeringsgatan I-3
103 20 Stockholm
Sweden

Mr R. Shreffler
P.O. Box 5713
Santa Fe
New Mexico 87502
USA

Dr O. Šuković
Institut za medjunarodnu politiku i privredu
Makedonska 25
11000 Belgrade
Yugoslavia

Acronyms and Abbreviations

ABM	Anti-Ballistic Missile
ACDA	Arms Control and Disarmament Agency
ACE	Air Command Europe
ADM	Atomic Demolition Munitions
AEC	US Atomic Energy Commission
AFAP	Artillery Fired Atomic Projectiles
ASROC	Anti-Submarine Rockets
ASM	Air-to-Surface Missile
ASW	Anti-Submarine Warfare
BAOR	British Army of the Rhine
CCD	Conference of the Committee on Disarmament
CEP	Circular Error Probability
CESC	Conference on European Security and Co-operation
CLGP	Cannon-Launched Guided Projectile
CENTO	Central Treaty Organization
DDR&E	Director of Defense Research and Engineering
DOD	Department of Defense
DPC	Defence Planning Committee
ERW	Enhanced-Radiation Weapon
ENF	European Nuclear Force
FBS	Forward Based Systems
GAC	General Advisory Committee
GSP	General Strike Plan
ICBM	Inter-Continental Ballistic Missile
IISS	International Institute for Strategic Studies
IP	Immediate Permanent Incapacitation
IRBM	Intermediate Range Ballistic Missile
IT	Immediate Transient Incapacitation
JCAE	Joint Congressional Atomic Energy Committee

LL	Latent Lethality
LRCM	Long Range Cruise Missile
MAD	Mutual Assured Destruction
MARV	Manoeuvrable Re-entry Vehicles
MBFR	Mutual Balanced Force Reductions
MIRV	Multiple Independently Targetable Re-entry Vehicle
MFR	Mutual Force Reductions
MLF	Multilateral Force
MRBM	Medium Range Ballistic Missile
MURFAAMCE	Mutual Reduction of Forces and Armaments and Associated Measures in Central Europe
NATO	North Atlantic Treaty Organization
NDAC	Nuclear Defense Affairs Committee
NGA	NATO Guideline Area
PAL	Permissive Action Links
PGM	Precision Guided Munitions
PSP	Priority Strike Programme
QRA	Quick Reaction Alert
SAC	Strategic Air Command
SACEUR	The Supreme Allied Commander in Europe
SALT	Strategic Arms Limitation Talks
SAM	Surface-to-Air Missile
SHAPE	Supreme Headquarters, Allied Powers in Europe
SIOP	Single Integrated Operations Plan
SLBM	Submarine-Launched Ballistic Missile
SLCM	Submarine-Launched Cruise Missile
SRBM	Short Range Ballistic Missile
SSM	Surface-to-Surface Missile
TAC	Tactical Air Command
TH	Theatre of Hostilities
TNDV	Tactical Nuclear Delivery Vehicle
TNF	Tactical Nuclear Force
TNW	Tactical Nuclear Weapon
TSP	Tactical Strike Program
USAF	United States Air Force
WTO	Warsaw Treaty Organization

Part One. Basic data on tactical nuclear weapons

1. Background materials in tactical nuclear weapons (primarily in the European context)

M. Leitenberg

Introduction

This chapter is designed as an introductory compendium of information on tactical nuclear weapons (TNWs) in general and their distribution worldwide, but by far the greatest part of its contents pertains to tactical nuclear weapons in the context of Europe. Most of the information that it contains derives from US sources. As a source of descriptive information concerning such aspects of the weapons as their numbers, operational characteristics, location etc., the data chapter deliberately omits any comprehensive survey or summary of the greatest volume of the literature concerning tactical nuclear weapons, those sources which discuss the strategy, tactics or doctrine related to their use. A bibliographic guide to much of this other literature is presented on page 342. Information from such material was however drawn on where it might have been the only available source to inferences concerning technical characteristics or actual national policy of a particular nation regarding its tactical nuclear weapons.

I. The USA and NATO

Tactical nuclear weapons – definitions and distribution

The phrases 'tactical nuclear weapons', and more recently 'theatre nuclear weapons', are dependent for their meaning on customary usage. They are defined largely by exclusion, and to be set aside from the category of 'strategic nuclear weapons'. This latter group is again usually defined arbitrarily in the West as those nuclear warheads delivered by ICBMs (Intercontinental Ballistic Missiles), SLBMs (Submarine Launched Ballistic Missiles), long-range heavy bombers, fractional-orbital space-borne systems, and probably ABMs (Anti-ballistic Missiles). 'Tactical nuclear weapons' are then intended to include all other kinds, and they are often also intended to convey in the public mind the notion of battlefield use, although many have quite other targets. These other 'non-strategic' categories of nuclear weapons include:

(a) Air-dropped free fall bombs and glide bombs
(b) Air-to-surface missiles and air-to-surface stand-off missiles
(c) Airbreathing cruise missiles
(d) Surface-to-air missiles
(e) Shorter-range surface-to-surface missiles
(f) Air-to-air missiles
(g) Artillery
(h) Depth charges
(i) Torpedoes and rocket torpedoes
(j) Atomic demolition munitions, or nuclear land mines
(k) Ocean mines [1]

In the United States' nuclear weapon stockpile, there were as of 1973 some 26 types of strategic and tactical nuclear warheads with more than 57 different versions or modifications providing varying yields, security control features, or other adaptations [2]. Equivalent details concerning Soviet, British or French nuclear weapons are not available. However, there are no absolute criteria which distinguish 'tactical' nuclear weapons from 'strategic' nuclear weapons.†
If one examines such criteria as:

— Weapon yield
— Weapon range
— Location of delivery system
— Location of target
— 'Alert' or readiness status

one discovers a continuous range of values, with much overlap, between tactical and strategic nuclear weapons. The concept of tactical nuclear weapons did not derive from certain quantitative limits, or parameters. Nor were the nuclear weapons so designated or assigned ever subsequently required to fit such quantitative limits or parameters. Thus, there are some tactical nuclear weapons – air-dropped bombs with yields of over one megaton – that have very significantly higher yields than some strategic nuclear weapons. Part of the NATO tactical nuclear weapon force, both aircraft and some Pershing missiles, has been part of the Quick Reaction Alert (QRA) force on constant readiness, as are strategic nuclear weapons. The arbitrariness of the designations tactical and strategic as regards nuclear weapons was demonstrated with unusual clarity some years ago in two US Congressional Hearings.

> *Senator Symington:* You say this weapon is a tactical bomb despite it having a few hundred kilotons.
> That is many, many times the Hiroshima 14. As I understand it, our forward-based aircraft can carry two of these – actually more than that. They could carry two with extra gas tanks. Isn't the word

---

† Nevertheless, for the remainder of the paper, the quotation marks will be omitted from around the word 'tactical', the point having been made in this opening section that the designation is an arbitrary one.

'tactical' a little misleading? They would be performing a strategic mission if they destroyed a city.

At the end of World War II, when our P-51s took off from Iwo Jima, they could barely get to the southern coast of Japan. When they got there, if they knocked out a Japanese plant, they were performing a strategic mission. But in the Battle of the Bulge, we took the biggest bombers in Europe to support the troops in an effort to contain the Germans. Obviously that was a tactical mission, despite the use of a strategic bomber.

I don't understand why this is a tactical nuclear bomb if it has a capacity for exceeding the Hiroshima bomb [deleted] times as big and can reach Soviet targets.

General Giller: Yes sir; you certainly have described the situation.

Senator Symington: I can remember being in Turkey back in 1959 where the commanding officer said, 'Every one of my pilots has a town in his mind he knows from the air better than you know your own hometown on the ground'.

Those weapons were no doubt more powerful than Hiroshima, but nothing like the bomb in question. So I don't quite understand why it is always classified 'tactical'. Will you comment?

General Giller: Certainly. As you say, the target system describes the purpose of the bomb perhaps better than the yield. I think it is the jargon of the trade, perhaps, that those bombs that are targeted through SAC and which are in the Triad are normally called strategic bombs whereas, those that are assigned, with fighter-bombers, are normally called tactical bombs.

Quite a few of these delivery systems and bombs are capable of being employed in either tactical or strategic roles. [Deleted]. The Russians as I understand it define strategic as anything that can strike the homeland. Therefore, their argument is any system which can strike Russia is strategic and should be counted in with the ICBMs.

On the other hand, IRBMs and MRBMs pointed at NATO are not strategic. At this point, I think I find it rather confusing. I think it is confused in people's minds, one versus the other. When we are in NATO, part of NATO, we should look at the IRBMs as strategic, also.

Senator Symington: Dr Kissinger, when on public television, in Moscow, was asked, 'How do you justify our 1064 against their 1618 ICBM launchers?' we getting the lesser figure; and 'How do you justify 710 against 950 in submarine launching tubes?'

He replied, 'You must not forget our strategic bombers and our forward-based aircraft'. We can't have it both ways, as I see it.

Obviously, this is going to be a matter of serious discussion in any future SALT talks.

General Giller: The same airplane is capable of taking that same bomb and [deleted] and going after a railyard 100 miles away. How we are going to solve that problem in SALT, I don't know.

Representative Hosmer: You are talking specifically – you are answering questions relative to the B-61 bomb?

General Giller: The high-yield version.

Representative Hosmer: It has [deleted]. Is that correct?

General Giller: Yes, sir.

Representative Hosmer: Are those yields [deleted]?

General Giller: Yes.

Representative Hosmer: If the [deleted] the employment would be in a tactical mode. If [deleted] you would be in a strategic mode if you were in a strategic war; is that correct?

General Giller: The answer is yes and no. The way you described it, it would be yes. All I can do is comment on the fact that I could find it inconceivable of expending several hundred kilotons on a railyard. Whether that is strategic or tactical, I think, depends on how you view it.

Representative Hosmer: The end use to which it is put is the definition of the weapon?

General Giller: Yes, sir.

Representative Hosmer: This is a dual-purpose weapon. You can use it for strategic purposes or tactical [3].

The same question was treated in a second Congressional Hearing in 1973 with the same result. Whether a particular nuclear weapon was 'tactical' or 'strategic' 'would depend on the target against which it was used'; '. . . the definition, you know, depends on the situation under discussion' [4]. In recent years, the very same air-dropped bomb may be deployed both with strategic bomber aircraft and with 'tactical' nuclear strike aircraft. For many past years one of NATO's air-dropped weapons had a yield of 1.4 Mt. (See Appendices, p. 109). Some Poseidon submarines are now assigned to NATO 'tactical' nuclear targets. Thus, it has been the common presumption that 'tactical' nuclear weapons were reserved for a category of targets that did not overlap with targets designated for 'strategic' weapons that permitted uncritical use of the term. In terms of the real deployments of tactical nuclear weapons, there has always been a rather extraordinary amount of divergence between the concept as publicly discussed, the actual weapons and yields subsumed or allotted under it, and their targets. The inconsistencies are clearly not all on one side. The USSR defines any nuclear weapon which impacts on the USSR from a West European launching point as strategic, but does not so designate, at least not publicly, those of its medium and intermediate range missiles, launched from within its own borders, which would impact on West European targets. It is probably very much for those reasons that the phrase Theatre Nuclear Weapons has recently come into use, simply to indicate that these were weapons intended to be used in some particular area.

Numbers and locations of tactical nuclear weapons. Estimates of US tactical nuclear weapons have ranged from 50 000 and 'tens of thousands' in 1966 [5] to

22 000 in 1975 [6]. The 1975 publication indicates the following breakdown for the locations of US tactical nuclear weapons:

Europe	7 000
Atlantic Fleet	1 000
Asia	1 700
Pacific Fleet	1 500
United States	10 800
Total US tactical nuclear weapons	**22 000**

A subsequent publication by the same group indicated an estimate of from 661 to 686 US tactical nuclear weapons in South Korea in January 1976 [7]. President Carter referred to '700 atomic weapons in Korea' [8] Congressman Dellums later indicated a precise number of 666 [9]. These are for Sergeant and Honest John missiles, 8-in nuclear artillery and F-4 aircraft [10]. Other sources have estimated US nuclear weapons in the Pacific area at between 6000 and 12 000. (This included an estimate of 3156 weapons stored in Hawaii, which is part of the continental United States, and 500 strategic weapons on Polaris–Poseidon weapons. The high estimate derived from a Yomiuri claim of 4000 US nuclear weapons on land and 8000 on ships 'in the Pacific area' [11].) Official US government statements have indicated the number of US tactical nuclear weapons in Europe (chapter 2), but have never given numbers for their location in any other area.

Figure 1.1. Stockpile mix of US weapons

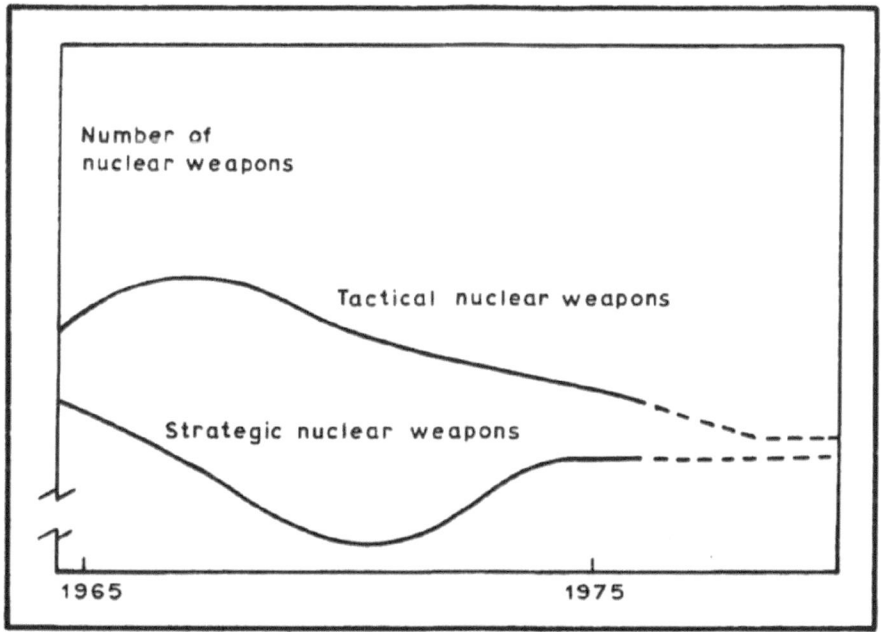

The US Governmental security policy regarding nuclear weapons locations is that it will neither confirm nor deny the existence or location of US nuclear weapons located anywhere. In part, this is at the request of the nations where the weapons are deployed, since in most nations the existence of US nuclear weapons within their borders is a difficult internal political issue. Thus they generally have requested that the United States not declassify the fact that US nuclear weapons are located in their specific nation – even though the evidence that they are there is obvious and generally known by their population [12].

Without specifying particular numbers, the US Department of Defense recently released information which indicates that the numbers of US tactical nuclear weapons reached a peak around 1967 and has been declining since then [13]. This is corroborated by the Congressional Hearings of 1976 which indicated that tactical nuclear weapons in several specific older categories, such as land-based and ship-based air defense weapons, had been or were being retired. In 1973, Senator Stuart Symington still indicated that 'The United States has now deployed in foreign countries, on our ships, and in this country some tens of thousands of nuclear weapons . . .' [15].

The question of secrecy regarding the overseas location of US tactical nuclear weapons has produced exceedingly unusual Congressional testimony. In a year and a half long investigation, the Senate Committee on Foreign Relations had sought answers to

> . . . questions on both the political and military implications involved in the placement of those weapons in foreign countries.
>
> In addition, [we] asked about government-to-government agreements or understandings under which such weapons had been placed in each country, including questions of how they could be used or withdrawn. In some countries, however, not all this information was available at the Ambassadorial level. Even high-ranking military officers in certain countries where such weapons were located did not have precise answers.
>
> It was clear that many years had passed since the political implications of the placement of these weapons had been thoroughly considered, if, in fact, they had ever been so considered. One example: As our published hearings on Taiwan show, the ranking United States Army officer in that country testified he was unaware whether or not nuclear weapons were located on Taiwan.
>
> In more than one country, the American Ambassador stated that he professed not to know whether nuclear weapons were there. In several cases where they were located, the American Ambassadors in question said they did not know what understandings with the host country had been arrived at with respect to their possible use [16].†

One of the Committee's staff personnel subsequently published an article with two pieces of even more significant information [17].

> In one instance, nuclear bombs were stored in a foreign country without the knowledge of that country's president. One US Army officer in command of a nuclear weapons depot in the Far East

---

† One wonders whether to accept at face value all the testimony given in open hearing.

described joint exercises with the host country army using dummy nuclear weapons, although we had no agreement with that country on sharing such weapons. What was being done violated provisions of the Atomic Energy Act.

The second point demonstrated a rarely seen and even more rarely understood aspect of the mechanistic basis, the 'operational procedure' derivation, of even the most important defence policy decisions.

> . . . in two Far East countries US-piloted F-4 fighter bombers armed with nuclear weapons stood at the end of runways on 15-minute alert. Against whom? we asked. At a secret Pentagon briefing we were told the host countries initially had permitted us to station F-4 squadrons; that F-4s could carry both conventional and nuclear weapons; that F-4s were most cost effective when nuclear armed and the most efficient mode to be so armed was on 15-minute alert. In just that off-hand manner serious nuclear weapons policy was made.

The numbers and locations of tactical nuclear weapons manufactured and deployed by the USSR are unknown. An estimate for all Soviet nuclear warheads of 5000–10 000 was made in 1966, but this did not indicate any breakdown between strategic and tactical warheads. It has become customary to refer to 3500 as the number of Soviet tactical nuclear weapons deployed with its forces in WTO nations. This has come about through the constant repetition of this number in the press and in the academic literature. As will be indicated below, there is no publicly available evidence to indicate that Soviet tactical nuclear weapon *warheads* are located in WTO nations along with their delivery systems. The number 3500 appears to be derived from available Soviet delivery systems with Soviet forces in WTO nations, but offers no direct evidence as to the actual presence of tactical nuclear warheads with those delivery vehicles, nor even the degree to which tactical nuclear weapons may be assigned to such systems, if they are present.

The numbers and location of French and British tactical nuclear weapons are also not known. It is possible that each country has several hundred or more of such weapons. British nuclear weapons were apparently based overseas in the past, but no good evidence one way or the other seems available on this point.

An historical review: the development of tactical nuclear weapons and their entry into NATO Europe

The previous section dealt with questions of definition regarding tactical nuclear weapons. We should also like to know something of where the notion of developing tactical nuclear weapons came from, and about their entry into NATO Europe. This section will therefore look at the significant dates and events regarding that process, and at the decisions and policy for their deployment and control. The following pages will be somewhat sketchy and far briefer than the available information on the subject would permit. Particulars regarding specific weapons, launchers and warheads, their numbers etc., will appear in the data sections (pages 109–136).

The notion of developing tactical nuclear weapons – nuclear weapons for tactical missions, at first on land and then at sea and in the air – developed quite early. The first theoretical studies in the area of 'limited nuclear war' began in 1948 in Project Vista, at the California Institute of Technology. Several sources of interest and support for their development can be identified. In fact they were a major issue in the 1949–50 internal US government debate on whether or not to initiate a crash programme to develop a thermonuclear bomb. This is because the alternative goal which was recommended by the General Advisory Committee of the US Atomic Energy Commission (AEC), led by Drs Oppenheimer, Fermi and Rabi, stressed instead an increased development of tactical nuclear weapons [18]. A second source of support for development of tactical nuclear weapons was one of the five individual Commissioners of the Atomic Energy Commission at the time, Edward Murray. Murray also opposed the development of thermonuclear weapons, and urged the development of TNWs instead. The third source of support was the nuclear weapon development laboratories. Research and development on warheads for cannon, and for smaller air-dropped weapons for the US Navy and the US Air Force (USAF) started quite early, and designs were available by 1949. The problem at this time was scarcity of fissile material, and the USAF strongly opposed any diversion from the strategic bomb programme. For the USAF in the period 1950–52 'tactical' just meant dropped by USAF aircraft on a target other than a major Soviet city, perhaps on a 'battlefield', on troops. Weapon yields were actually going up during this period, due to the emphasis on weapon design improvements, so it was not a matter of lower yields, only of making a bomb smaller in size so as to be carried by aircraft smaller than long range bombers. The nuclear cannon was ready in 1952, and was written about by Ralph Lapp in the following year [19].

The fourth source of support for the development of TNWs was the US Army. As early as 1950 General Gavin wrote an article in *Combat Forces Journal*, a US Army publication, on 'tactical' nuclear weapons, and another by General Omar Bradley had appeared in October 1949 in the *Saturday Evening Post*, a general circulation magazine. On 4 June 1950 Army Chief of Staff General Collins stated that atomic weapons could be developed for Army use, specifically mentioning artillery shells and guided missiles. Two days later the Army Secretary, Frank Pace, also referred to the 'revolutionary new weapon' as the best way to stop masses of heavily mechanized ground troops. This interest was greatly stimulated by the US Army's unhappy experience during the Korean War. Perhaps the key for the Army was that the USAF was proposing that it be the service assigned the 'mission' of dropping tactical nuclear weapons in support of Army combat operations. The Air Force already had the strategic mission. The Army wanted to be 'out from under' the USAF; it clearly wanted to have its own tactical nuclear combat support arm [20]. Then there was the question of the USAF accumulating all the most significant 'missions', which was more important in its secondary and derivative effect, the effect on that services' shares of the annual military expenditure appropriation. The only way to stop one service from having a disproportionate share of the appropriated funds was to stop it from gaining all the 'mission' assignments that required more extensive outlays for the weapon systems related to those missions. The US Navy was soon to take on the USAF over precisely the same issue – nuclear weapon delivery roles – in the bitter 1953–54 'carrier–B-36 controversy'. (In the 1950 USAF–US Army confrontation over 'tacticals' the Navy role was minimal, since it was seeking a portion of the 'strategic' mission for its carriers, not the 'tactical' nuclear one). This was one of the first, but certainly not the last of the US interservice disagreements over nuclear weapon policies. (When 'massive retaliation' was announced, with the prominent support of the Chairman of the Joint Chiefs of Staff, Admiral Radford, the Army Chief of Staff General Bradley opposed it and resigned due to its slighting of the role of the Army. General Maxwell Taylor took a similar position five years later. Other later debates concerned the 'limited deterrent' and other aspects of nuclear weapon policy.) An incidental outcome of yet another such conflict in 1956 is the present reason that there are no US ground-based 'tactical' nuclear missiles with ranges greater than that of the roughly 400 miles of the present Pershing missiles in Europe. The debate between the USAF and the Army over Army development of both Medium-Range Ballistic Missiles (MRBMs) and Anti-Ballistic Missiles (ABMs) resulted in the decision that the Army would retain the ABM role, but give up any missile development in the MRBM range [21]. The range set in that decision for Army short-range surface-to-surface missile development was actually 200 miles, which was doubled for the Pershings, but when the Army subsequently asked for an 'extended range Pershing' as well, the request was turned down. (When the 1200 mile range Mace was deployed in F.R. Germany with US forces, as it was for some years, it was staffed by USAF personnel.)

Paradoxically, the successful testing of a thermonuclear fusion device in 1952 gave the notion of TNWs impetus, and raised 'the possibility that very large (megaton) weapons would be used tactically on the battlefield'. In 1953, the Army revealed that it was studying the applications of nuclear weapons in infantry warfare, and by 1956 it had activated its first division intended to fight with nuclear weapons. By December 1959 all the five American divisions stationed in Europe had been remodelled to the Pentomic organizational structure.

During this period the notions of the Administration underwent a process of change as both weapons development programmes for tactical nuclear weapons and notions of 'limited (nuclear) warfare' evolved. Along with the New Look the Administration approved the creation of a varied arsenal of tactical nuclear weapons. Throughout the New Look period, however, this goal usually had been linked closely with stress on air-atomic forces in general. In 1953 and 1954 Administration spokesmen did not clearly differentiate between different types of nuclear weapons and the different purposes to which they might be put. However, as the weapon development programmes progressed, tactical nuclear weapons began to be linked more and more with 'limited' war. The appearance of the weapons in sufficient numbers for battlefield use in 1954 and 1955 helped to legitimize limited nuclear war in the thinking of the

Administration. Once again one has an example of doctrine, and the years of its subsequent evolutions, *following* from a particular achievement in weapon development.

The final domestic US influence on tactical nuclear weapons development and deployment was the Joint Congressional Atomic Energy Committee (JCAE). (The two main European influences, the NATO allies and the office of the Supreme Allied Commander, Europe (SACEUR) were not to play a role until later). This influence probably did not play a role until some years later, in the late 1950s, but when it did, it was attributed to the most parochial motivations possible.

> ... tactical nuclear weapons ... found ready support in the Joint Congressional Atomic Energy Committee, whose members saw, among other things, greatly expanded production of nuclear weapons, which consume relatively large amounts of fissionable materials, as a way of expanding Atomic Energy Commission activities in their home districts.

Similarly, in 1974 testimony, Secretary of Defense Schlesinger said there was 'some measure of truth' to the charge that once fissionable materials began to be produced in large quantities, the Joint Chiefs of Staff wanted to use all the materials available, irrespective of any previous plan or warfare doctrine. To illustrate his point, Schlesinger noted that the eight-inch nuclear shell development 'reflected the availability of U-235' and not any real military need. A former Assistant Secretary of Defense, A. Enthoven recalled that the 'nuclear artillery rounds ... for land warfare in Europe were never subject to the same kind of rational analysis' brought to bear on conventional weapons. 'It was simply a race to equip everybody – even the infantry – with nuclear weapons' [22].

It is not commonly understood how early nuclear weapons were introduced into Europe. General Norstad indicated that 'In fact, the first nuclear elements were introduced into the alliance in the early summer of 1952. . .' [23]. It is possible that this refers to the 280 mm nuclear cannon, which were not considered very useful, and were few in number. The date otherwise often given for the introduction of these is 1953. In October 1953 a US National Security Council decision authorized the US Joint Chiefs of Staff to base their plans on using tactical and strategic nuclear weapons against conventional attacks whenever this was militarily advantageous. (The President retained the exclusive authority to order their use in any particular contingency.) By 1954, US aircraft carriers in the Mediterranean apparently were able to deliver nuclear weapons as well. These moves as well as earlier ones involving strategic nuclear weapons were known to the USSR.

> The historical responsibility for transferring atomic weapons to Europe belongs to the US. . . . They were the ones who, in July 1948, stationed their bombers armed with nuclear bombs in England, and then in the fall of 1953, sent atomic warheads to the European continent and unfolded the nuclear weapons race. The Soviet Union was forced to take responding defensive measures [24].

12

However, several years would pass before the USSR had any 'tactical' nuclear weapon capability. In early 1955 US forces in F.R. Germany included Honest John missiles and 30 atomic cannon [25]. In January 1957 the USA announced that it would set up six 'atomic support command' task forces in the coming fiscal year (June 1957 to June 1958). However, it was reported that the first of these SETAF was already in prototype operation in the North Italian theatre, and was armed with Honest John and Corporal missiles. Elsewhere in Europe the US reportedly already had Honest John, Corporal, Matador, and Little John missiles and the 280 mm atomic cannon [26].

Once tactical nuclear weapons were introduced into Europe, a new group of influences on their numbers and deployment began to have increasingly important roles. These were SACEUR, NATO's Supreme Commander, and what will be used for want of a better designation, 'european political leaders'. The Lisbon meeting of NATO in 1952 had set a force goal of 96 divisions, *not* including the manpower contributions of Greece and Turkey. After the USA introduced the 'new look' into its army in Europe, involving nuclear artillery and missiles, NATO reduced its conventional force goals to 48 divisions, including those to come from Greece and Turkey. The European political leaders did not want to raise conventional armies. They also did not want to fight another *conventional* war in Europe, and so they became devout believers in the concept of 'deterrence'. That deterrent was SAC, the US *strategic* nuclear force. *Tactical* nuclear weapons *in Europe* were an overt, explicit way to invoke escalation, a 'chain' to hold on to SAC. The order of events in the years from 1952 to 1955 seems to be critical, and would be worth a good deal more research. The question is whether tactical nuclear weapons were, in fact, introduced to counter presumed massive numbers of Soviet troops, which is the conventional wisdom. A first impression is that this may not be so at all. Tactical nuclear weapons were introduced, Western Europe did not want to raise troops, and after the introduction of the tactical nuclear weapons, NATO dropped its troop goals and began the process of doctrinal rationalization. As indicators of the US policy changes, SHAPE (Supreme Headquarters Allied Powers in Europe) held its first manoeuvres involving tactical nuclear weapons in the *Sagebrush* and *Carte Blanche* exercises in 1954 and 1955. European political leaders applied substantial pressure on the Eisenhower administration to constantly increase the numbers of TNWs in Europe. Once the weapons were in Europe in relatively large numbers, by 1959, the pressure to raise the numbers increased even more, from SACEUR, from individual services, and from European political leaders. As will be indicated, the weapons were intended and planned for use, which led to NATO's unending contradictions over warfighting, deterrence, and command and control over the TNWs.

In 1951, during the Korean War, President Truman agreed to the transfer of a limited number of non-nuclear weapon components of the stockpile to the armed forces for deployment to a few specific overseas locations. By September 1952, another agreement was reached, giving complete weapons to the military and specifying that the Department of Defense would state its needs for numbers and types of weapons, and the AEC would propose production rates before the

President determined the actual production schedule. By 1956, virtually the entire stockpile had been transferred to the military, although custodial responsibility remained in the AEC. In 1959, the full military custody of nuclear weapons was formalized.

On 2 March 1955, the NATO Council approved an agreement between the NATO partners for cooperation regarding atomic information. This agreement committed the United States 'from time to time', and subject to the restrictions of the Atomic Energy Act, (which was amended in 1954 to permit dissemination of some atomic information), to furnish NATO with atomic information needed to develop defence plans, train personnel in the employment of, and defence against, atomic weapons, and evaluate enemy capabilities. The strategy adopted by NATO Council in 1956 envisaged that, except in the case of a very small-scale attack, the period of conventional or tactical nuclear combat would be short and would rapidly escalate into an all-out nuclear exchange. Finally, at the NATO Heads of Government meeting in Washington in 1957, President Eisenhower formally proposed that the United States stockpile tactical nuclear weapons in Europe. Two weeks after the Washington meeting the USA informed the NATO Council about the TNW delivery systems that it intended to supply to NATO countries under the mutual aid programme for fiscal year 1957 [27]. These included the Honest John missile, Matadors with a range of 800 km, and Nikes. Both the Nike Ajax and Nike Hercules had been equipped in the USA with nuclear warheads at that time. In March and April 1957 General Norstad urged that missiles be installed in all NATO command zones, that allied forces learn how to use them, and that NATO allies develop measures, 'to ensure that these arms can be used effectively in case of emergency' [28]. In July of 1957 US Secretary of State Dulles publicly announced that the USA was considering the establishment of a nuclear weapon stockpile for NATO in Europe [29]. The proposal was by that time obviously under discussion with the NATO allies of the USA. In December 1957 the Council decided to establish stocks of nuclear warheads, 'which will be readily available for the defense of the Alliance in case of need', and to equip the allied forces with systems to deliver the warheads. IRBMs were also placed at SACEUR's disposal. Prior to 1957 the United States had also deployed three strategic nuclear weapons capable systems in Europe, – the Air Force's Thor missile, the Army's Jupiter missile, and B-47 aircraft. Seven squadrons of 15 rockets each had been deployed, four in Britain, two in Italy and one in Turkey. It is stated that there were no formal agreements with the host NATO nations regarding these forces, and that there was no stockpiling of nuclear weapons prior to 1957 [30]. However, in view of what has already been indicated regarding US deployments between 1952 and 1957 it is not clear whether 'stockpiling' was meant here to refer only to weapons intended for NATO allies of the USA, or some other condition regarding US forces, (i.e. that tactical nuclear warheads were not deployed with their delivery systems). In 1956 Roger Hilsman wrote, referring to US forces in Europe, 'the Ground Forces have been supplied with nuclear weapons' [31]. There is therefore at least some ambiguity about the meaning of stockpiling. Information in the literature on dates concerning this topic are often poor, or vague, or confuse *US* deployments and *NATO* deployments. The customary phrase that tactical nuclear weapons were deployed in the mid- or late 1950s may intend to indicate either the first *substantial* deployments, or may be a sign of just such confusion. Throughout, no one ever seems to refer to entry of TNWs on aircraft carriers in the Mediterranean.

In 1957 the first NATO 'Guidelines' were written, at the same time as the first equipping and training of non-US NATO forces with TNWs. Little information is available concerning these first Guidelines. Limited war was 'recognized', but the use of nuclear weapons was called for in response to any major aggression. 'SACEUR thenceforth was to base his forward planning on the assumption that a large range of nuclear weapons gradually would be introduced into both NATO and Soviet bloc armories' [32]. Between 1957 and 1962 the USA negotiated bilateral agreements with Canada, France, Greece, Italy, Turkey, the UK, F.R. Germany, Belgium, and the Netherlands, for cooperation 'in the operation of atomic weapons systems for mutual defense purposes'. (The agreement with France was subsequently revoked.) The Executive Branch then entered into two kinds of technical implementing agreements: a formal Program of Cooperation covering one or more specific nuclear weapons delivery systems, and a (military) service-to-service stockpile agreement. Programs of Cooperation and stockpile agreements have been concluded with each NATO country where the USA has nuclear weapons deployed. The stockpile agreement states that the location of nuclear weapons storage sites is to be determined by SACEUR in agreement with US military authorities and the government of the country concerned, that the costs are to be borne by the user nation unless NATO agrees to fund the site costs under its infrastructure programme, that US personnel will have custody of the weapons and that external security will be the responsibility of the user government. By January 1961 Honest John missiles were deployed with the forces of the UK, F.R. Germany, the Netherlands and Greece. In November 1965 it was publicly stated that US nuclear warheads were mounted on QRA aircraft of nine NATO allies [33].

While stockpile agreements refer to broad categories of weapons, they do not identify individual systems. In 1967 the USA agreed to inform countries of the details of the nuclear weapons stockpiled in their territory or stockpiled elsewhere for the use of their forces. That information is conveyed in an annual letter from the US Secretary of Defense to the minister of defence of the country concerned, enclosing an annual report which lists the type, quantity, yield and location of all US warheads stockpiled in the country (and where appropriate, stockpiled elsewhere for the forces of that country) and the types of weapons in which these warheads can be used.

During the period 1960–68 (the Kennedy–Johnson Administration) Secretary of Defense McNamara made a series of statements indicating either precise numbers of tactical nuclear weapons in Europe, or percentage increases over previous levels in some number of months. These statements enable one to make a reasonably good approximation of the NATO TNW stockpile deployed in Western Europe over the years, particularly after 1960. In 1973 and 1974 further information on the numbers of weapons became available:

- 1960, end of the Eisenhower Administration, approximately 2500 TNWs
- August 1963 ('... in the past 24 months alone ... a 60 per cent increase ...') also reference to 'several thousand in Europe'; around 4000 TNWs
- October 1965, 5000 TNWs
- November 1965 ('... more than 5000 ... currently; would be increased by 20 per cent in the next 6 months. This will represent a doubling of nuclear weapons strength in Europe over the past five years' [34].
- 1966, 6000 TNWs
- December 1966, 6000 TNWs
- and finally in an October 1968 statement by Secretary of Defense Clark Clifford, 7200 TNWs [35].

A very large proportion of these weapons are in the Federal Republic of Germany. In early 1963 Secretary of Defense McNamara had also stated that:

> In the field of tactical nuclear weapons. ... The US at present has in stockpile, or planned for stockpile, tens of thousands of nuclear explosives for tactical use on the battlefield, in anti-submarine warfare, and against aircraft. They include warheads for artillery, battlefield missiles, demolition munitions, bombs, depth charges, air-to-air missiles, and surface-to-air missiles. The consensus is that the US is presently substantially superior in design, diversity and numbers in this class of weapons [36].

In 1973 the Senate Foreign Relations Committee report gave the following synopsis of the numbers of TNWs as of that date.

> Secretary McNamara did not so state, but the 7000 warhead figure referred only to land-based tactical warheads for support of NATO stored in European NATO countries. Thus, the figure did not include Strategic Air Command warheads [deleted] and warheads in ships afloat some of which [deleted].
>
> The number of land-based nuclear warhead "authorized" to be deployed in Europe for support of NATO rose steadily in the period after Secretary McNamara's statement and by fiscal year 1970 had reached [deleted]. The authorized figure has since dropped to [deleted] for fiscal year 1974, and the projected figure for fiscal year 1975 is [deleted]. According to the Department of Defense, these fluctuations in "authorized" deployment figures have resulted from an annual "re-examination of military requirements".
>
> As of 31 July of this year, we were told there were in fact [deleted] land-based nuclear weapons in support of NATO Europe stored at over 100 special ammunition sites in [deleted] European countries – [deleted]. About one-third of these sites are for weapons to be used by US forces. The remaining two-thirds are for weapons to be used by allied forces, although all nuclear weapons in Europe are kept in the custody of the United States whether they are for the use of US or foreign forces [37].

It would appear that the number 'authorized' have not always been delivered to Europe. One knowledgeable Congressional authority referred to 10 000

TNWs (in Europe) in 1972, but the '7000' figure is still universally used, and it would appear from the following that it is in that general neighbourhood, with the caveats indicated above (land-based, stored in Europe). In 1974, Secretary of Defense Schlesinger provided new evidence as to the numbers, and stated that the totals had been reduced.

> There are approximately 7000 US nuclear weapons in Europe, including naval weapons. All of these US weapons are available for NATO missions; none is a SAC weapon. . . . The weapons range from subkiloton to megaton yields. . . . The number rose from zero in the mid-1950s to slightly above 7000 in the early 1960s and has remained relatively constant since then. The mix of weapons within that total has changed to provide for modernization. . . . We withdrew without replacement the Mace, Matador, Lacrosse, Thor, Jupiter, Bullpup, Falcon and the Davy Crockett battlefield weapons. Our stockpile has been reduced, in overall totals, from over 7000 to less than 7000 since the mid-1960s [38].

In view of the stated far larger numbers available and with logistic replenishment possible, it is not always clear why these specific numbers get the emphasis that they do. As was indicated, the 'tens of thousands' as the number describing the total available US TNW stockpile is still valid today.

Leaving the numbers of weapons behind, and returning to 1960 and the new Democratic Administration, one finds the period of the 1960s presenting an extremely confusing web of processes involving tactical nuclear weapons in NATO Europe. There was the simultaneous large *increase* in the numbers of weapons deployed into Europe† at the same time as the Kennedy–McNamara doctrine of 'flexible response' was enunciated – over great NATO resistance. This doctrine increased the stress on limited (tactical) nuclear war at the same time as it attempted to improve NATO's conventional defence capability and end the notion of immediate recourse to tactical nuclear weapons in the event of conflict in Europe. One had the increase in participation, planning and operation of TNWs by non-US NATO nations. One had the development of 'two-key' decisional processes – a phrase intended to denote the combined necessary decisions of the USA and a second NATO nation on the question of TNW use, not two actual keys involved in the operation of a particular weapon. This however produced a puzzling two-sided situation: was the joint 'control' meant to facilitate TNW use when the USA wanted to hold back, or to prevent US use when the allied NATO nation did not want it to take place. All of these processes seemed to be going on at the same time, and the public record did not always make it readily apparent which motive produced which deployment or doctrine.

As part of its NATO policy throughout the 1950s, the USA sought to discourage the acquisition of nuclear weapons by other members of the alliance.

---

† At the start of the Kennedy Administration in 1961, Secretary of State Dean Rusk wrote a secret but nonetheless publicized letter to the Secretary of Defense, McNamara, suggesting consideration of the removal of nuclear weapons from Europe [39]. Instead, as indicated, the number was built up considerably.

17

However, within five weeks of taking office President de Gaulle informed Dulles of his determination to develop such weapons, with or without US assistance. He refused to allow US Jupiter or Thor missiles not directly under his control to be stationed on French soil. Six months later, in June 1959, he forced the removal of more than 200 NATO fighter bombers from French bases because they carried nuclear weapons that France did not control. As long as France remained a full partner in NATO, it blocked US efforts to move the alliance's formal doctrine away from quick resort to nuclear weapons. De Gaulle rejected the US concept of 'a pause' in any European armed conflict to avert an immediate escalation to nuclear war. Thus France repeatedly professed its determination to unleash an immediate strategic retaliation to any attack. After 1954, the German military in NATO had insisted on the position of forward defence, holding the Elbe instead of the earlier, more flexible 'fallback' defence. The Federal Republic of Germany had demanded an atomic detonator placed close to its frontiers as part of the forward defence. When the Kennedy Administration took office in 1960, it found a proposal awaiting its decision for installing electromechanical devices on nuclear weapons located in West Europe that would prevent their being fired without authorization. The decision was taken in 1961. Its purpose was to ensure that no US-made nuclear weapon could be set off by the NATO allies until authorized by the US President to do so. The practical effect of these new controls, however, was to eliminate the prospect anticipated in Europe that the Supreme Allied Commander would use US nuclear forces in Europe to meet a Soviet attack even without clear authorization from the President. With one stroke, the Supreme Commander's role had been radically reduced. Previous uncertainties about custody and command arrangements before the weapons were locked had assured the US nuclear commitment to its allies by entrusting the Supreme Commander with a possible decision to use. The perspectives of the USA and of its NATO allies were thus strikingly different even on the 1960–61 question of security locks. There was a contradictory welter of interactions, of motives, of the weapons and their disposition and operation. At times, European critics of the USA feared that the USA would start a nuclear war needlessly. On the other hand, they wanted to rely on the US nuclear deterrent rather than to increase conventional defence efforts. Early 1960 US proposals in the Eisenhower Administration for NATO nuclear 'sharing' maintained a US veto on TNW use but extended a similar veto to the allies: they could *prevent* the use of nuclear weapons assigned to their forces. However, they could not prevent the US use of nuclear weapons assigned to its own forces, nor could they force US use of TNWs when they chose. The USA would still maintain a veto over the use of shared nuclear weapons. The proposals were rejected.

In 1962, the USA proposed new 'Guidelines' for NATO, embodying the notions of 'flexible response'. They de-emphasized immediate use of TNWs, but the numbers deployed climbed at the same time. From the public record, the status of the new guidelines appears ambiguous, since they were not formally accepted by NATO until the very end of 1967. General Norstad had suggested the idea of 'the pause', urging NATO to resist conventional attack by conventional means. Nevertheless, testifying in 1967, Norstad indicated that the application of the conventional response contingency was quite limited; dependent upon the ability of the local commander, in a relatively narrow sector, to resist. For several years early in the 1960s, the US promoted a scheme to develop a NATO multilateral nuclear force (MLF). The idea ultimately failed because the US was unwilling to delegate to individual allies or to NATO itself the power to decide when to initiate the use of nuclear weapons, and because European countries, with the exception of the Federal Republic of Germany, showed no enthusiasm for an MLF. After the collapse of the MLF proposal, and following an amendment to the US Atomic Energy Act, Secretary of Defense McNamara suggested the establishment of two new permanent NATO organizations for nuclear planning. The Nuclear Defense Affairs Committee (NDAC), made up of NATO defence ministers, and, under it, the Nuclear Planning Group (NPG) were established in 1966. The NPG meets semi-annually and is intended to go deeper into the details of subjects raised by the NDAC. The USA, the UK, F.R. Germany and Italy are permanent members of the NPG, and there are three or four rotating members who serve eighteen month terms. All NATO countries participate except France, Iceland, Luxembourg and, at present, Portugal.

In 1967, the Nuclear Planning Group, in the 'Healey–Schroeder exercise', finally formally adopted the US strategy of 'flexible response' [40]. As far as the European theatre was concerned, the concept of a massive-interdiction strike against the hinterland of the invader, erroneously called massive retaliation, had finally been abandoned. The emphasis was now on the tactical use of nuclear weapons.

General NATO strategy, with respect to both conventional and nuclear forces, was set forth in a document entitled 'Overall Strategic Concept for the Defense of the NATO Area', known colloquially by its serial number as '14/3' and dated 16 January 1968. The basic elements in the doctrine set forth in 14/3 are nuclear deterrence, forward defence and flexible response, elements which NATO countries had presumably accepted informally a number of years before adopting 14/3. 14/3 replaced 14/2 which had provided for a strategy based on massive retaliation [41]. Whether 14/2 had been considered operative in the years from 1962 to 1968 is unclear. In 1968, US Secretary of Defense Clifford presented to NATO's Nuclear Planning Group 'a detailed outline of how demonstration type nuclear bursts could be used to show an invading enemy the West was not afraid to resort to such weapons'. 'One or two well-placed nuclear detonations could be considered in the early phase of a massive, uncontrollable assault by enemy forces over NATO borders' [42]. However, the material discussed in this paper in the section on 'Targets and Plans' would seem to indicate that in 1973–74, NATO's actual plans for use of TNWs were relatively little different from those before 1967–68. The following extended excerpt from the 1973 Senate Committee on Foreign Relations Report is intended to provide some insight into the situation regarding 'Command and Control' of tactical nuclear weapons reached after the 20 years of NATO integration.

In 1969, NATO drew up agreed general guidelines for consultation procedures on the use of nuclear weapons. These guidelines proceeded from three decisions that had already been taken. The first decision was taken at the Athens meeting of the North Atlantic Council in 1962 and produced what were called the 'Athens Guidelines'. The second decision was taken at The Hague in April 1968 by members of NATO's Nuclear Planning Group. The third was taken in London in May 1969 at a ministerial meeting of the Nuclear Planning Group.

In brief, NATO doctrine is that in the event of a Soviet attack with nuclear weapons in the NATO area, the alliance would respond with nuclear weapons on the scale appropriate to the circumstances. Consultation would [deleted]. In the event of a full scale Soviet attack with conventional forces, indicating the opening of general hostilities in any sector of the NATO area, the forces of the alliance would, if necessary, respond with nuclear weapons on the scale appropriate to the circumstances. Again consultation would [deleted]. In the event of a Soviet attack which did not fulfill the conditions described in the first two cases, but which nevertheless threatened the integrity of the forces and the territory attacked and which could not be successfully held with the existing conventional forces, the decision to use nuclear weapons would be subject to prior consultation in the North Atlantic Council. In all cases, special weight would be given to the views of the NATO country most directly affected – that is, the country on, or from, whose territory nuclear weapons would be employed; the country or countries providing the nuclear warheads; and the country or countries providing or manning the contemplated means of delivery.

As far as consultation procedures are concerned, any request for the use of nuclear weapons in the defence of NATO either from a member government or from a major NATO Commander, and any possibility for the use of nuclear weapons in defence of NATO by a nuclear power, would be communicated immediately to the NATO governments and to the Defence Planning Committee (composed of all NATO members except France). The normal forum for consultation would be the Defence Planning Committee where member governments would be able to express their views, in particular on the political and military objectives of the proposed use of nuclear weapons, the methods of use and the possible consequences either of use or non-use. These views would then be communicated to the nuclear power concerned, and the decision of the nuclear power would be conveyed to the allied governments, the North Atlantic Council and the major NATO commanders. US officials estimate that this consultation procedure could be accomplished in [deleted].

SACEUR would thus not be permitted to use nuclear weapons unless there were consultations with NATO member governments directly and with NATO itself through the Defence Planning Committee. But the converse does not apply, because no NATO body has the authority to order SACEUR to use nuclear weapons. The release of nuclear weapons can only be authorized by the President of the United States (or, for British weapons, the British Prime Minister). Before releasing or ordering the use of nuclear weapons in Europe, the President is bound to consult if time and circumstances permit.

In a technical sense, the President cannot order SACEUR (who is simultaneously the Allied Commander responsible to NATO's

Defence Planning Committee as well as the Commander, US Forces in Europe) to fire a nuclear weapon; he can only release the weapon to him (although he can unilaterally direct the same commander, in his national capacity as commander of US Forces in Europe, to employ nuclear weapons). SACEUR would then regard the President's decision to release a nuclear weapon to him as a valid reflection of NATO's collective interest and will, although the release is not a command so that SACEUR would still retain discretion as to whether or not to fire the weapon. The NATO guidelines do not explicitly cover [deleted]. Nor do they provide guidance for situations in which [deleted].

The agreed NATO guidelines state that in times of crisis the procedures for general consultation should be set in motion at the earliest possible stage in the crisis – [deleted]. We were told at SHAPE that in most NATO procedural exercises the decision to use nuclear weapons is usually reached [deleted] [43].

In practice, things have apparently been both more flexible and more difficult. The flexibility was expressed in unusual testimony by General Lauris Norstad in 1966 in which he attempted to give a general notion of the latitude and informal arrangements of SACEUR, and of his own personal relationship as SACEUR with two US Presidents, Eisenhower and Kennedy, and with the heads of state of other NATO governments [44]. The difficulties are demonstrated by the contradictory positions, or the constant contradictory requirements of warfighting and deterrence, in the positions of the Federal Republic of Germany on TNWs and their use. The Federal Republic of Germany displays an extraordinary degree of ambivalence concerning the first use of TNWs, and in the late 1960s had sought a veto power over the firing of nuclear weapons from its territory. Germany has apparently never agreed to the 'prechambering' of Atomic Demolition Munitions and, as late as 1973, there still was no formal agreement between the FRG and the USA on the use of nuclear shells for 8 in and 155 mm artillery to be fired from FRG territory, after these weapons had been deployed with US forces in the Federal Republic of Germany for many years, and were the TNWs mostly closely involved with the forward battle zone. Such an agreement was apparently signed in 1974 [45]. Nevertheless, the West German annual defence 'white paper' of 1976 stressed – as it had in past years as well – 'the Western alliance's readiness if necessary to make first use of nuclear weapons against a WTO' [46].

An anticipation of how events might actually transpire in the event of a European conflict indicates the remaining difficulties. Secretary of Defense Clifford's 1968 presentation to the NATO–NPG took care to indicate that even demonstration use of a very few TNWs risked escalation by both sides. A Brookings Institution analysis, (written in 1971 before Secretary of Defense Schlesinger's further and particular emphasis on selective use), was even more pessimistic in attempting to analyse the possible degree of Presidential control of the use of TNWs in the field.

The characteristics of a weapon system and tactics for its use also affect the degree of presidential control over this form of warfare. Some disturbing questions arise in this connection:

– Deployment of weapons to a large number of units near the front means that once the President authorizes their use, control over firings would be diffused in the very areas where communications are likely to be disrupted early in a war and where confusion could most readily occur. Thus, local commanders might well have to make decisions about what weapons to use and when, and the President would have difficulty in controlling moderating, or terminating their use once the tactical nuclear battle had begun.

– The problem arising from this diffusion of weapons is compounded because we provide these weapons, under US guard, to allied units. In the confusion of a war situation, allied units might want to use tactical nuclear weapons in defence of their territory earlier than we would, and might try to take control of them for this purpose.

– The diffusion of control once tactical nuclear war has begun would also reduce the range of options open to the President. In practice, diffusion means that the President might have little choice but to authorize using all the weapons or to decide not to use them at all. Thus, he might be inclined in a crisis to postpone the decision even if that means risking the destruction of most of the weapons. Of course, the other side of this situation is that, if he decided on use, events could rapidly get out of control [47].

The context of the European conventional military balance

The reason customarily given for the introduction of tactical nuclear weapons by the USA and NATO into Europe, and for their maintenance there, is to counter an alleged Soviet and WTO superiority in manpower and in conventional military weapons. With time, other rationales have also become prominent, to deter Soviet use of tactical nuclear weapons or to deter any Soviet conventional incursion at all, but the first reason indicated is still the one most often heard in various public statements or analyses. It would seem sensible to include at least some comment concerning this claim, given that it serves as the dominant public rationale for the deployment and maintenance of tactical nuclear weapons with NATO, and the central context in which this deployment takes place.

No thorough survey of this question is attempted here, and no attempt has been made to survey assessments of the conventional balance in the 1950s. Other studies are available which have looked at the NATO–WTO conventional military balance in more detail [48].

In 1961 and 1962, the US Department of Defense under Secretary of Defense McNamara carried out a thorough 'systems analysis' of NATO and WTO conventional forces. This study took in all aspects of land and airborne weapons, their subcomponents, operational characteristics and capabilities, manpower, mobilization, logistics etc. The conclusion was that the WTO did not have a conventional superiority, but quite the opposite, that NATO did. Only three public references were made to this work, which was considered extremely sensitive, all in 1963. They were made by Secretary of Defense

McNamara [49], by Assistant Secretary of Defense Nitze (ISA) [50], and by Assistant Secretary of Defense Enthoven (Systems Analysis) [51]. However, these references were relatively brief, and contained no specific numbers or analyses. The difficulty, and the cause of the sensitivity, was that extensive publicity of the conclusions would have been inconsistent with Secretary McNamara's simultaneous effort to get European NATO leaders to *increase* their conventional military efforts, and to decouple 'tactical' nuclear weapons in the European theatre. Since it was decided (in 1962) to give this latter effort priority, nothing more was said publicly concerning NATO conventional superiority – or capabilities – until 1968.

In 1968, the kernel of the study was released to the House Committee on Armed Services by Assistant Secretary of Defense Enthoven in a heavily censored report entitled, *Review of a Systems Analysis Evaluation of NATO vs. Warsaw Pact Conventional Forces* [52]. After leaving office later in 1968 at the end of Secretary Clifford's tenure, Enthoven put this material into a series of papers which appeared in 1969 [53–54], 1974 [55], 1976 [56], and into a book in 1971 [57]. It should be noted that the US Department of Defense has maintained this perception of NATO's conventional power by and large until the present day, or at least through 1975 through the tenures of Secretaries Clifford, Laird, Richardson and Schlesinger, with three of these Secretaries making public statements to that effect. For example, on 7 June 1973, it was reported that 'a major new Pentagon study . . .' had come to the same conclusions again [58] and Secretary of Defense Schlesinger reiterated this position to the NATO defence ministers in July 1973 [59] and in January 1973. When publicly pressed on the question in Congressional testimony, US defence and military spokesmen chose to emphasize in their reply that, although the study had been proceeding since October 1971, it was not yet completed and its conclusions not yet final [60].

Enthoven states unequivocally that 'for 20 years, the WTO forces have been persistently exaggerated by our military, by our media, and therefore in the popular mind' [61]. It is also interesting that George Kennan indicates that such overestimates had become established by the time he wrote a memorandum (which concerned US reliance on nuclear weapons as a deterrent in Europe and the notion of first use) with contrary implications just before leaving official office in 1950:

> I cannot remember that this paper was ever seriously considered or discussed. . . . The views put forward here conflicted with what was already established military policy. They conflicted with the ideas we had formed as to where the essentials of our own defense were to be found. They conflicted with the reactions of Congress, the military establishment, and the public to the news of the detonation by the Russians of an atomic weapon. They conflicted with the views we had communicated to our European allies, and with the manner in which we had taught these latter to see the needs and the guarantees of their own defense. (Some of them would have been the first to react with alarms and indignation at the very idea of attempting to live in a world where *our* atomic weapons did not stand as the guarantee of

their security.) These views conflicted, furthermore, with the highly inflated estimates of Russian conventional military strength that had already become ingrained (we will not at this point ask why) in the official assessment of NATO defense needs, and with the resulting brief on our part that we could never meet the Russians successfully on non-nuclear ground [62].

It is clear that numbers have both been grossly misrepresented and used in misleading ways.† For example, when weapon systems as complex as tactical aircraft are evaluated for two opposing forces by military planners, they are not simply compared by total deployed numbers on the two sides. Aircraft are divided into groupings by missions and then they are compared by the amount of ordinance capable of being carried, ('payload'), combat radius, air time aloft over the battlefield area, weapons complement, mission-related electronics suit, electronic countermeasures, basing, pilot training, refueling capability, and yet other characteristics. In 1974, when it suited US military and defence planners to stress NATO's tactical aircraft superiority, it was pointed out that 'In the past the emphasis was on the numerical superiority of the Warsaw Pact' [65]. That was for the public. *No* military leadership is so grossly incompetent. It should be assumed that Western military planners have made the proper and relevant evaluations for the purposes of their defence planning. Similarly, a recent report which quite accurately pointed out recent improvements in Soviet tactical air capability stated that 'In 1968, Soviet tactical air forces were oriented primarily toward the defence of air space over Pact territory, and accordingly were composed largely of short-range, low-payload aircraft possessing only incidental ground attack capabilities' [66]. However, neither this report nor the reporting of it in the public press raised the question of what purpose was served by reporting the NATO–WTO tactical aircraft balance in all those fifteen previous years solely in terms of allegedly higher numbers of total Warsaw Pact aircraft [67]. The question of why NATO military and civilian leaders publicize misleading statistics which reflect to their disadvantage is seldom, if ever, addressed. The situation has led to the involuted irony in which a sophisticated strategic analyst could write in 1976,

---

† Examples are legion and could be culled from 20 years of the international press and defence literature. Perhaps a classic instance is illustrated by the following. ' "There are 175 Red divisions on a war footing; but that isn't all that counts", declared General Lyman M. Lemnitzer, Army Chief of Staff at the Pentagon from October 1960 to October 1962, "You musn't be overwhelmed by mere figures" ' [63].

Such gross misrepresentations have not only appeared in the public media but over the years in various official pronouncements, such as major addresses by a former British Defence Minister, Healey [64], or the Federal Republic of Germany's government 'Whitebooks'. A prime problem throughout the 1960s was the inaccuracy of data in this area printed by the Institute for Strategic Studies in London. ISS data on Soviet tactical nuclear weapons over the years has been both inaccurate and misleading, regarding numbers, delivery systems (types) and disposition. The same holds for their data on NATO–WTO conventional balance. The conventional weapon comparisons were examined in the 1969/70 SIPRI Yearbook [48a].

Common sense would suggest that the national leaders of sophisticated European nations could hardly make an error so crude as to compare units that were quite unequal. But against this presumption, there is a mountain of evidence which demonstrates beyond doubt that the terms of the comparison are almost always much closer to those suggested by simple divisional counts than, say, manpower counts [68].

The author was deploring the ostensible political benefits the USSR obtained from West European 'perceptions' of Soviet conventional superiority, without ever troubling to point out that the statements which generated those public perceptions were all made by NATO spokesmen – and more critically – were by and large misrepresentations.

The only sensible answer that anyone has been able to supply for this situation is that it is apparently dependent on budgetary allotment struggles in Western nations. One recent statement which did address this question directly, though it related primarily to the overall US–Soviet military balance, rather than to the NATO–WTO one, was so rare – and blunt – as to seem shocking. (The speaker was General George Buis, former director of the French Armed Forces Centre for Superior Military Studies, and current president of the French Foundation for National Defence Studies.)

It's part of an alarmist campaign of intoxication that just does not reflect the reality of ongoing American superiority. . . . The facts don't back up the statement of alarm. They happen, though, every year when they're about to vote the US military budget and right on schedule with the NATO annual general meeting [69].

Nevertheless, the repercussions of NATO's self-depreciation over the years are both serious and widespread, not the least of which is its role as rationale for the necessity and function of tactical nuclear weapons. For example, one of the more frequent claims for the utility of tactical nuclear weapons on the European battlefield would be that these could counter the Soviet preponderance in tanks and armour, and that with such armour Soviet–WTO forces could cross the European continent in a very short period of days or weeks.

Nuclear cannon artillery contributes to deterrence of both nuclear and conventional attacks by providing an important capability for deterring the massing of Pact artillery and armor, and substantially blunting a Pact exploitation attack by destroying armored units and their supporting artillery [70].

The conventional balance in Europe is such that in the event Warsaw Pact forces are able to mass secretly and to apply armored pressure to any given point in the defense line, NATO ability to defend with conventional forces would be greatly weakened. The ability to employ nuclear weapons selectively under all-weather conditions against armored thrusts greatly contributes to theater deterrence and is an intermediate option between conventional warfare and general nuclear war [71].

To support this conception, NATO spokesmen have recently pointed to the massive use of armour by the Arab nations in the October 1973 Middle East War.

> Alliance Defense experts, aware that tanks are crucial to the East–West defense equation in Central Europe, note that the Arabs sent 2500 tanks against Israel compared with 2000 tanks used by the Nazis in invading Russia in World War II [72].
>
> In an address to the French Academy of Political Sciences, on November 12, NATO Secretary-General Joseph Luns said Egypt, Syria and Iraq had deployed 5000 tanks and 900 missile batteries at the start of the war. In all, he said, 'the Arab armies lined up as many tanks and armoured vehicles as did the Nazi armies when they attacked Russia in June 1941 . . . Syria launched its attack against the Golan Heights with 1200 tanks. It was a real miracle that these armoured forces failed to reach Haifa, only 65 kilometres away, in 24 hours' [73].

Moreover, the October 1973 Middle East War is an analogue of the scenario that NATO planners speak of as particularly threatening, the concentration of great amounts of armour on a very short front line. Yet, if anything, the example seems to demonstrate precisely the opposite lesson from the one that Mr Luns intends to convey. All those tanks, and on so short a front line, were stopped before they got more than ten miles, by perhaps one-hundredth the number of troops that would be available to NATO forces, and with the loss of around 1000 fatalities in the first two days of the war. If the defences of Israel could hold against a tank attack of that magnitude, why shouldn't NATO defences be able to?

More recently, criticism of NATO's conventional capability to withstand a WTO conventional attack has switched to a new basis. This is variously described as the 'short war–long war' controversy, or the 'surprise attack' controversy. This position claims that the real issue is not the measures of the conventional military balance, even using more relevant indicators and more accurate numbers. These would show NATO superiority in some categories and WTO superiority in others. Rather, it is suggested that the military doctrinal and organizational concepts of the WTO and of NATO forces are so different as to provide a dangerous imbalance to the advantage of the WTO. It is suggested that Soviet forces are organized for offensive *blitzkrieg* tactics, a quick breakthrough and a short war, while NATO forces are organized for a long and defensive war [74]. (There is a contradiction between this thesis and the rapidity with which NATO planning presumes a transition to its use of tactical nuclear weapons, but this will be discussed later in this chapter (page 35).

Targets and plans

Something can be said regarding NATO targeting of its tactical nuclear weapons. In a 1973 discussion of targets for tactical nuclear weapons in Europe the following were referred to: airfields, rail yards, the Soviet IRBMs, and underground sites [75]. Rail yards were said to imply the wagons in them, and airfields the aircraft on them. Surface-to-air (SAM) missile sites, POL-logistic lines of communications, 'rail choke points', the railway rails themselves, and underground command and control facilities were then added in. It was indicated that roughly half, 52 per cent, of these targets were 'hard', and half were 'soft'. Elsewhere the Pershing missile in particular is referred to as 'a counter nuclear weapon, to wipe out airfields from which enemy planes might attack, and other major targets' [76]. These targets are integrated in what is known as the General Strike Plan (GSP) which is, in turn, a part of the Single Integrated Operations Plan (SIOP) of the US. Two long excerpts from the 1973 Senate Committee on Foreign Relations report are included at the end of this section. They provide some description of the GSP, the Priority Strike Program (PSP) within it, and in particular, the use of QRA aircraft.

Some information about targets is of course given by the ranges of various delivery systems for tactical nuclear warheads. The targets of nuclear SAMs, torpedoes, depth charges, air-to-air weapons etc., are clearly set. Similarly, ADMs are intended for bridges, passes and the like. Nuclear artillery with a range of from eight to ten miles is restricted to use in the front before it, against personnel or armour. However QRA aircraft, carrier-borne aircraft, and missiles such as the 400-mile-range Pershing and the even longer range earlier Mace and the NATO-assigned Polaris warheads are clearly intended to reach far into WTO territory and into the USSR. In the early years of US strategic targeting plans, the Tactical Air Command (TAC) was assigned coresponsibility with the Strategic Air Command (SAC) for targeting of the USSR (and East Europe). It was estimated in the 1960–62 period that TAC and SAC could together deliver 18 000–20 000 megatons against the USSR in a 24 hour period [77]. In 1970, the West German magazine *Stern* published a series of three articles which released portions of a compromised US targeting manual entitled *Nuclear Yield Requirements*, for NATO air-dropped weapons [78]. These ranged from 2.5 kt to 1.4 Mt, and the targets included airports, bridges, railroads, military headquarters and troop concentrations, ammunition and weapon depots, sluices, dams, and harbours, which were all specifically identified by location. The targets were in WTO states and in NATO states allied to the US, with the latter group presumably to be considered if they were overrun by WTO troops. The West German NATO authorities subsequently confirmed the authenticity of the manual, but stressed that it dated from 1962, 'an outdated strategic period'.† West German spokesmen also stressed that 'not one NATO atomic bomb would be exploded . . . without the political consent of the NATO Council', and portrayed the target planning as 'an element of intimidation' [79].

---

† The manual and other material had been turned over to Soviet authorities by a subsequently convicted US military courier. The Soviet KGB subsequently released the material to the press of the Federal Republic of Germany. On another occasion Soviet authorities have released what was purportedly another compromised document, a *Nuclear Target Study* of the Baghdad Pact Organization, dated 3 February 1958.

In 1972, a RAND study which had described a number of scenarios for the use of tactical nuclear weapons in an entirely different theatre, in Laos and in North Vietnam in Southeast Asia, was publicly released [80].

The 1975 report to the Congress by the US Secretary of Defense supplied several capsule summaries of NATO TNWs, which supply targeting information.

Surface-to-Surface Missiles. NATO's SSMs consist primarily of Pershing, Sergeant and Lance, with Lance currently being deployed to replace the older Sergeant missile and Honest John rocket. The primary role of Pershing is attack of fixed targets; Lance, Sergeant and Honest John provide tactical support to the battlefield.

Some Pershing missiles are on peacetime Quick Reaction Alert (QRA) at fixed locations, QRA missiles are designated against specific WP, high priority, time sensitive targets.

Nuclear artillery. Artillery's high accuracy, low yield, rapid responsiveness, and ease of control by local commanders should provide for effective attacks against targets in proximity to friendly troops. Because of its relatively short range, confining nuclear effects to the immediate battle area, it is judged that use of nuclear artillery in limited nuclear conflict probably has less chance of resulting in escalation to theatre-wide nuclear war than longer range SSMs or tactical aircraft.

Nuclear-capable tactical aircraft. Some of NATO's tactical aircraft are completely nuclear-capable, that is, configured to carry nuclear weapons, supported by nuclear weapons, and with crews designated and trained for nuclear missions . . . Nuclear-capable tactical aircraft will continue to have a place in the NATO theatre nuclear posture. They provide a means of rapidly concentrating nuclear firepower anywhere in the area of NATO operations. Against nonfixed targets well beyond the front lines, the manned aircraft has a potential advantage over current missiles in that the pilot could make last minute changes in his aim point, to correct for target movement, providing in effect a form of terminal guidance.

Atomic Demolition Munitions. ADMs are nuclear demolition devices which are manually emplaced and detonated by timer or command. They can be used to destroy bridges, cave in tunnels or defiles, cut roads, and otherwise create barriers to slow enemy movement or induce concentrations of his forces. These actions could produce lucrative targets for attack by conventional or nuclear forces and buy time for conventional re-inforcements. Being defensive weapons and most likely to be used on NATO territory, they probably have lower escalation potential than most other theater nuclear weapons, often without direct casualties.

Studies are underway to examine alternatives in the form of earth penetrators delivered by missiles or aircraft.

Submarine-Launched Ballistic Missiles. Currently the United States is committed to share with NATO a portion of its sea-based strategic nuclear deterrent system in support of the Alliance. The highly survivable Poseidon RVs provide high confidence that they will be available under all conditions of war initiation. Since these RVs are relatively ineffective against hard targets, other systems are required, such as Pershing with its higher yield and tactical aircraft with a higher yield capability and greater accuracy. Because of its relatively low yield, Poseidon will produce a low level of collateral damage except when employed against military installations colocated with urban areas. Here, weapons with lower yields and greater accuracies such as those currently deliverable by tactical aircraft would be used [81].

Elsewhere, the Secretary of Defense described

> ... the desirable characteristics of tactical nuclear forces which are necessary to maintain a credible theater nuclear deterrent. Basically, these forces should have the capability for the following:
> – Quick dispersal to match a Warsaw Pact surprise dispersal.
> – Survivability and controllability while dispersed.
> – Denying the enemy his military objectives without excessive collateral damage.
> – Providing for selective, carefully-controlled employment options.
> – Penetrating enemy defenses [82].

The material summarized in this and in several other sections of this paper indicates that over a 20 year period, the stockpile of NATO TNWs was built up in Europe with the presumption that if need arose to save Europe against a Soviet invasion, virtually all the warheads would be used at the same time – at the minimum, that hundreds of weapons would be delivered against targets in the rear support areas of the WTO forces. The degree to which these estimates were actually reduced by the 1968–69 Guidelines, and by discussion of very low numbers of 'demonstration shots' is unclear from the public record.

In 1966, General Norstad had made the following observations regarding possible use of tactical nuclear weapons in the European theatre:

> The enemy may not necessarily be using atomic weapons, but if he indicates in a big way that they will be coming this would really make the decision for us. But there are some areas, perhaps short of that, where a decision would have to be made.
> ... Assume there is a small attack in a certain area and forces of a certain size move across and this is the beginning of an effort to spearhead a major attack, and intelligence supports this. You might, under these circumstances, decide you can't meet this with conventional forces. One of the several plans for dealing with the situation would certainly contemplate the use of some limited number of small atomic weapons against carefully selected targets. This might be five weapons. Maybe that would be satisfactory. You might have 100 or 200. We are talking in that range. You would find a starting point in that range [83].

In November 1969, the NATO Nuclear Planning Group, at a meeting in Virginia, drafted the first agreed NATO political guidelines covering possible initial use of tactical nuclear weapons by the alliance. (The Nuclear Planning Group consists of seven of the NATO defence ministers serving in rotation.) Though these were not made public, they said in effect that in response to a conventional attack which could not be stopped with ordinary forces, NATO might use tactical nuclear weapons. But it would do so in ways designed to minimize the chances of automatic escalation. As one NATO planner interpreted the agreement, low yield weapons would be used, and they would be used in 'tens rather than hundreds'. The enemy would be told of the intent to use restraint. The enemy command and control system would be deliberately spared, making it possible for the enemy commanders to control their forces. In subsequent meetings, NATO has taken up the question of how nuclear weapons might be used *after* their initial introduction. It is not clear from the public record what state of definition is arrived at in these consultations, or how often or to what degree changes are made in actual NATO operational plans. Beyond that, the relation of such operational plans to NATO political control is an entire additional issue. The communique of the May 1973 Nuclear Planning Group included the following:

> Concluding the first phase of a comprehensive work programme started by the Nuclear Planning Group in 1970, the Ministers considered the last of a series of studies dealing with possible defensive tactical employment of nuclear weapons by NATO in deteriorating situations after initial use has occurred. This study, sponsored by the Netherlands and the United Kingdom, examined questions of NATO nuclear defense in a maritime context. The next phase of the work programme, now in progress, involves the comparative analyses of the various studies for the purpose of gaining deeper insight into policy matters pertaining to the defensive tactical use of nuclear weapons. This step, in turn, is expected to contribute to the further refinement and elaboration of existing political guidelines [84].

At the end of the June 1975 Nuclear Planning Group meeting, the statement released sounded much the same. The defence ministers had reviewed studies 'dealing with the defensive tactical deployment of nuclear weapons in cases where initial use has not achieved its purpose'.

Indications that actual US planning probably had not changed that much with time were given by the following exchange in Congressional testimony in 1973:

> *Senator Symington:* Last month, Dr Walske, then Assistant to the Secretary of Defense for Atomic Energy, testified before this subcommittee. In this connection, he sent me a copy of a classified talk he had given in which he had discussed theater nuclear war. In developing a scenario to stop a Soviet attack against NATO, he said that there was a possibility of using nuclear weapons on the battlefield to stop a conventional Soviet attack.
>
> In the talk, he said, and I quote: 'It can be done with a limited expenditure of nuclear weapons – I'll shock you by saying a limited expenditure is [deleted] on military targets'.
>
> Do you agree with Dr Walske that it would take [deleted] tactical nuclear weapons to stop a Soviet attack? If not, how many do you envisage would be required?
>
> *General Goodpaster:* There are certainly certain types of Soviet attacks that would require expenditure of [deleted] to stop. If you are thinking of heavy concentrations and massive attacks, many of the weapons that would be employed, however, are those of very small

yield and generally when I discuss this, I try to discuss numbers of weapons together with their aggregate yield, because we are talking of very small weapons in relation to the bigger tactical weapon and certainly, in relation to the strategic weapons.

For example, one survey that I have made of the aggregate of our small battlefield weapon yield in Europe showed that the total yield is less than [deleted].

Now, these are distributed, of course, and discretely, applied against the targets of an attacking force. But you would have to recognize that against these heavy attacks, you could very well be required to use weapons in that quantity [85].

The number of individual warheads which Dr Walske had given was obviously rather high. General Goodpaster supplied further details:

Is there a finite number of responses and degrees of retaliation?

General Goodpaster: No, Senator. In our planning, our concentration is on developing the widest possible range of responses from which we could draw, depending on how the situation would develop, and that extends from the conventional end of the spectrum to add force right on up to whatever is required.

Of course, there are many established modes of operation. The covering operations, the initial delaying operation, the direct defense on the main battle positions, the mobile defense in that area. But then, over and above that, we would add, if needed, and if approved, the controlled battlefield use of nuclear weapons, possible air defense use of nuclear weapons, possible use at sea, going from there into use in a more extended battle area which might include use of close-in interdiction purposes, extending even to counter-nuclear, counter-air always in a tactical role against military targets and beyond that we have plans, prepared plans and capabilities for conducting the full-scale use against fixed targets of military significance and the threats against our command extending throughout the WTO area.

Senator Symington: Does NATO preplan responses?

General Goodpaster: The categories that I have just described, starting from the top down [deleted] are very fully preplanned.

Below that, our operations are partially preplanned because, in many cases, they would be directed against nonfixed targets, against mobile targets, or against logistic support facilities that had come into play because of the way in which the attack had been developed.

The remainder of that planning would have to be done in the situation as it developed. So, we have a range that extends from fully preplanned to partially preplanned options.

Senator Symington: Do preplanned responses contemplate the use of nuclear weapons?

General Goodpaster: Yes. The preplanned responses that I have described do that. Now, we do not use the term 'preplanned' in relation to the operations of our conventional forces, for example, on their main battle positions but those, too, are planned in our general defense plan and in the general defense plans of the subordinate echelons.

Senator Symington: Which nuclear weapons do you plan?

General Goodpaster: For the full range of nuclear weapons available to us. We have done preplanning at the higher end of the spectrum. [deleted] weapons, PERSHINGS, SERGEANTS, that type of thing, against [deleted].

Then, we have the ADMs, of course, that have been planned on a contingency basis, if directed, against identified locations for the creation of obstructions.

Then, for the small nuclear weapons, we have the partial pre-planning for their use against mobile targets.

Senator Symington: What is the purpose in the NATO stockpile of high yield weapons deliverable to long distances by missiles or air-craft?

General Goodpaster: One of the purposes is for use against enemy air-fields. Another for use against enemy nuclear delivery means, rail concentrations. [deleted].

Senator Symington: Is the role of the high yield weapon considered to be defensive, retaliatory, punitive, or preventive?

General Goodpaster: In terms of the use that we would make of it, it is a defensive use in that we are trying to destroy the enemy's military capability that he is either employing or that we calculate he may employ against our forces [86].

All this of course runs directly into the question of 'first use', the phrase used to denote the first initiation of use of tactical nuclear weapons in the European theatre. There will be no extended discussion of the question or quotation of sources here, but it is clear that in the years both before and after approved NATO political guidelines, NATO and the USA have always stated that they would initiate the use of nuclear weapons if this were felt necessary [87]. Even one of the most moderate and reluctant statements – quite different from the ones which can be quoted from military sources – puts this unequivocally. In 1963, Secretary of Defense McNamara, after stressing the need to upgrade NATO's conventional defences, set forth the basic US policy governing the use of tactical nuclear weapons in the defence of Western Europe which holds to this day.

> This does not mean that the NATO forces can or should do without tactical nuclear weapons. On the contrary, we must continue to strengthen and modernize our tactical capabilities to deal with an attack where the opponent employs such weapons first, or any attack by conventional forces which puts Europe in danger of being over-run [88].

More recently, the policy was summarized by Secretary of Defense Schlesinger.

> We deploy nuclear weapons to Europe for three major reasons.
> First, maintaining nuclear capabilities is essential to deterrence as long as the Warsaw Pact maintains roughly comparable theater nuclear capabilities. These weapons help to deter use of nuclear weapons by the Warsaw Pact and, along with the conventional and strategic nuclear forces, provide a general deterrent across the entire spectrum of possible aggression.

Second, should deterrence fail, our tactical nuclear capabilities provide a source of nuclear options for defense other than the use of strategic forces.

Third, in keeping with the flexible response strategy, we do not rule out the use of nuclear weapons by the US and its Allies if necessary to contain and halt major conventional aggression [89].

This question plunges one back once again into the particular NATO maze concerning tactical nuclear weapons – nearly all of it composed of contradictory elements – concerning declaratory policy and doctrines, the weapons, their numbers, design and location, planning and training for their use, targeting plans, motivations and interests of the USA and its allies in introducing the TNWs in the first place and in maintaining them since, the reluctance of particular NATO countries to actually have any used on their soil, and involving the connection with the strategic deterrent, warfighting conceptions versus those of deterrence etc.

One of the more interesting aspects of NATO thinking has apparently been a consistent notion that it could use tactical nuclear weapons without the USSR replying in kind.

> *General Goodpaster:* In the context that we are talking about, a large-scale non-nuclear attack against Western Europe, if we were to apply in a controlled way limited numbers of nuclear weapons sufficient simply to stop the attack and impose costs and losses on their attack echelons, my own feeling is that the probabilities would be much less than even that they would immediately carry that to all-out nuclear exchange involving their own homeland.
>
> *Senator Symington:* In other words, your testimony before this committee is that we could use nuclear weapons in Europe and that the Soviets in all probability would not respond to nuclear weapons?
>
> *General Goodpaster:* I would not go so far as to say in all probability. I think you have to think probabilities but I think there is an appreciable probability that they would not [90].

Similar opinions were expressed a year before in testimony by other military spokesmen [91].

Estimates of how many people would be killed in a tactical nuclear war in Europe depend on assumptions made as to how the war unfolds. Such estimates have ranged from a matter of thousands, assuming the war stops immediately after a few nuclear weapons have been used; to from two to twenty million, assuming extensive use of nuclear weapons with some restraints; to 100 million if there are no restraints at all. 'Collateral damage', the term so often used in discussing the 'modernization', 'improvement', or 'mini-nuclear' TNWs, is the term used for the killing of civilians and the destruction of cities as concomitant results when using TNWs on military targets. In discussions of the present stockpile and its possible mode of use, it is often suggested that warheads would ostensibly be fused to explode in the air at a height sufficient to at least reduce fallout. Such estimates date back to the first tactical nuclear weapon exercises conducted by SHAPE in 1954 and 1955. The first, *Sagebrush*, took place in

33

1954 in the USA, but in a manoeuvre area intended to simulate the European battlefield. The use of nine TNWs, five bombs and four shells, was simulated. The second exercise, *Carte Blanche*, took place on 20–28 June 1955, and is much more well known [92]. It took place in the areas of Belgium, the Netherlands, F.R. Germany and eastern France, and involved the simulated delivery of tactical nuclear weapons by aircraft. Altogether 335 nuclear weapons were 'dropped' within 48 hours, very many on airfields, and nearly half on each of the two 'opposing' sides in the exercise. The figures that were released for West German mortality (1.7 million) and casualties (3.5 million) are likely to have been ridiculously low in view of the number of weapons used and the probable high yields at the time. Since then, numerous tactical nuclear war games have been played utilizing computer simulations. Quite often nuclear weapons are invoked relatively early, and equally often damage to NATO territory, particularly F.R. Germany, is extensive. An excellent and voluminous study of such possible outcomes was published in 1971 [93].

As has been indicated, NATO tactical nuclear weapon warheads are kept in stockpiles. Since these are in a relatively limited number of locations, and in some cases various security measures used in conjunction with these locations make them obvious, it is very probable that the USSR knows where these sites are. It has therefore been argued that in a time of extreme crisis, the USSR would have an incentive to try and destroy these stockpiles before they could be used. The same point has been made concerning the airfields, at which NATO QRA aircraft are located, and even where other NATO nuclear-capable aircraft are. These two factors, which might make such sites themselves prime nuclear targets, are two of the concerns which have often provided the source for suggestions for restructuring or modifying NATO's own nuclear forces. The interactions are suggested in the 1971 Brookings report.

> Our tactical nuclear force structure is based on the 'discrete fire' concept; namely, that tactical nuclear weapons will be fired against specific or known enemy targets, as in conventional warfare, and that they will be controlled and fired from forward positions. This structure is another factor contributing to vulnerability.
> . . . Hence, a major fraction of our launchers would be in a belt within one hundred miles of the front. Both systems would be well within range of the Pact weapons and thus would be destroyed in an initial attack. Aircraft-delivered nuclear bombs are located on fixed, known airbases . . .
> Once the nuclear threshold were crossed, both sides would be under pressure to use their nuclear weapons quickly before they were destroyed, and to use them on targets far beyond the front lines in order to attack the enemy's nuclear launchers, as well as its reserve troops, supplies, airfields, communications, and supply routes. These circumstances would compound the problems of using these weapons in a controlled or measured way – and in particular of limiting exchanges once they had begun [94].

Two last relevant items should be included in the discussion of this section. Quite often, one sees reference to one or another NATO estimate of how long a conventional battle could be sustained before NATO would 'have to' make use of tactical nuclear weapons. Very often, these are dependent on assumptions of a particular depth of penetration or particular strength of attack, or the crossing of certain logistic positions. Quite often, they make reference to official NATO planning and briefings. The estimates range from less than a day to several (two to three) days, with frequent mention of 72 hours [95].

The final point is that one sometimes finds reference in the strategic literature to a NATO TNW capability as a deterrent against a possible WTO use of a chemical warfare capability, or in retaliation for such a use.

The following excerpt is taken from the 1973 Senate Committee on Foreign Relations report, on aspects of NATO TNW targeting [30].

The Nuclear Activities Branch of SHAPE formulates NATO's operational plans for the use of nuclear weapons. It operates under both US guidance and NATO guidance contained in a document entitled 'Concepts for the Role of Theatre Nuclear Strike Forces in ACE' (Allied Command Europe), originally adopted by the North Atlantic Council in 1970 and last revised in 1972. [Deleted].

On the basis of US and NATO guidance, SHAPE has drawn up what is known as the General Strike Plan (GSP). Under the GSP, the weapons systems available to conduct nuclear strikes within the European theatre are the land based aircraft, sea based aircraft, fleet ballistic missiles and Pershing missiles which will be assigned to SACEUR by the NATO countries in times of war [deleted].

NATO provides for two broad categories of nuclear warfare – selective use and general nuclear response. While both require consultation within NATO as well as US Presidential approval, the GSP provides SACEUR with the capability for deliberate escalation with which he can expand the scope or increase the intensity of combat in order to strengthen his defensive effort or to insure the survival of his forces within the scope of authority the President has authorized him to use. The overall intent of the NATO nuclear concept is to provide for the selective and controlled application of force.

NATO strategy specifies that selective employment would be used on a controlled or limited scale either for demonstrative or tactical purposes. While the objective of selective use is to confront the aggressor with the risk of escalation with the aim of making him halt the attack or withdraw, NATO guidance also [deleted].

In the course of a Defense Department briefing, we were given hypothetical illustrations of selective nuclear use. One illustration envisioned a limited Soviet penetration in the central region of Germany which could not be contained by conventional NATO forces in the immediate area. Under such circumstances, the briefers said, SACEUR [deleted]. Another hypothetical illustration of the selective use of nuclear weapons was referred to as the theatre nuclear warfare option. Its objective would be to [deleted]. Targets would include [deleted].

Theatre nuclear warfare is the ultimate level of nuclear warfare which SACEUR would be authorized to conduct, subject of course to Presidential authority. A decision to resort to theatre nuclear warfare would not be made unless [deleted].

The second major category of nuclear warfare for which the GSP provides is general nuclear response which contemplates massive nuclear strikes against targets [deleted] in order to [deleted]. The general nuclear response would not be undertaken by theatre nuclear strike forces alone but only in conjunction with the execution of the Single Integrated Operations Plan (SIOP), the US Joint Chiefs of Staff plan for the widespread synchronized use of US nuclear weapons in an all-out war. Accordingly, responsibility for carrying out NATO's general nuclear response falls on US strategic forces based outside Europe.

Targets to be struck in the general nuclear response are part of what is known as the 'Scheduled Program' and include the Priority Strike Program (PSP) and the Tactical Strike Program (TSP). The PSP is comprised of those strikes of highest priority to SACEUR. They include [deleted]. Another group of targets covered by the PSP are those included in the Allied Command Europe's Critical Installation List. There are approximately [deleted] targets on the list of which theatre nuclear forces would strike about [deleted] under the PSP [deleted]. The TSP is a list of about [deleted] nuclear strikes against [deleted]. These are [deleted]. All PSP and TSP strikes have been coordinated, or 'deconflicted', with the US SIOP target list maintained by the Joint Strategic Targeting Center at Omaha.

We were told by Defense Department officials that a general nuclear response might be required under two contingencies. The first would be [deleted]. The second would be [deleted].

D. QUICK REACTION ALERT

Some weapons, both American and foreign, with nuclear warheads are kept loaded at all times on quick reaction alert (or 'QRA', as they are known). The number of aircraft, missiles and submarines on QRA depends on what is called the 'force generation level'. There are three such levels established by SACEUR's General Strike Plan.

At the normal peacetime QRA level, there are enough weapons on QRA to hit about [deleted] percent of the targets on SACEURs Priority Strike Program, or PSP, comprising the highest priority strikes of concern to SACEUR. The guidelines for NATO theatre nuclear forces specify that 'about [deleted]' land-based aircraft [deleted] Pershings and [deleted] Polaris and Poseidon missiles are to be kept loaded to meet the normal peacetime QRA requirements.

At the advanced readiness level, there are enough weapons on QRA to hit [deleted] of the targets on PSP. About [deleted] land-based aircraft, [deleted] Pershings (only the US and the Germans have Pershings, a weapon which has no conventional capability) and [deleted] Polaris and Poseidon missiles assigned to SACEUR are kept loaded.

At the maximum posture level, all remaining land-based dual-capable aircraft and all sea-based strike/attack aircraft are placed on QRA bringing the totals to [deleted] land-based and [deleted] sea-based aircraft plus [deleted] British Vulcan medium bombers. Thus, at the normal peacetime QRA level between [deleted] percent and [deleted] percent of the nuclear capable aircraft in Europe are on QRA, at the advanced readiness level that percentage rises to between [deleted] percent and [deleted] percent, and at the maximum posture level it is [deleted] percent.

Foreign aircraft are among those kept on QRA loaded with US nuclear weapons and ready to be launched immediately. As of the time of our visit, there were [deleted] US and foreign aircraft on QRA – [deleted] US and [deleted] foreign. By country, in [deleted]. This total of [deleted] did not include [deleted].

Also on QRA at the time of our visit were [deleted] batteries of Pershing missiles with [deleted] launchers. These are maintained by US forces although foreign forces which have Pershings have been trained in the employment of nuclear weapons using dummy warheads.

1974: An attempt to reduce US tactical nuclear weapons in Europe

Pressures began to develop around 1970 for a rationalization and reduction of the US tactical nuclear stockpile in Europe. The motives of various interested groups were different. One of the earliest considerations introduced was the cost of the force and its attendant US personnel [96]. More significant arguments which followed were:

— Vulnerability of such elements as QRA aircraft and forward based nuclear artillery,
— Number and yields of the TNW stockpile were too high to be really functional,
— Command and control of certain elements, such as forward-based nuclear artillery might prove very difficult once use had been initiated,
— The general Soviet nuclear capability had grown to the point where the original rationales for the weapons were no longer supported by an overwhelming US strategic nuclear superiority,
— Security of the stockpile,
— The original largely political impetus for the large growth in numbers of the stockpile, both from European sources and from various domestic US pressures, was for reasons largely irrelevant to European military requirements.

A report prepared by an *ad hoc* US Congressional group in 1971 [97] and a Pugwash meeting on the subject later in August 1971 both produced no response whatsoever [98]. During the next four years in which this subject was under active discussion the USSR showed no interest in the subject at all. An influential Senate Committee on Foreign Relations report appeared in December 1973 [99] and the Committee held very influential Hearings in March and April 1974 [100]. Legislation passed in August 1974 required the Secretary of Defense to provide a report on the subject, which was presented to Congress in April 1975 [101]. Finally, a study on the subject was released by the Brookings Institution in December 1974 [102]. The Brookings report suggested that US tactical nuclear weapons in Europe could be reduced from 7000 to 2000. In the Senate Committee on Foreign Relations Hearing, former Assistant Secretary of Defense Enthoven suggested that the force could be reduced to 1000 weapons. Pressure continued well into 1975, long after the issue was actually decided [103].

The internal debate within the government and the US Department of Defense took place in early 1974. Most significant was the indication that Secretary of Defense Schlesinger and at least some sections of the Department of Defense and of the military services favoured at least partial reductions. There had been earlier indications that at various times influential segments of NATO and of its Nuclear Planning Group had also been prepared to give up nuclear strike aircraft and certain other segments of the NATO nuclear forces that were the most vulnerable or had the most drawbacks from one or another point of view. Starting in the spring of 1974, there was a very well reported series of articles in the public press indicating that Secretary of Defense Schlesinger was considering reducing the number of tactical nuclear weapons deployed in Europe. The stories spanned a spread of six months. All were specific. They were clearly written with access to persons in the government. The Secretary of Defense stated explicit reasons why the reductions would be desirable.

> Defense Secretary James R. Schlesinger is planning on reducing the sizable stockpile of nuclear weapons in Europe as well as cutting back on the number of atomic-armed missiles and planes kept on alert.
>
> To the European allies and the Pentagon, according to associates, Mr Schlesinger has begun pressing the view that the United States has more nuclear weapons in Europe than it can effectively use. He also is known to believe that the United States is in effect encouraging an atomic exchange by keeping so many planes and missiles on a nuclear alert.
>
> At Mr Schlesinger's direction, therefore, the Defense Department is headed for the first major revision in its nuclear posture in Europe since it started stationing nuclear weapons there nearly 20 years ago.
>
> The United States has about 7000 nuclear warheads in Europe – a figure that high-ranking Defense officials believe in retrospect was an arbitrary reaction to military and political pressure.
>
> Starting in the mid-nineteen fifties, the Eisenhower Administration . . . came to the conclusion that tactical nuclear weapons provided a way of offsetting Soviet superiority in Soviet manpower.
>
> This concept found ready support in the Joint Congressional Atomic Energy Committee, whose members saw, among other things, greatly expanded production of nuclear weapons, which consume relatively large amounts of fissionable materials, as a way of expanding Atomic Energy Commission activities in their home districts.
>
> As analysed by Mr Schlesinger, the original military considerations that led to the build-up of nuclear weapons in Europe have been overtaken by events or were based on mistaken assumptions.
>
> Mr Schlesinger also believes that while the nuclear shield remains an essential ingredient in the strategic posture of the Atlantic alliance, it has become increasingly unlikely that either side will resort to nuclear warfare now that the Soviet Union has reached rough parity with the United States.
>
> Even in event of a resort to nuclear warfare, Mr Schlesinger is making clear to associates he does not foresee large-scale use of atomic weapons as likely.

To a large extent the military requirement for 7000 weapons was based on the assumption that if the Western allies were driven to a nuclear response, hundreds of weapons would be delivered against targets in the rear support area of the Warsaw Pact forces.

In challenging this assumption, Mr Schlesinger argues that if the United States were to drop a hundred or so nuclear weapons, the natural Soviet response would be to retaliate with an all-out nuclear attack [104].

However, by November and December 1974, it would appear that the issue was foreclosed. There would be no reductions in the US tactical nuclear weapons stockpile in Europe [105]. Rather, it was announced that a *larger* number of Lance missiles would be placed in F.R. Germany, and it appeared that at least some of these were to be a true increment, and not a replacement. This finally drew a Soviet comment, of protest [106]. The reasons which determined the maintenance of the *status quo* were political and not military, and indicated the dominating influence of the US Department of State and of the other NATO allies on the question. *Any* reduction, no matter what its nature, and *regardless* of the substantive questions involved, were seen as upsetting to 'the perceptions' of NATO allies on the degree of US 'committedness' and US military support. Any actual analysis of relative US–USSR military balance, or of the tactical nuclear weapons themselves was largely irrelevant. At the same time, a Dutch proposal to a NATO defence ministers' meeting that tactical nuclear weapons be de-emphasized in European defence, and that such proposals be part of MBFR negotiations, apparently met with ambiguous response judging from publicly available reports. It was reportedly rejected by the NATO defence ministers. Nevertheless, Secretary Schlesinger simultaneously indicated that such a proposal might be offered at the MBFR talks [107]. NATO subsequently offered a reduction of 1000 TNWs, composed of some aircraft and some surface-to-surface missiles, in exchange for the proposed withdrawal of a Soviet tank army and its complement of tanks – 68 000 men and 1700 armoured vehicles – from the German Democratic Republic [108]. The USSR and the WTO – which had all along been pressing that Western tactical nuclear forces be drawn down as part of any conventional weapon reductions negotiated in the MBFR talks – rejected the proposal.

The degree to which the US Department of State maintained control over the issue, and its formalistic nature was coincidentally demonstrated again within a few months, in mid-1975. It was again indicated that some months earlier the Department of Defense had presented President Ford with a plan to reduce and rearrange the TNW stockpile in Europe. 'Defense Department circles regard the stockpile as too large for military requirements and composed of too many obsolete weapons' [109]. In this case, the proposal was specific and quite limited: to remove all the nuclear warheads for Nike-Hercules anti-aircraft missiles from Italy and Greece by 1 July 1975. In both cases, the proposal was blocked by the US Department of State for reasons involving domestic political considerations in Greece and Italy. Even legal considerations could be brushed aside.

At the time of the Greek–Turkish confrontation over Cyprus a year ago, the United States removed atomic warheads from Greek and Turkish fighter-bombers and transferred atomic warheads from Nike Hercules units manned by Greek forces.

When Greece withdrew its forces from the NATO military command last year a legal question arose as to whether the United States could station atomic warheads there for use by Greek forces. Under the 1954 Atomic Energy Law, atomic weapons may be stationed in a foreign country only under arrangements for mutual defense.

Again largely for political reasons, State Department officials have tended not even to raise this legal question and the atomic warheads have remained in Greece. In both Greece and Italy, the warheads, while officially under US control, are assigned for use by the forces of the two nations [110].

The only changes that could be made were substitutions and rearrangements in the TNW stockpile assignments, such as the replacement of some QRA aircraft by additional US Poseidon missiles which took place at this time.

The next time the issue of possible reductions in US tactical nuclear weapons was raised, by then Secretary of Defense Rumsfeld to a January 1976 NATO Nuclear Planning Group meeting, it was apparently indissolubly linked with the simultaneous replacement of older TNWs by a new generation; 'replacing old weapons with smaller and more accurate arms', and to replace 'big, dirty, and inaccurate' TNWs with 'clean, neat, and accurate ones' [111].

Another instance of difficulty in altering the status quo in a situation involving the overseas basing of tactical nuclear weapons developed in 1977 concerning the new US administration's efforts to remove such weapons from South Korea. In this case, plans to remove tactical nuclear weapons under US Army and US Air Force control in South Korea drew protests from both US military and US Congressional sources, (as well as from South Korea) [112].

In 1970, a US Senate committee report stated the following:

> The placement of these weapons abroad was undertaken at a time when relationships between the super powers was one of inequality: when the vast preponderence of nuclear power rested with the United States . . .
>
> Most of the weapons were deployed initially during the mid- or late-1950s. Often the original missions of the weapons have been changed. In most countries over the past few years the overall level or number of the warheads maintained has increased; and there is no case where warheads have been totally removed from a country.
>
> In but one known case – this because of a change in delivery systems – has the number of such nuclear weapons been reduced [113].

The safety and security of nuclear weapons in European or other overseas locations

In the early 1960s, the question of nuclear weapon security became one of particular interest and concern to US government policy makers. This occurred at the time of the new Democratic administration of J. F. Kennedy and the tenure of Secretary of Defense Robert McNamara. Coincidentally, the next few years would see an increase of several fold in the number of US tactical nuclear weapons in Europe. However, at this time, security and safety of nuclear weapons meant only two things. In the order of their perceived danger and of the concern exhibited about them, these were, first, the problems of unauthorized use, command and control and catalytic war and, second, physical accidents that might occur to nuclear weapons in their storage, handling, shipment, or while deployed.

In the 1970s, two additional kinds of concerns were added to the category of nuclear weapon safety and security. The first was a real one, in that such attempted incidents had already historically taken place. This was the danger of capture of nuclear weapons by dissident military groups during coup d'état attempts or similar situations. The second was still a theoretical concern, as it was an anticipation of possible dangers which might take place in the future, but which, as far as publicly available evidence indicated, had not yet occurred. This was the theft or capture of nuclear weapons by terrorist groups.

Regarding the earlier concerns, fears of unauthorized use or of command and control failures led to the development of positive control systems, electronic locks, and various other technological devices and administrative procedures. (See section on PAL systems, below.) The purpose of these efforts was to prevent unauthorized use either by what is referred to as an aberrent or demented individual, or by a military commander during wartime conditions either as a result of a communications failure or, against orders, on his own judgement. To some degree, these same concerns had also contributed to the interest which led to the earlier and abortive 1958 'conference of experts on surprise attack' held between several NATO and WTO nations (primarily the USA and the USSR).

Concern over the effects that might follow a nuclear weapon accident had produced even earlier efforts at control. The US Atomic Energy Commission had been sufficiently concerned to carry out a series of nuclear safety experiments to determine the safety of nuclear weapons in case of accident [114]. Data concerning nineteen tests at the Nevada Test Site 'which resulted in a measurable nuclear yield' were released. (There may, therefore, have been additional tests as well which resulted in no measurable yield.) 'These experiments were designed to provide data about the behaviour of the various weapons and devices under conditions like those which might occur in case of fire or accident' [115]. The experiments were begun in November 1955, and the nineteen announced safety tests were completed in October 1958. Weapons were purposely dropped from aircraft, exploded with dynamite, set on fire, and involved in vehicle crashes. Reviews of nuclear weapons accidents, as comprehensive as the available public evidence permits, have appeared in other SIPRI publications [116, 117]. The more recent of these surveys indicated that there have been at least 113 nuclear major accidents (more serious) and minor accidents (less serious) since 1950. The available information pertains primarily to accidents and incidents of US nuclear weapons and secondarily to those of the USSR. Evidence concerning France and the UK is ambiguous. It is clear that accidents and incidents involving nuclear weapons are frequent, occurring on a worldwide average of perhaps one every few months. They occur to probably all the different kinds of nuclear weapon systems while these contain nuclear warheads, and in probably every kind of activity in which these weapon delivery systems take part: in silos, in the air, on the sea and, in port, under the sea-surface, on land, and so forth. However, the available data deals for the most part with accidents of strategic nuclear weapon systems, and it is not possible to provide any meaningful breakdown of nuclear weapon accidents and incidents to indicate how many of them might have involved tactical nuclear weapons in particular, of the USA, the USSR, France and the UK.

However, it is the two newer concerns – capture of nuclear weapons by dissident military or by terrorist groups – which have been both the focus of recent interest and the motive for increased efforts at upgrading security of nuclear weapons, most particularly in Europe. For it was incidents that took place in Europe which literally forced consideration of both of these kinds of situations.

The first of the two problems to receive any public attention was that of the attempt to seize nuclear weapons by a dissident military group. In 1968, D. G. Brennan reported on an incident which developed involving dissident French military forces and a nuclear weapon that was at the then French test site at Reggan, Algeria [118]. The academic literature on strategic problems or on nuclear proliferation took absolutely no notice of the question, and as far as the public record would indicate, neither did any government circles. In December 1970, a US Senate Committee on Foreign Relations subcommittee which was looking at US security agreements with other nations, at overseas commitments resulting from these, and at the overseas location of US tactical nuclear weapons, reported that 'The Subcommittee Staff received reports on problems that had developed during periods of crisis within a number of countries where our nuclear weapons are stored' [119]. In the event, despite reference to 'a number of countries', the information regarding only one incident in one country was released. The incident took place in Greece, in April 1967, in the course of the coup being carried out by a military junta. Military units under the command of the junta surrounded a depot of US nuclear warheads and were withdrawn after a period of US protest [120, 121]. It was reported that 'the Defense and State Departments give little importance to the incident' [122]. The official US position was that even if the weapons had been seized, they could not be activated without US help – ostensibly because of PAL (Permissive Action Link) or other systems associated with them. Two other incidents were subsequently described by a staff member of the Committee.

> . . . In a country that had recently experienced a revolution, Americans in nuclear custodial units recalled how a tank from the host country army positioned itself in front of a QRA aircraft piloted by a host country air force pilot. In the uprising the army and air force were on opposite sides of the battle. In another country, a local general asked

his American counterpart if he could use a tactical nuclear weapon in a fight then going on with a neighboring country [123].

QRA aircraft are normally prevented from taking off by being positioned at the end of runways and being screened off by a wire fence, at times a double wire fence. There is also an armed US guard present to prevent unauthorized seizure of the unarmed bomb. In addition, for the bomb to be armed, US personnel would have to adjust the combination lock command and control devices on the weapon. Attention was again drawn to this situation by Greek–Turkish hostilities accompanying the July 1974 Cyprus crisis. US nuclear warheads were reportedly removed from Greek and Turkish aircraft that were assigned to Quick Reaction Alert (QRA) missions. Stand-by preparations were made to fly in US Marine detachments in helicopters, if necessary, to remove the nuclear warheads entirely [124]. In a short report soon after, Congressman Long reported that

> It is a little known fact that the United States deploys nuclear weapons at some foreign locations hundred of miles from the nearest American installation. These are NATO bases run by the host nation, and the host nation is responsible for security. At these NATO bases the United States stations a small team of security guards to perform custodial duties for the nuclear weapons [125].

A member of the US armed forces had indicated similar details in a letter to Senator Stuart Symington.

> I am aware of the existence of [US] tactical nuclear warheads . . . in Greece . . . and . . . in Turkey . . . No more than four–six US soldiers guard the bunkers which store the nukes! Most of the troops (about 40 per detachment) are housed about a quarter of a mile from the bunkers and could easily be isolated from the warheads [126].

A Joint Congressional Atomic Energy Committee report referred to the same stimulus.

> . . . the Greco-Turkish war over Cyprus in July 1974 and the consequent overthrowing of the Greek Junta highlighted renewed Congressional concern over the security of weapons which might be deployed in vulnerable and outlying areas in countries where the political situation may become unstable [127].

Finally, in 1975, the US Department of Defense in its Nuclear Weapons Security Primer officially corroborated this situation in general terms.

> (d) Host Nation Takeover. Some US nuclear weapons have been earmarked for use by our allies. Physical security of these weapons is provided by the host country although custody is maintained by small US detachments. In addition, weapons are stored in the sovereign territory of allied nations for use by US forces, if required. Thus,

changes in attitude of allies must be evaluated carefully and on a
continuing basis so that appropriate action can be taken as warranted
should this threat arise [128].†

The number of these weapons vulnerable in this way, as well as to terrorists, is
of course sizable. As was indicated, about two thirds of the over 100 nuclear
weapon storage sites in Europe are for weapons to be used by Allied forces.
Congressman Long indicated that Quick Reaction Alert (QRA) aircraft
'nuclear-loaded-ready-to-fly-aircraft . . . are deployed by the hundreds around
the world . . . In one small country I have looked into, the US deploys nearly
700 nuclear weapons. At one Air Base in this same country we keep 82 nuclear
weapons . . . At another overseas location I have found that over 200 nuclear
weapons are in storage' [129]. The NATO theatre was not the only area to
which Congressional concern was directed. Attention was also drawn to South
Korea.

> The nuclear weapons, known here as 'special ammunition' are
> considered particularly worrisome in South Korea because of the
> unpredictable and sometimes bellicose governments on both sides of
> the 38th Parallel and the close proximity of the weapons depots to the
> dividing line. The depots are reported to be within 35 to 50 miles of
> the Demilitarized Zone.
> According to informed sources, nuclear warheads have in the
> past been flown almost routinely to the edge of the DMZ by helicopter
> for training exercises [130].

It is clear that US administration policy has changed in the last few years on
these issues and to some degree has been responsive to Congressional pressures.
A 1975 Senate report indicated that earlier pressure by the Joint Committee on
Atomic Energy had resulted in nuclear weapons being removed from QRA
aircraft at a particular US base in Europe by March 1973 [132]. The earlier
1970 Senate subcommittee report had indicated much less administration sym-
pathy for any examination of previously disregarded issues involved in overseas
basing of TNWs.

> In at least one instance, security classification has been used to
> prevent legitimate inquiry by the proper committees of the Senate
> into matters which the Executive Branch did not want to discuss.

---

† In 1974, a retired high-ranking US military officer offered a rather puzzling piece of testimony:

> Now, on the matter of the Government, this lesson that we have learned from the
> Greeks, our NATO ally, should tell us something about the vulnerability of any of our
> military forces in a foreign country. Even when we think of them as our staunch and
> true friends, if they tell us we cannot take our planes off the ground, we cannot let our
> ships leave, we wouldn't be able to get them out of a foreign country even if the Govern-
> ment of that country did nothing physically to restrain us. We might have to fight our
> way in to get back our nuclear weapons [131].

The statement seems very implausible – that US aircraft or ships would not in fact leave and take
their nuclear weapons with them – but it was never rebutted by any Department of Defense or
administration spokesman.

In one striking instance, the Executive Branch, by means of the use of classification, sought to prevent any discussion in closed session by the Foreign Relations Committee, apparently because of concern over the context in which the information in question would be discussed.

At issue was the placement of United States tactical nuclear weapons in foreign countries. The Committee's interest and concern was directed toward the foreign policy aspects stemming from the placement of nuclear weapons in a foreign country.

The Subcommittee sought information with respect to the understandings made with the host governments having to do with the introduction of United States nuclear weapons, the manner in which they would be used, the manner in which they were to be protected, and the manner in which they could be removed from those countries.

Inevitably, any such discussion led to the military reasons behind the introduction of such weapons, also knowledge about those weapons on the part of citizens of the host country.

There can be no question that the placement of nuclear weapons in a foreign country automatically creates – in itself constitutes – a new relationship between that country and the United States.

The stationing of nuclear weapons in foreign countries represents a special kind of commitment between the United States and the host country. In almost every one of these countries a veil of secrecy hides the presence of such weapons. Nowhere is this veil stronger than in the United States.

Most people here are unaware of the fact that United States tactical nuclear warheads have been and are stationed in countries all around the world [133].

If nuclear weapons continue to proliferate to additional countries and if the phenomenon of the military coup continues at its present rate, it seems a fair assumption that these two events will interact with increasing frequency in future years [134].

The second of the two more recent concerns, the potential attempt to seize nuclear weapons by groups identified as 'terrorist', has received by far the greater amount of both public and governmental attention. Perhaps this is paradoxical, or perhaps it is because it is a less sensitive political issue, the identified actor not being an established government, ally, or a segment of its military forces. The report of former Secretary of Defense Schlesinger on tactical nuclear weapons in Europe stated that 'The potential threat to nuclear weapons by terrorist elements has been of serious concern since the start of international terrorism in 1970' [135]. Congressional and other US governmental sources have often referred to the 1972 Munich Olympics episode as a critical incident prompting re-evaluation.

Ever since the terrorist action during the 1972 Olympics at Munich, the threat of US deployed weapons due to terrorist attack became of increased concern to the Congress [136].

It is very likely that these same events prompted additional security measures and planning for depots of tactical nuclear weapons in France and the United

Kingdom as well – and perhaps even in the USSR. It might be useful at this point to indicate in their entirety the categories of threat to the security of stored nuclear weapons as spelled out by the US Department of Defense in 1975.

(a) *Covert Penetrators*. This threat would arise from the desire of a hostile element to infiltrate personnel into a nuclear weapon site for the purpose of gaining intelligence information on our weapons or sabotaging them.

(b) *Unauthorized Use*. Only the President can authorize use of nuclear weapons. Our security measures must be designed to preclude use of a nuclear weapon by our own people until Presidential approval has been granted.

(c) *Psychotic Attack*. This threat could result from one of our own people becoming deranged and attempting destruction or unauthorized detonation of one or more nuclear weapons.

(d) *Host Nation Takeover*. Some US nuclear weapons have been earmarked for use by our allies. Physical security of these weapons is provided by the host country although custody is maintained by small US detachments. In addition, weapons are stored in the sovereign territory of allied nations for use by US forces, if required. Thus, changes in attitude of allies must be evaluated carefully and on a continuing basis so that appropriate action can be taken as warranted should this threat arise.

(e) *Terrorist*. International terrorism, during the past few years, has demonstrated that it is a force to be reckoned with. Because of the violent, efficient, and rapid manner by which terrorist acts have been executed, terrorism poses a potential threat to our weapon stockpiles and is driving most of the new security upgrade efforts [137].

We have already discussed the second of these, 'Unauthorized Use', and the fourth, 'Host Nation Takeover'. The third category, 'Psychotic Attack', essentially falls within the same problem of unauthorized use. This paper does not go into the rather extensive material available from US sources on its 'Personnel Reliability Program', which removes individuals *already serving* as nuclear weapons personnel for various infractions, such as drug use, alcohol abuse, medical–psychiatric reasons or other infractions. Data from two recent reports indicate disqualification of 5128 out of 119 625 US personnel in 1975, a rate of just over 4 per cent, and 1868 of 23 450 personnel in Europe in particular, around 8 per cent [138]. When one realizes that these are *annual* figures, the numbers of individuals disqualified is sizable. The duties and knowledge of these personnel are unknown, but it raises some troublesome questions. It is possible to assume roughly similar rates of disqualification in French, UK and Soviet nuclear weapon personnel. (However, because the total numbers of nuclear weapons personnel in these other countries are liable to be very much lower, tactical nuclear weapons far fewer and less dispersed, it is possible that original personnel selection criteria might be stricter and subsequent disqualifications fewer.)

United States Congressional concern on the subject of nuclear weapon security has been extensive [139], and the public literature is growing as well [140].

Congressional investigations had little trouble demonstrating numerous examples of lax and faulty security in an area in which official statements for over ten years had stressed nothing but the great care which is taken and the security with which weapons are guarded. In practice, things often turned out to be a bit more careless and not as uniform as the official pronouncements would have had one believe [141]. In addition, there was the problem that certain aspects of those security measures that have been in force for years make the storage sites quite visible. At some sites, no changes had been made a year after a first site visit in 1972. An indication of the added attention is the record of budget expenditure for nuclear weapon security since 1970.

Fiscal Year	$, millions
1972	3.8
1973	9.6
1974	12.0
1975	29.6
1976	93.9
(projected)	
1977	135.8
1978	76.7 [142]

In addition, the USA estimated that an additional $146 million might be spent by NATO for upgrading nuclear storage sites in its 'five year infrastructure program' that was being planned in 1975 [143].

Congressional suggestions for improvements focused on:

— Improvement of the physical security at existing sites.
— Withdrawal of weapons from more vulnerable areas to more secure areas.
— Total withdrawal of weapons from nations where political conditions have become unstable [144].

Some nuclear weapon storage sites were in fact closed down and consolidated. It was reported that the US Department of Defense closed 97 nuclear weapons storage sites in the calendar years 1974 and 1975 [145]. There were no indications of how many, or if any, of these were in Europe, but apparently at least several were. However, extensive consolidation would produce a situation considered unstable in event of a NATO–WTO military crisis, ostensibly inviting pre-empting attacks if NATO nuclear weapon stockpiles were in a very limited number of locations. This runs counter to NATO interests for strategic reasons, which would in fact dictate maximum dispersion consonant with control, and the two conflicting motives have begun to appear in the most recent policy statements. Nevertheless, one now sees nuclear weapon security mentioned in nearly all official statements regarding NATO and tactical nuclear weapons.

> . . . it has become evident that preparations must be made to defend the weapons against an overt, violent attack by a larger group using sophisticated guerrilla tactics.

This potential threat has caused a comprehensive reexamination of our storage site security. Both short and long term strengthening and restructuring of procedures and requirements have produced actions such as revised security standards; reduction of weapons movements; consolidation of storage sites; increased site defense and training of security forces; improved physical layouts to include lighting and road barriers; and improved weapons security devices . . .

If NATO is to improve its deterrent posture for the future, the following major conditions must be met for theatre nuclear forces:

First, we must reduce their vulnerability to sabotage, seizure and conventional assault. Measures are already underway to ensure this condition in cooperation with our Allies . . .

Future Goals. (*a*) Theatre nuclear force improvements which are under review include the following: . . .

— Continued improvement in security of nuclear weapon storage sites and, where militarily sound and economically advantageous, consolidation of sites [146].

The nuclear weapons laboratories, under the auspices of the Energy Research and Development Administration have done extensive advanced development in safety, security and command and control . . . Improved disablement, nonviolent destruction, PAL sophistication and automated code handling systems exemplify improvements available to our weapon stockpile [147].

Additionally, we are making every effort to ensure that our theatre nuclear forces and their support are structured to provide a high level of security against surprise attacks, terrorists and unauthorized use [148].

Peacetime Security and Storage. The security of nuclear weapons in peacetime is continually being reviewed. In planning and implementing nuclear storage on a world-wide basis, we must balance such factors as survivability of warheads in case of a surprise attack, security of individual sites under a terrorist attack, capability for weapon dispersal in a crisis, and cost. As a result of the latest site-by-site review, we have closed or plan to close a number of sites for a savings of $20–30 million annually. We will further upgrade the physical security of the remaining nuclear storage sites at a cost of about $1.5 million per site.

We are also making significant advances in the safety characteristics of and security devices attached to individual warheads [149].

However, in addition, nearly every mention of 'modernization' of tactical or theatre nuclear weapons in recent years has referred to the provision of additional safeguards in new weapon designs. The problem is, as the above example indicates, that these new designs are often simultaneously – but not always, as in the case of air-dropped TNW – radiation-enhanced weapons, and the budget appropriation is for the entire design, safety features *and* radiation enhancement. For example, 'The new 8-inch nuclear projectile, XM753 . . . design incorporates improvements in command control, safety, and security, so that it is more secure in stockpile storage and more responsive to control on the battlefield' [150]. 'The installation of the new electronic safeguards involves virtually rebuilding a weapon, General Giller said, Only those weapons earmarked for replacement with advanced models are involved' [151]. Nevertheless, at least some present tactical nuclear weapons appear to have several additional security safeguards in addition to PAL systems. One is a small removable portion which makes the weapon inoperable and which can be removed or even removed to another location. There are also a number of different devices, some in the weapons, some in the possession of the personnel in custody of the weapons, that would make the weapon inoperable or actually destroy them [152]. New research has tended to focus on automatic tamper-proof and self-destruction devices, to be contained inside the weapon. If someone unfamiliar with the method of handling the weapon were to attempt to operate it, the weapon would be disarmed, and perhaps destroyed. These systems would replace current combination lock command and control devices.

Modernization of US tactical nuclear weapons, the new debate, 1970–77

The review of material in this section concerns development of a more or less new generation of US tactical nuclear weapons. However, over the years, dating back to 1958, this subject matter has gone under a series of different descriptive labels. The prospective nuclear weapon designs or munitions have been variously called 'neutron bombs', 'clean weapons', 'enhanced-radiation weapons' or 'mini-nukes'. In the period 1974–76, when the request for appropriations for these were meeting resistance in the US Congress, they were nevertheless what was being referred to when proponents spoke of 'modernizing' or 'improving' tactical nuclear weapons. At times, they were inherent in the appropriations request for a particular weapons, for example, the AFAP (Artillery Fired Atomic Projectiles).

A precise definition of these weapons is not available. A 1973 report to the US Senate Foreign Relations Committee used the following:

> . . . a new generation of tactical nuclear weapons which combined low and variable yield possibilities with enhanced radiation character-istics and which could be used with dual capable artillery and later guided or other 'smart' bombs [153].

The possibility of newer and more accurate guidance systems is a recent addition, fortuitous and coincidental. The critical element is the warhead design. The concept for them was there well before laser or other guidance possibilities came along which could be added to the delivery vehicle to increase the accuracy of the bomb, missile, or artillery shell. It is important to understand that the suggestion for such warheads goes back to the period 1958–60. The question of improved guidance, which was a part of the requests for US authorization and appropriations for these kinds of warheads in 1973 and 1974 has in some cases been put aside and in others received less emphasis in the more recent requests in 1976 and 1977. In defending the concept and the requests for approval for development and procurement of these weapons former Secretary of Defense Schlesinger repeatedly stressed that 'There is no accepted definition of the term "mini-nukes" '.

It should be noted that smaller low yield tactical nuclear warheads are not new. We had the Davy Crockett, a very low yield nuclear weapon system, in the stockpile from 1961 to 1970, and currently have very low yield Walleye air-to-surface missiles, ADMs, artillery rounds, and bombs. Thus, the concept of the 'mini-nuke' that some writers have portrayed as representing a radically new family of new designs of extremely small size does not exist as such. We continue to seek more accurate systems and more lethal effects from lower yields to improve military effectiveness and decrease unwanted damage. Such improvements could increase the credibility of our tactical nuclear stockpile through better warfighting capability, thereby enhancing deterrence . . .
As indicated earlier, a new 'mini-nuke' program does not exist . . .
There is no accepted definition of the term 'mini-nuke' [154].

Back in 1963, Dr Alan Enthoven, then Deputy Assistant Secretary of Defense for Systems Analysis had revealed that there were small nuclear weapons whose yield was less than the largest conventional explosive weapons [155]. It was recently indicated that the 'Davy Crockett' warhead had been of 'a few hundredths of a kiloton'. And in 1971, Scoville and York reported its yield as 'tens of tons of TNT, not kilotons' [156]. The USA is presently reported to have at least four kinds of subkiloton tactical nuclear weapons, 155 mm artillery shells, some of the 8-inch artillery shells, some of the bombs, some of the Lance warheads, and some of the ADMs [157]. It is perhaps this situation that enabled the USA to make what seemed to be an important statement in 1974, and which at the time was nearly universally misread and misunderstood to mean precisely the opposite of what it said. The statement was made by Ambassador Joseph Martin at the CCD in Geneva [158]. A newspaper report of it stated that 'The United States gave assurances today that it would not develop a new generation of miniaturized nuclear weapons that could be used interchangeably with conventional weapons on a battlefield' [159]. On the same day, Dr Iklé, then head of the US Arms Control and Disarmament Agency said, 'We have no intention to move in a direction that could blur the distinction between nuclear and conventional arms' [160]. What the US statement at the CCD made clear was that the development of these weapons would continue:

I wish to turn now to a subject introduced last year by the former leader of the Swedish delegation, Mrs Myrdal, and raised again by Mrs Myrdal's successor, Mrs Thorsson, at our 633rd plenary meeting. In her statement on 9 August last, the Swedish representative referred to newspaper accounts of the development of a 'new generation' of tactical nuclear weapons, so-called 'mini-nukes', which, in her words, 'would blur the present distinction between conventional and nuclear weapons'. Such a development, she said, would drastically aggravate the nuclear threat against non-nuclear weapon States and 'affect the very premises on which adherence to the NPT is based' (CCD/PV.620, p. 14). I should like to explain why such fears are wholly unwarranted.
The term 'mini-nukes' is misleading in two important respects. First, the coinage of this new catchword in itself conveys the false impression that we are talking about a radically new and futuristic family of weapons. Secondly, the diminutive element of the term

'mini-nuke' falsely suggests some miniature nuclear device which can be handled and used in the same manner as conventional weapons. I would like to correct both of these misimpressions.

We are not now, nor have we been in recent years, at the brink of some qualitative breakthrough in tactical nuclear weapons development. Instead, as Secretary of Defense Schlesinger has stated, we have been engaged over many years in a gradual process of moderately upgrading our tactical nuclear stockpile. There is nothing new about the existence of tactical nuclear weapons of very low explosive yields. Indeed, it is public knowledge that nuclear weapons with explosive yields measured in the sub-kiloton range were introduced in the United States inventory many years ago. No decisions have been made to produce or deploy any new systems.

In response to speculation that further development of low-yield tactical nuclear weapons would blur the present distinction between conventional and nuclear weapons, I wish to state categorically that the United States Government has no intention whatever to treat such tactical systems as interchangeable with conventional arms. We fully appreciate that the distinction, or 'firebreak', between nuclear and non-nuclear arms is a major factor in preventing nuclear warfare, and we will not act to erode this distinction. The very special command-and-control and safety arrangements that apply to nuclear weapons in general have of course always applied to small-yield nuclear weapons as well [161].

The press report correctly indicated that '. . . the process of moderately upgrading our tactical nuclear stockpile . . . *will continue*'.

Thus, under the new policy, according to officials, the Defense Department could continue development of a smaller, more precise warhead for its Pershing missile, which is now equipped with warheads ranging up to 400 kt, and the Army would be permitted to produce a new family of atomic artillery shells for its 155-MM Howitzer and 175-MM rifle [162].

Thus the 'new' policy was the same policy as before, and the references to not 'blur(ring) a distinction', as well as the rest of the official statements, were essentially facile evasions of the central issue. Similarly,

. . . when the subject came up at the NATO Nuclear Planning Group meeting in Ankara in 1973, the United States produced a public statement which said that the United States had not developed or deployed a new family of precision guided miniaturized nuclear weapons and that the United States would only undertake such a decision after full consultations with its allies. In a subsequent press conference, when the German Defense Minister was asked about mini-nukes, he replied, drawing on the US statement, that the US was not thinking about equipping its troops with these weapons [163].

Initial NATO and West German response had, in fact, been negative, but more importantly both the US and West German statements stood against all the available evidence. The weapons had been under development for nearly fifteen years by then.

The concept for 'clean', 'neutron' or 'enhanced-radiation' nuclear weapons was developed, as far as public knowledge would indicate, in the US in the period 1958–1960. A short but rather unusual description of the events, individuals and forces involved in a major weapons innovation – involving this particular case – recently appeared as part of the June–July 1977 public debate on this subject in the USA [164]. It includes most of the relevant pertinent events of the 1958–1960 period with one exception. This was the visit of Dr Edward Teller to then President Eisenhower, with the request – or promise – that if President Eisenhower would hold off his approval of a comprehensive nuclear test ban treaty then under negotiation with the USSR, the nuclear weapons laboratories would be able to develop 'clean' nuclear weapons within a relatively short period. One recent press account indicates that Dr Teller's visit took place in 1959, and that he asked for a period of eighteen months† [165]. The pressure continued on into the Kennedy administration and one of the reasons pressed on the new President to end the nuclear test moratorium (before the USSR reinitiated testing) was in order to test new 'clean' weapon designs.

Table 1.1. Yield distribution of US, Soviet and UK nuclear weapons tests, until January 1971

Numbers of tests

	Low	Low-inter-mediate	Intermediate	High	Other
	Less than 20 kt	20–200 kt	200 kt to 1 mt	Greater than 1 mt	(No yield indicated)
USA†					
Before PTB‡	160	25	28	14	71
After PTB‡	120	65	10	4	0
USA total	**280**	**90**	33	18	71
USSR§					
Before PTB‡	20	12	32	43	17
After PTB‡	5	24	9	1	5
USSR total	**25**	**36**	**41**	**44**	**22**
UK	**6**	**12**		7	0

† Omits all Vela, safety and Plowshare tests.
‡ The Partial Test Ban Treaty came into force on 10 October 1963.
§ These numbers are Soviet tests announced by the US Atomic Energy Commission: they are therefore lower than SIPRI figures.

Source: World Armaments and Disarmament, SIPRI Yearbook, 1972, p. 409.

† The visit to President Eisenhower in 1956 by Dr. Teller and two other scientists, also on behalf of 'clean' thermonuclear weapons, and which was publicly reported at the time, did not refer to low yield weapons [166].

Senator Dodd, a vociferous supporter of continued US testing, raised the claim (as so often occurs in strategic weapon development issues) that in the absence of US testing, the USSR would develop clean weapons before the USA, with dire consequences [167]. From publicly available data, even after the 1963 Limited Nuclear Test Ban Treaty and into the early 1970s, the USSR had carried out relatively few nuclear weapon tests in the lower yield ranges (Table 1.1).

Despite the fact that the data in this table include only publicly announced US and Soviet nuclear tests, and only as of January 1971, the remaining approximately 30 or 40 per cent of additional underground tests would not change the relative US–USSR proportions within the test yield categories to any significant degree. The US reportedly carried out its first test of an 'enhanced-radiation' weapon for tactical nuclear use in 1963 [168]. The programme to this point had been carried out not as a response to a weapon 'requirement' produced by one of the military services, but under the direction of the US Atomic Energy Commission. It would develop that in 1973 at least a part of the US Army was still hesitant about deploying low yield 'enhanced-radiation' weapons. It was reported that in 1964 US Secretary of Defense McNamara cancelled all further Department of Defense funded research and development regarding low-yield nuclear weapons, work which had been in progress for several years. But in 1967, the US Atomic Energy Commission still reported it was doing further work on 'radiation-enhanced' weapons and the first indications of a new push for their development by the military appeared in the same year.

> A major effort is under way to perfect small precise tactical nuclear warheads, including some that are practically free from fallout . . . 'that could make nuclear weapons more usable in Europe' according to a top American military planner . . . Thus the major program to develop what one general called 'cooky-cutter nuclear weapons'. 'We're fast developing small warheads offering very precise predictable effects' [169].

Work by the Department of Defense on these weapons was apparently reauthorized at a very early date in the Nixon administration. It was known to have the active support of Dr J. Foster, Director of Defense, Research and Engineering; Melvin Laird, the Secretary of Defense (and later James Schlesinger, his successor); and Henry Kissinger. An important front page article appeared in the *Washington Post* in 1969 attributing to Nixon and Kissinger a 'new' reliance on tactical nuclear warheads in Europe, a 'lowering of the threshold', a greater strategic reliance on nuclear weapon use, harkening back to Eisenhower administration tendencies (with the traditional rationales of saving money and shortage of troops etc.) – and the authorization of the new tactical warhead weapon designs [170]. In his 1971 review of foreign policy, the President, in referring to NATO, indicated that the USA would develop concrete programmes to ensure the provision of 'modern and sufficient' theatre nuclear weapons [171]. There is no reason to doubt that the President supported the subsequent moves. They were consistent with a broad programme of developments and deployments in the Nixon administration, under both Secretaries of

Defense Laird and Schlesinger, to enhance the 'warfighting' capabilities of nuclear weapons, both strategic and tactical, and to make them more 'usable'. Soon after these decisions by the new Nixon administration, publicity began to appear in support of the programme, in the military and strategic literature [172], and in the public press [173]. The first European appreciation appeared, though this report had only a very limited distribution [174]. The proponents hoped that:

> From the political point of view, if the NATO stockpile were modernized along the lines described above – that is, toward essentially clean enhanced-radiation weapons having yields predominantly in the low sub-kiloton range up to 1 kiloton – the Enthoven 'Firebreak' philosophy would no longer apply, and the credibility of a tactical nuclear defense of Western Europe would be restored [175].

Beecher's 'inside' article made clear that:

> For nearly a decade a debate has been under way, principally behind closed doors, over whether the United States should stay with, or fundamentally transform, the stockpile of some 7200 relatively large and 'dirty' tactical nuclear weapons deployed in Western Europe.
> Tactical nuclear weapons are once again being developed into a front-line deterrent in Europe after a decade in the doldrums when it was feared that their refinement would invite too-easy employment [176].

The year 1972 saw the surprising statement of Secretary of Defense Laird that 'the Nixon administration is considering deploying smaller, cleaner tactical nuclear weapons in Western Europe' [177]. This report carried three important additional items of information. New tactics and doctrines for the use of the cleaner weapons were being worked out by the Army and the Air Force. The services were to devise improved command and control procedures so that if the President authorized the use of a few of the weapons, others would not be fired. Finally, there was the first mention of more accurate delivery systems for artillery shells, air-to-surface missiles and glide bombs, all involving homing on laser illumination of the target.

In 1973 came the first appropriation request for new nuclear artillery shells of the new types, the XM-517 for the 155 mm gun, and the XM-673 for the 8 inch gun [178]. There were said to be 360 US 155 mm guns in Europe, and 324 in the hands of NATO allies. In the case of the 8 in guns, there were 326 for both US and NATO nations. The Army was requesting 'several thousand new nuclear shells' [179]. The then Secretary of Defense, Elliot Richardson, stated that he had no knowledge of the request, and that it had not been brought to his attention. Procurement and funding requests for nuclear munitions are directed to the budget of the US Atomic Energy Commission, not the US Department of Defense, and are reviewed by the Joint Committee on Atomic Energy and the appropriations subcommittees on Water, Power and Atomic Energy of the Senate and House Appropriations Committees of the US Congress. The Appropriations request for the new shells was turned down. It is quite important to understand that this occurred for a combination of three reasons, *none of which was directly related to the substantive question of 'mini-nukes' or enhanced-radiation properties* per se. The first of these was the cost of the new nuclear shells. The cost of each shell apparently ran to $400 000 and the total request for the shells was for $1.3 billion, though military spokesmen later claimed that most or all of this cost would be regained by salvage of the nuclear material in the older nuclear shells that would be retired. The second reason was that the Army had requested the first of the newer enhanced-radiation warheads, or 'mini-nukes', particularly for *artillery*, rather than for another kind of tactical nuclear weapon, for example, for Pershing missile warheads, or for tactical nuclear bombs. Many members of the Subcommittee on Military Applications of the Joint Committee on Atomic Energy had strong and particular reservations concerning any and all nuclear artillery. These reservations concerned risks of two parallel sorts inherent in a tactical nuclear weapon of any type placed very close to NATO–WTO borders, and US and NATO nuclear artillery was positioned in that way. In a potential combat situation, tactical nuclear weapons so close to enemy forces were in risk of being overrun. Hence, there was a good deal of concern for their security in a combat situation. Related to that was the fear that their forward positioning would force a decision on their use far sooner than might otherwise be desirable or necessary. Since one of the general considerations in possible restructuring of NATOs tactical nuclear forces was the drawing back of exposed, forward and 'soft' TNWs, and since the members of the Joint Committee had strong reservations about nuclear artillery for the reason indicated, they were reluctant both to provide so large a cost for a new generation of nuclear artillery warheads and to give nuclear artillery a new lease of life in doing so. The third reason was that Senator Stuart Symington had replaced Senator Henry Jackson in 1973 as chairman of the Military Applications subcommittee of the Joint Committee [180]. Senator Jackson had held that post since 1955. The Committee as a whole had long been a promoter of ever more atomic weapons, and the subcommittee had played an instrumental role in the late 1950s in pushing for thousands of tactical nuclear weapons. Senator Symington opposed any further increases in the tactical nuclear weapon stockpile. It was the conjunction of these three factors which defeated the first request for a 'mini-nuke' replacement – a new lower yield, enhanced-radiation artillery warhead – and *not* a discussion of the pros or cons of lower yield or radiation-enhanced weapons. This was not to take place until three more years had passed.

There were two other particularly noteworthy aspects that became evident in these Hearings. The first concerned the apparent dissatisfaction of the Federal Republic of Germany in particular with the proposals for new low yield weapons. In reply, it was stressed that 'Those matters have not been formally proposed for approval [181]. Perhaps more striking was the second aspect.

> *Senator Symington:* Would you comment on the Army's reluctance to consider the use of the [deleted] as a primarily kill mechanism against battlefield personnel?

General Goodpaster: I would rather comment on the issue, itself, and say that I think that these are possibilities that are very important and I am hoping and I know that the Army is in the process of evaluating this. I cannot foretell the outcome but my people are in touch with others who are looking into this very possibility.

I will say that I, for one, hope that if the results of our analyses and operational studies come out as I anticipate, I hope that the Army will be prepared to go forward with us in supporting weapons of the kind [182].

These US Army evaluations were taking place while the Atomic Energy Commission and SACEUR were making clear their unqualified support for the weapons. The public record does not indicate the nature of the disputes, compromises and compensations that go into a final administration decision on such an issue. Any doubts are seldom displayed before a Congressional appropriations body. In the very same fiscal year, another US Army spokesman made it explicitly clear that the service intended to come back again the next year with the same appropriations requests. 'We intend to continue the program next year' [183]. Doubts were certainly not in any evidence in SACEUR's request:

NATO defence-only requirements (with short-range weapons)

(a) *Nuclear Weapons* – The development of [deleted] low yield tactical nuclear warheads would not appreciably decrease the number required for the defence of NATO. [Deleted] lower yield weapons, delivered more accurately, can be employed against a larger variety of targets than larger weapons with high residual contamination in most limited conflict situations we can envision.

The military aspects of lower yield, smaller delivery errors, and dominant battlefield kill mechanisms, however, are not the only reasons for seeking modernized warheads. The employment of tactical nuclear warfare in the populated areas of Europe would likely inflict a number of civilian casualties. Lower yield warheads, with [deleted] could allow the use of these weapons in a given situation with greatly reduced collateral damage and casualties. The overall military and political advantages to be gained by such modernized weapons is that their use in plausible scenarios in Europe would appear more suitable and credible to our allies and potential enemies alike. Our allies would find these weapons as more acceptable modes of defence because of the expected resultant reduction in civilian casualties. The enemy's knowledge of our deliberate steps to reduce collateral damage cannot fail but suggest to him our determination to use tactical nuclear weapons if necessary. The establishment of a modernized tactical nuclear weapons stockpile in Europe enhances the flexible response doctrine of NATO.

The receipt and installation of the latest PAL devices would not affect the required number of nuclear weapons for defence of NATO. These devices are designed to preclude the unauthorized use of nuclear weapons – either by design or by accident. The latest PAL devices will provide better security and more flexibility [deleted] [184].

It was at this time that the requests for these weapons also began to indicate that they would simultaneously embody new security devices [185]. 'Current combination lock command and control devices would be replaced with [deleted]. A security container would have a tamper-proof system that if activated and [deleted] and prevent unauthorized use' [186].

In addition to the 1973 NATO Nuclear Planning Group meeting in Ankara already referred to, the US Department of Defense also brought up the subject of 'mini-nukes' at a conference with its allies held in Paris in May 1973. It informed them that the new tactical nuclear weapons could be ready for deployment to Europe in three to five years. This produced the first public European responses [187]. Their development was particularly opposed by the French. Criticism of the new warheads from NATO allies was derived from a different set of considerations than those which arose from the domestic US arms control community. It was not the question of the reduction of the 'firebreak' and the greater possibility of initiation of use and subsequent escalation. Quite the opposite. It was that it was taken to imply 'a *reduced* US nuclear commitment'. By making the weapons more usable *in Europe*, it implied a greater decoupling of NATO tactical nuclear weapon use from the US strategic nuclear deterrent. It was that connection that the Europeans had always wanted, above all else. With supreme irony, this was *precisely* the opposite of the argument made by the US proponents of the new warheads: the greater likelihood of use of the newer warheads, due to their reduced collateral damage, increased the deterrent value of the TNWs in Europe, and the entire escalatory chain. The European discussion also provided the context within which Sweden's delegate to the CCD criticized the USA for the prospect of the introduction of the new warheads into the European theatre [188]. However, by December 1974, the British Defence Minister, Mr Mason, stated that 'Britain would go along with any American reassessment of the tactical nuclear stockpile in the interests of greater accuracy and economy of warhead yield' [189].

In 1974, the Department of Defense pressed its request again [190]. Press response was limited [191]. Secretary of Defense Schlesinger and General Andrew Goodpaster, SACEUR, were particularly strong in their support for the weapons. This time it was announced that the Department of Defense sought development funds for a new Pershing missile warhead [192]. The existing Pershing warheads ranged from 60 to 400 kt in yield, with most of the missiles carrying the higher yield. These were exceedingly high levels, higher than the yields in Poseidon SLBM and Minuteman ICBM strategic nuclear MIRV warheads. They were an open invitation to rationalization and reduction. The proposed new Pershing II warhead was to go down to the range of 1 kt and get an additional radar guidance system to increase its accuracy. In 1977, the following description of the Pershing II was given.

> The Pershing II system, now in advanced development, is to be a modular improvement to the Pershing Ia. The improvements involve replacing the warhead and guidance and control section with new nuclear warheads and a terminally guided reentry vehicle. Pershing II will have the same range but a radar area correlation terminal guidance system, now being developed, will provide greater accuracy. Several warheads are being considered for the system. The program

57

cost estimate is $1 billion. In June 1978 the Defense Systems Acquisition Review Council (DSARC) II is scheduled to consider whether the system should enter engineering development [193].

It was also reported that laser guidance packages had been tested for shorter range missiles and that a laser homing version of the Lance missile warhead might be developed [194]. Atomic artillery had a most usual range of from 8 to 15 miles, and in some cases was positioned even closer to the WTO border so that its shells could reach across the border, thus raising the problems discussed earlier. The Pershing missile had a range of from 60 to 450 miles. General Goodpaster asked Congress to approve the entire new generation of 'mini-nuclear' weapons. Not all the new generation tactical nuclear weapons were to be in the 'subkiloton' range; some were to be in the 'kiloton range', (a designation which had previously been used by the Pentagon to indicate anything between one and several hundred kilotons). This left some ambiguity. Secretary of Defense Schlesinger was unequivocal that the new weapons were desirable.

> Increased accuracy permits the use of lower yield weapons to achieve the damage desired on a target. At the same time, the use of a more accurate, lower-yield weapon decreases the probability of unwanted damage. In this sense, it is generally accepted that an improved nuclear stockpile would provide a greater capability for planning against a massive conventional or nuclear attack. The combination of improved military effectiveness and increased allied acceptability enhance the deterrent effect of our tactical nuclear stockpile.
> . . . Our nuclear weapons in Europe are present for deterrence, and deterrence is made credible by a credible warfighting capability [195].

His April–May 1975 report to Congress on TNWs stated that,

> . . . If deterrence fails, weapons with low collateral damage would reduce civilian casualties and perhaps reduce the risks of uncontrolled escalation . . . SACEUR's targeting is intended to limit the collateral damage . . .
> Further reductions in collateral damage can be made by improvements in weapon systems (e.g., reduced yields, special warhead effects such as enhanced radiation, improved delivery system accuracy) [196].

But at the same time, there was much playing with words. Secretary Schlesinger's simultaneous statement that 'a new "mini-nuke" program does not exist' depended on the listener's understanding of the word 'new', and the US statements in 1973 in Ankara and at the CCD in May 1974 were largely misleading.

In June 1975, NATO defence ministers met in Monterey, California. Their joint communique said that their meeting had reviewed 'possible improvements in theatre nuclear force posture, including possible steps to enhance the physical security of nuclear weapons stored in Western Europe, and the implications of new weapons technology upon the alliance's nuclear posture' [197]. Press reports indicated that 'There have been some changes in the composition of the stockpile, with older warheads replaced by smaller and presumably more discriminate weapons' [198]. Though these may have been lower yield weapons, perhaps newer air-dropped weapons such as the B-61 bomb which had come into the stockpile, with lower yields than earlier air-dropped bombs, these were not yet the 'mini-nukes' or enhanced-radiation wearheads. There had as yet not been a second procurement request, or approval, for these. It was also reported that the USA had assigned additional Posiedon missiles to NATO. These apparently took over the targets which had formerly been allotted to some NATO QRA aircraft. In addition, F-4 aircraft were to be replaced by F-15s and the latter did not have the F-4's nuclear capability. It is likely that such a replacement would also produce some lowering of yields. Coincidentally, it also increased the link between NATO's tactical nuclear weapons and the US strategic deterrent, reinforcing the convergence which the USA's NATO allies feared would be weakened by the introduction of newer low-yield warheads. Earlier, in 1973, the US Senate Foreign Relations Committee report had indicated that,

> The North Atlantic Treaty allies have agreed upon a strategy providing that any large-scale use of nuclear weapons by forces within Europe should also be accompanied by United States forces stationed outside of Europe [199].

Department of Defense budget requests for FY 1977 related to 'mini-nuclear' or enhanced-radiation TNWs received minimal attention [200]. In its funding request for continued development of the XM-753 8 in nuclear artillery projectile (AFAP), the Army stated that 'After completion of testing, long lead-time procurement will start to support low-rate initial production of fuses, training projectiles, and test equipment in FY 1978' [201]. This drew no Congressional or press attention at all.

One other important event relevant to the question of procurement of a new generation of enhanced-radiation tactical nuclear weapons took place in 1976 and early 1977, before the virtual explosion concerning the question in June and July 1977. In 1975, the US Arms Control and Disarmament Act had been amended to require the executive branch to provide 'arms control impact statements' to the Congress with requests for authorization or appropriation for certain programmes. The Law required that the analysis to be provided 'shall include a complete statement analysing the impact of such programme(s) on arms control and disarmament policy and negotiations'. The first group of these statements, sixteen in number, were submitted in August 1976 under the Ford administration. Eleven concerned Department of Defense programmes, and five were from ERDA and concerned new nuclear warheads. The Department of Defense provided a statement on the XM-753 nuclear artillery shell, and ERDA provided one on the W-79 8 in nuclear artillery shell. The statements provided to Congress by the Administration, after some delay, were considered a charade and a disaster [202]. The eleven statements on DOD programmes averaged four sentences each. Statements on a large number of programmes that were originally promised were not submitted at all. Finally, the sixteen skimpy statements were nevertheless all classified. The appropriate Senate committees replied that the statements provided 'do not comply with the law and are unacceptable'. (Nevertheless, no attempt was made by Congress to hold back the appropriations for the programmes in question, to which the statements applied.) Twenty-six statements were resubmitted for FY 1978 programmes, 21 by the Department of Defense and again five for ERDA. These were unclassified and were published [203]. However, they were again judged to be unsatisfactory, being too short (most ran to a page) and for not addressing the appropriate questions. This time, three of the 26 statements concerned enhanced-radiation warheads. They were for:

— The XM-785 improved 155 mm nuclear projectile.
— The XM-753 improved 8 in nuclear projectile.
— The W-79 warhead for the XM-753 [204].

(Both the 155 mm and 8 in nuclear projectiles were also variously referred to as the AFAP, Artillery-fired Atomic Projectile.) The Committee on Foreign Relations of the US Senate and the Committee on International Relations of the House then requested the Congressional Research Service of the Library of Congress to prepare a set of model Arms Control Impact Statements. One of these was an analysis of the XM-753 8 in nuclear projectile [205]. The evaluation of the three impact statements was by and large critical, for the most part pointing out the large number of questions and considerations that had not been addressed in them. The model impact statement was ten pages long, setting out a large number of points under three categories,

— factual information and issues related to military considerations,
— arms control policy,
— arms control negotiations,

and addressing each in turn. It was indicated that the Army's planned artillery modernization of both the 8 in and 155 mm nuclear projectiles would cost between $2 and $3 billion over a 10 year period. If further nuclear testing were necessary for the warheads, it might impede the possibility for a Comprehensive Test Ban, which President Carter had proposed in a 24 January 1977 press interview.

All of these materials were made public in April 1977. They produced no Congressional or press response at all.

A scant two months later, the FY 1978 budget requests for ERDA, the testimony for which had been given in March, were released [206]. These contained the funding requests for several enhanced-radiation tactical nuclear warheads. The primary request was for the 'W 70 Mod 3 Lance enhanced-radiation warhead'. There was also a request for the W-79 8 in artillery fired atomic projectile. It was also indicated that before the end of FY 1978, the DOD might also ask for funds for the Pershing II warhead, with a terminal guidance system and lower yield enhanced-radiation warhead, and/or for a replacement for the 155 mm nuclear artillery shell. Still further ahead, research and development funds were asked for an enhanced-radiation air-dropped weapons. The only funds deleted by the Committee were for the 155 mm nuclear artillery shell, on the apparent understanding that if the Army got its 8 in nuclear shell, it would not ask for the 155 mm one as well. The request – which should not have been so very surprising in view of the history reviewed here – had several significant points to it:

— That this time the Lance warhead was stressed first (earlier requests had emphasized first the nuclear artillery, and then the Pershing II),

— That four new enhanced-radiation warheads were anticipated, with the likelihood of yet a fifth, the air-dropped bomb, to follow,

— It was indicated that the first new Lance warheads could start being delivered in two years, i.e. 1979,

— It was indicated that the number of dual-capable 155 mm howitzers given in the Senate Foreign Relations Committee report in 1973 was incorrect. There were a total of 2196 dual-capable 155 mm howitzers and 450 dual-capable 8 in howitzers in Allied Command Europe.

The series of events which followed in June and July 1977 must be considered astonishing and was entirely unexpected. A reporter for the *Washington Post*, Walter Pincus, wrote a news story on the ERDA warhead budget request for the *Post*. Some years earlier, Pincus had been a Joint Committee on Atomic Energy staff assistant for Senator Symington, and had a long interest in tactical nuclear weapons. The nation and the press were in the midst of the President's decision on the B-1 strategic bomber, and the strategic cruise missile. Nevertheless, further articles on 'the neutron bomb', as the subject came to be called, appeared in the *Washington Post* and spread to other major US newspapers, and the press response grew and grew†. Under the leadership of several dedicated US Senators, Congressional pressure began to build up, and it culminated in a secret session of the US Senate on 1 July to debate the question [207]. Though more acute, and of shorter duration, in some ways the debate ranked with those that had taken place on the ABM in 1969 and, more recently, on the B-1 bomber. It became known that President Ford had approved the production request for a Lance enhanced-radiation warhead in his annual stockpile review on 24 November 1976. Nevertheless, no arms control impact statement on the warhead had ever been sent to the Congress, as the law required, (although one was sent for the *non*-nuclear Lance warhead which the administration had also sought to have put into production). At the end of June, the Senate now asked the administration for such a statement. When one was quickly prepared by the US Arms Control and Disarmament Agency, President Carter permitted it to be quickly forwarded to the Senate for use in their debate. It became evident that the warheads for the new weapons had already been tested, at least all or in part, but probably completely, and that they could be deployed to NATO within eighteen months of a decision to put them into production (i.e. by 1979). US Department of Defense officials indicated that 'NATO had been endorsing the development of the weapons over the past three years (i.e. since 1974) 'and

---

† Because of the very rare and unusual nature of the public response to this issue, a rather comprehensive list of references is appended at the very end of the reference section and individual articles are not referenced here.

61

that these were the warheads ambiguously referred to in NATO communiques as "improved weapons for the alliance" '. There were other reports 'that almost all the 15 member countries (of NATO) have said they would like the weapon'. The Senate debate, and its reflection in the public press, was quite a good one, precise to the critical issues. Both opponents and proponents generally agreed that yields on most NATO tactical nuclear weapons were far too high, especially those on the Pershing and for air-dropped weapons. Reports attributed a yield of 10 kt (or up to 50 kt) to the Lance warhead that would be replaced and the new warhead to be 1 kt or less, so that the difference was less in this particular case than it would have been for the Pershing. Because the neutron radiation was short-lived, it was stated that an area targeted with enhanced-radiation weapons could be re-entered within hours and the radius of severe blast and fire damage would be much smaller. There were claims for its utility against tank forces, but questions about its general battlefield utility, since neutron doses must be very high to be lethal, and even then there is some delay before death occurs. But the crucial issue was one on which the opponents and proponents begin with agreement on a common understanding and then see the consequences in precisely opposite fashion. Both sides agree that the reduction in collateral damage would make the use of the new warheads more feasible in any eventual conflict, and thus more likely. The proponents present the same argument as former Secretary of Defense Schlesinger had given earlier: such increased likelihood of use increases the credibility of the deterrent against any Soviet incursion into NATO territory. Opponents say that if any nuclear weapon use is initiated, no matter how small, it will lead to an escalation to full-scale nuclear war. The question then is whether deployment of smaller yield enhanced-radiation warheads for tactical nuclear weapons serves to deter war, or to deter it any more than the presently deployed warheads of far larger yield, and therefore more dangerous warheads taken individually, or whether such deployment increases the likelihood of initiating use of nuclear warheads that are of lower yield and less dangerous taken individually, but more likely to initiate escalation. The question is whether 'tactical' or 'limited' nuclear war can actually take place, either with present NATO (and presumably WTO) targeting arrangements, or even under any imaginable circumstances. Since the only answers that one can give to these questions are entirely matters of judgement and 'faith', and since the questions bring us directly back into the verbal gymnastics of all tactical nuclear warfare doctrine, it is clear that one can arrive at nothing more than such verbal formulations by way of answers. The primary support for 'mini-nuclear' weapons have been the weapon development laboratories, NATO's SACEUR, after 1968 the Secretaries of Defense and Director, Defense Research and Engineering (DDR&E) of the Department of Defense, and all who argue for more *useful* warfighting nuclear weapon capabilities. These proponents claim that escalation would not be inexorable and though there would be no obvious 'firebreak', larger or strategic nuclear exchanges would not follow. There were also two important tangential questions, as to whether a possible Comprehensive Test Ban would be delayed, or again sidetracked, and what effects US acquisition of a new generation of tactical nuclear weapons would have on the US efforts to prevent further proliferation of nuclear weapons to other nations. Finally, there was the question of whether a renewed NATO commitment either to general notions of a tactical nuclear deterrent, or more specifically to the idea that a small nuclear exchange can be conducted without serious risk of expansion to general nuclear war, would short-circuit any NATO efforts to improve its conventional forces.

The effort to delete production funds for the new Lance warhead failed in a Senate Appropriations Committee by a 10–10 tie vote. The 1 July vote left the Senate 43 to 42 against deleting funds. A final vote was to be held when the Senate reconvened on 13 July after a recess. In a press conference on 12 July, President Carter indicated that he had not known that the production request for the new warheads was in the appropriations bill until it became a public issue. There was some ambiguity in the subsequent phrasing.

> Before I make a final decision on a neutron bomb's deployment I would do a complete impact statement analysis on it, submit this information to the Congress. But I have not yet decided whether to approve the neutron bomb, I do think it ought to be one of our options, however [208].

It was not clear from this if the 'option' was to be one for a decision-making process *before* production, or for *deployment*. However, it was reported that the President had privately informed the Senate the previous day, on 11 July, that he wanted the Senate to approve the funds for production as they appeared in the ERDA budget request. On 13 July, the Senate voted 58 to 38 not to remove the funds for the weapons from the bill [209].

There was some comment from the Soviet Union on this process, and it remains to indicate the nature of these. In February 1974, a week after General Goodpaster's testimony on behalf of the new enhanced-radiation warheads was released, *Pravda* published a prominent commentary which stated that such smaller tactical nuclear weapons and the proposal for them violated the spirit of the agreement on preventing nuclear war signed by President Nixon and Party Secretary Brezhnev in June 1973 [210]. It said that General Goodpaster and NATO hoped to 'circumvent' the agreement. It did not explain how the new warheads would do that. (Perhaps ironically, the arms control impact statement prepared by the US National Security Council in 1976 for the new 8 in artillery shell offered the shell 'as a positive arms control contribution', specifically in the context of that same June 1973 USA–USSR agreement.) In May 1977, it was reported that the USSR had already protested against 'new types of US tactical radiological weapons . . . through the SALT channels and is calling for a ban on the weapon. . .' [211]. On 19 June, *Pravda* again referred to 'research into neutron based bombs' as one of several US programmes indicating a US 'intention to step up the arms race' [212]. The *Pravda* commentator stated that such weapons 'are practically a chemical warfare weapon' – which seems something less than accurate. Finally, on 9 July, Tass released a dispatch in which the proposed neutron bombs were said to be inconsistent with President Carter's stand on human rights. It was grouped with other new

US warheads in the strategic nuclear area which could 'only complicate the international situation and bring about a new, extremely dangerous round of the arms race', and whose development 'would complicate the SALT talks' [213].

For the sections on France and the UK – as for the USA and the USSR – the discussion is limited to their tactical nuclear weapons, despite the fact that their stockpiles of all kinds of nuclear weapons are smaller. Some confusion is produced by the fact that delivery vehicles may in some cases be the same for both 'tactical' and strategic weapons, but that is essentially also no different the USA and the USSR, although it in some way appears more 'noticeable' for France and the UK, again because of their smaller total forces.

II. Tactical nuclear weapons of France

France is known to have two forms of tactical nuclear weapons, air-dropped weapons and the warheads for its Pluton surface-to-surface missile. The air-dropped munitions are in turn carried by two elements, land-based attack aircraft, and French aircraft-carrier-borne aircraft. There is no evidence as to whether France has any nuclear ASW weapons. The land-based aircraft assigned to deliver air-dropped tactical nuclear weapons are the Mirage III-E and the Jaguar. Two squadrons of 15 Mirage III-E aircraft each reportedly began equipping with the AN-52 bomb in October 1972 [214, 215]. Altogether there would be 60 aircraft for the tactical nuclear weapon air delivery role, 30 of each type. The bomb is dropped at low altitude, with a retardation parachute [216]. As early as 1967, France announced that French forces would be equipped with a low-altitude nuclear weapon delivery system (by the summer of 1967) that would permit the weapon to be delivered from altitudes as low as several hundred yards [217]. Yields for the AN-52 warhead were variously given as 10, 15, 10–15, or 10–25 kt, and it is the very same warhead referred to as the *arme tactique commun* that is also used for the Pluton missile. More recently, it was stated that two-thirds of the tactical nuclear weapons for the Mirage III-Es, Jaguars and the Pluton were of 10 kt, and the remaining one third, apparently again for all three delivery systems, were of 25 kt [218]. Earlier it had been reported that the aircraft would probably deliver only the more powerful of the two warheads [219]. The warheads were developed during the 1971–75 French defence budget period [220].

The first report of a French aircraft-carrier delivered nuclear weapon capability appeared as early as 1963. This stated that '. . . the Dassault Etendard 4M shipboard fighter (is) being equipped for nuclear missions . . . Forty of the French Navy's 90 Etendard fighters, currently being delivered, will have similar capability' [221]. Since this same report also indicated that Mirage IIIs then fully operational had the same capacity, while tactical nuclear weapons reportedly did not go to the latter aircraft until their development later in the

1970s, it is unclear what bomb the Etendard 4M was to deliver and when it was deployed. Presumably, if it was also the AN-52, it was not until 1972 or after. When Jaguar Ms were planned for the two French aircraft-carriers, the Foch and the Clemenceau, it was stated that they would carry a nuclear air-to-surface missile, presumably the Martel [222]. Purchase of these aircraft was, however, cancelled by the French Navy and replaced by the Super Etendard. There will apparently be 12 Super Etendards on each aircraft-carrier replacing the present Etendard IV-Ms. The available evidence is ambiguous, but implies that nuclear weapons are already deployed on these, though not precisely when that deployment was first initiated [223]. A French report in late 1973 indicated that tactical nuclear weapons were not yet at that time deployed with the Etendards.

> 'Il est question de doter de cette arme atomique tactique les avions Super Etendard de l'aeronavale. Le problème de l'emploi de ces armes et de l'autorité qui décidera de leur utilisation ne semble pas encore résolu de façon définitif,' souligne le rapporteur [224].

The 1971–75 French Defence Plan also reportedly called for the development of a nuclear air-to-surface missile for the 1980s, presumably to equip the *avion de combat futur* [225].

In February 1975, the French surface-to-surface solid-fuelled missile Pluton became operational with French forces. There had been some delay, its earlier appearance having been predicted for 1972, and then 1974 [226]. However, the production order was not placed before early 1972. One hundred and twenty Pluton missiles were ordered. Their range is given as a maximum of 120 km or 75 miles, with a CEP of 200–400 m. The warhead is to explode at 300–400 m in altitude, and its targets were given as 'targets of opportunity, (mobile military targets, logistic concentrations, obstruction of points of passage), as opposed to tactical nuclear air bombs which are directed at more resistent targets, (airfields, weapons factories, etc.)' [227]. The Pluton missile is mounted on a modified AMX-30 tank chassis. The missile body is stored in a container on the tank, and the warhead section is carried on a second vehicle and attached shortly before launching [228]. There will be a total of 40 launcher vehicles, six 'regiments' of three batteries each, each battery with two launchers and two batteries for 'reserve' (or 36 plus 4). These are to equip five mechanized divisions, and with the tactical nuclear delivery air arm, comprise the *forces de manoeuvre*. As of 1976, only four of these regiments, with 24 Pluton missiles, were reportedly deployed. Pluton deployment is to be completed by 1978. All Pluton missiles and their launchers are located within French borders, in Eastern France, though half of the regiments to which they are assigned are based in the Federal Republic of Germany. However, up to the present time, France has resisted the Federal Republic of Germany pressure to move these into the Federal Republic of Germany and closer to the WTO borders [229]. Such a move would probably require some joint-use controls on the part of Bonn, which may explain the French refusal.

By 1963 and 1964, French interest in a tactical nuclear weapon programme had been reported, at least as regards air-dropped weapons [230], and by 1968, a reasonably accurate description of the programme could be given [231]. An article by the French Chief of Staff, General Fourquet, appeared in 1969, attaching great importance to tactical nuclear weapons [232]. In 1972, for the first time, a French Defence Minister very sketchily presented the policy for these weapons, that they are intended for use against enemy 'forces', presumably thus distinguishing them from strategic weapons [233]. In February 1975, when the Pluton became operational, then-Prime Minister Chirac devoted an entire address to it. He stated that its purpose was to define a specific geographic zone delineated by the deployment of the weapon and its operational firing range. This gave a potential invader a precise indication of at which point he would incur a nuclear strike, and that 'only a land-based' (and presumably relatively short-range) nuclear weapon system could 'establish such a zone of nuclear interdiction, and the Pluton makes it possible' [234]. This was soon further amplified by the new French Chief of Staff of the Armed Forces, General Guy Mery, in March 1976. In an address that explored the possible uses of French forces alongside other NATO forces, General Mery indicated that 'The possible use of our tactical nuclear weapons . . . is, as might be expected, the major problem. . .' [235]. Again, it was stated that these were 'counter-military weapons, i.e. intended for the battlefield and its surroundings'.

> But they are most of all intended to have a major effect on this battlefield – signalling to the enemy our determination to proceed, if necessary, to the point of bringing massive reprisals to bear if he should continue his present action.
> I do not think that the number of weapons is necessarily very important. The character of final warning which clothes their possible use excludes, in my view, any idea of a 'nuclear battle' and, on the contrary, militates in favor of usage which is as short and massive as possible.

The culmination of this sequence of policy statements regarding French tactical nuclear weapons came in an address by French President Giscard d'Estaing in June 1976. He spoke against previous French emphasis on the use of its strategic nuclear weapons in any circumstances, a position akin to 'massive retaliation' and offered instead that

> . . . far from what has been said or written (in France), diversity of means would reinforce deterrence if it were called into play . . . (rather than) the probability of our using, and of our determination to use, this sole means, which would be the ultimate means.
> I believe that the 'all-or-nothing' approach in defence matters might lack credibility. France might find herself in complex situations; they might take the form of very major disturbances in neighboring countries, or uncertainty as to the reaction of one country to a change in the political situation of another. If France can only speak or act in all-or-nothing terms in these situations, her position will lack credibility. Let me give two illustrations to prove this.
> (1) The first is the importance all our experts place on tactical nuclear weapons.

Even advocates of pure deterrence do not think that France can do without these weapons. Arms of this type, whether carried by rockets or airplanes, are undeniably nuclear and for this reason come under the control of the President of the French Republic. But these weapons are not only instruments of deterrence; they are also weapons for battle. They can be used as one of the many means of deterrence. They represent the opposite of 'all or nothing' [236].

In a very recent interview, the French Defence Minister, Yvon Bourges, sought to untangle some of the problems produced in the verbalizations of French policy by tactical nuclear weapons in particular.

This did not mean a switch from nuclear 'massive retaliation' to 'flexible response'. 'France has never adopted a strategy of all or nothing, for such a doctrine is little credible and irresponsible. She does not, for all that, lay herself open to the risks of a "flexible response" through the use of tactical nuclear weapons. The concept of use of those weapons is closely integrated with that of strategic weapons. For France, the use of tactical nuclear weapons would be the fearful means of signifying her determination to go to the bitter end – a kind of warning signal in effect – to an enemy who would have underestimated her determination. In short, from the moment of involvement of tactical weapons, the battle would have fundamentally changed in character', M. Bourges stressed [237].

It is useful to indicate this sequence of policy statements to make an important point. Something is known of the French decisions that brought about its tactical nuclear weapon programme. Whether the above rationales are intelligent or not, they in fact bear no relation at all to the sequence of circumstances and events that produced French tactical nuclear weapons. The decision was an example of what is called institutional or bureaucratic pressure in the process of major weapons acquisition. The French decision to make tactical nuclear weapons, particularly the Pluton, was a bureaucratic one, resulting from Army pressure, and was approved by President de Gaulle without much thought of what to do with them. Nuclear weapons had been acquired by France in the first place for political reasons not directly related to national military defence. The weapon systems for these had first been assigned to the French Air Force and the French Navy; strategic doctrines for these followed after. Pressure developed for a nuclear role for the Army, in which particular personal military advisers to President de Gaulle played an important role. Once the decision to produce Pluton and tactical nuclear weapons was made, several problems followed. Perhaps the most remedial was the need to find adequate personnel with which to staff the weapon system. Then, two schools of thought developed within the Army on what to do with them – how to propose using them – which directly involved questions of command and control and delegation of authority for initiation of use. To some degree, the differences of opinion represented those of two sectors within the Army: the infantry and the cavalry (i.e. armour). Some of the statements quoted above, for example M. Debré's, were indicative of particular stages or positions in this debate. The question of what to do with tactical nuclear weapons remained an unresolved issue when the government of President Giscard d'Estaing came to power, and to some degree is so still today. The question of command and control was resolved so that only the French President can decide on each individual nuclear strike. What we see in sum, however, is a situation in which a major weapon with the most serious defence implications is procured for reasons *unrelated* to and *before* any consideration of those very implications has taken place.

Something more can be said about French command and control of its nuclear weapons. The bombs for the French Mirage IV strategic strike force are reportedly mounted at all times on the aircraft, except when they are removed for checking [238]. The 'core' of the bomb is reportedly not in its place except when the bomb is 'armed', which takes place in flight. In order to drop the bomb from the aircraft in flight, one needs a 'key' which is given only during the mission, and the code for which the pilot does not know. The code is only given during the flight as well. In these aircraft, the firing circuits also involve cooperation between two individuals, the pilot and the navigator. Aircraft used for training flights do not carry the nuclear weapons. It is not known whether procedures similar to these are followed with the Mirage III-E, Jaguar, and Etendard aircraft equipped for tactical nuclear weapons, but there is no reason to assume that they are not very similar or the same. As regards the Pluton, it was already indicated that the warhead was kept on a separate vehicle from the missile body. Further,

> Le contrôle de l'engagement repose sur l'existence d'un boîtier de sécurité verrouillé sur chaque munition, qui interdit le tir du missile et la mise à feu de la charge en l'absence d'un numéro de code transmis par le chef de l'Etat au moment de la manoeuvre. L'ordre de tir vient au coup par coup, ou pour la totalité des véhicules lanceurs. Le missile ne s'arme qu'une fois en vol . . . [239].

As early as 1967, it was reported that only 'the Chief of State alone can give the order to act . . . If he is not available or unable to act, the Prime Minister, by strict protocol, can in turn assume the ultimate responsibility' [240]. These conditions were to hold for tactical nuclear weapons as well as for the French strategic forces.

It is interesting to note that the French Communist Party holds as its official policy that it will abandon 'strategic' nuclear weapons, but not France's battlefield, tactical nuclear weapons [241].

When France was still a full participant in the NATO military structure, she had two battalions of Honest John nuclear missiles, with US warheads, and also 155 mm nuclear artillery, also with US warheads in US custody. It was reported that in 1969 French Chief of Staff General Fourquet approached NATO 'to learn what kind of tactical nuclear support the French forces might expect from NATO in certain circumstances' [242]. In response, 'NATO apparently has made a vague agreement with the French to provide them with tactical nuclear weapons "when required" at some unspecified point after hostilities break out' [243]. French commanders in the Federal Republic of Germany had never ceased to cooperate with NATO forces in training for the use of these weapons and it is possible that in 1969 the French forces had still retained their Honest John missiles and launchers, minus the warheads. It is not known if this arrangement continued once France deployed its own Pluton and air-dropped tactical nuclear weapons.

III. Tactical nuclear weapons of the United Kingdom

Like France, Great Britain has tactical nuclear weapons in two components. Those for its land-based Honest John and Lance missiles carry US warheads under NATO bilateral arrangements. Those for delivery by its attack aircraft, and its V-bomber force now assigned to NATO, are of its own manufacture. Britain's nuclear attack aircraft were both land-based and aircraft-carrier borne. Britain apparently also produced its own ASW weapons. Some ambiguity is perhaps produced by the reassignment of the V-bomber force, as to whether to consider its weapons and targets 'tactical' or 'strategic'.

British stockpiles of nuclear weapons were built up much earlier than those of France, and probable in substantially larger quantities. As early as 1957, it was reported that

> Inasmuch as Britain exploded its first hydrogen bombs only this past spring, it may be presumed that its stockpile of these weapons is relatively small, but its stockpile of atomic bombs was stated by Mr Sandys to be 'substantial' [244].

and in 1960,

> Qualified sources said Britain's supply of nuclear weapons was sufficient to enable her to arm a V-bomber force, stock overseas arsenals and contribute to the (NATO) alliance's nuclear arsenal [245].

The same report indicated that Britain would disperse its V-bomber force to overseas bases and would stockpile nuclear weapons at overseas bases. Scimitar and Buccaneer aircraft on Britain's aircraft-carriers, as well as the land-based Canberra, also had nuclear delivery capability. Brown describes the Canberras as having constituted 'the Central Treaty Organization's nuclear strike force' [246]. (This implies an allotment of British nuclear forces to CENTO in the manner of nuclear weapons both allotted and supplied to NATO by the USA, and the joint nuclear weapon targeting, arrived at in NATO.) In 1964, Brown indicated that Britain had accumulated about 10 000 lb of uranium-235 and 3500 lb of weapon grade plutonium and that output of the latter was then being slowed down [247]. In the same year, Beaton estimated that Britain might have 'perhaps 1500' weapons [248]. In 1969, Brown wrote that 'Britain's nuclear stockpile is likely long to remain the most powerful Western European one' [249].

Britain's U-235 gaseous diffusion plant at Capenhurst, as were the United States installations and the French plant at Pierrelatte, was established to meet military requirements. Britain's U-235 production was terminated several years ago at Capenhurst. Press reports in 1969 indicated that production might be reopened, but this is contradicted by the following Parliamentary exchange in 1970, though its phrasing is a bit convoluted.

> Enriched uranium: Mr Brooks asked the Secretary of State for Defence whether he proposes to continue to obtain the total enriched uranium requirement for Great Britain's nuclear armament from gas diffusion plant.
> *Mr John Morris:* No, Sir. The assessment made in 1963 that supplies of fissile material already available or assured would be sufficient to meet all defence requirements for the foreseeable future is still valid. Production of U-235 for military purposes at Capenhurst Gas Diffusion Plant ceased in 1963, and there are no current plans to resume this production [250].

Britain has eight plutonium reactors in her weapons programme [251], and her weapons development and devices are concentrated on plutonium in contrast to the US programme which concentrated in uranium. Under exchange agreements with the United States, Britain has been able to obtain enriched uranium, particularly associated with the trigger mechanisms for thermonuclear weapons, in exchange for plutonium. Britain also is able to buy nuclear fuel for the reactors of her nuclear-propelled attack and Polaris submarines from the United States. However Britain fabricates her own nuclear weapons warheads.

The sequence of assignments of various British aircraft to the tactical nuclear strike role is a bit confusing due to the reassignment of various of these same aircraft from one role to another, from 'strategic' nuclear weapon delivery to 'tactical' nuclear or to conventional ordnance, and from carrier-based to land-based with possible changes in missions. It is often ambiguous if aircraft (F-4 Phantoms) which replaced former carrier-borne nuclear delivery aircraft (Buccaneers) nevertheless maintained that mission and capability.

Britain's V-bomber force of nearly 200 aircraft had carried megaton weapons [252]. Their strategic role was said to have been passed over completely to Britain's first two Polaris submarines on 1 July 1969. The V-bombers had been equipped with free-fall bombs and with the 200 mile range 'Blue Steel' stand-off bomb, or air-to-surface missile, which could be launched from high or from low levels [253–255]. The first units became operational with the stand-off missile in February 1963. A part of the Victor and Vulcan fleet were designed as aerial tankers for air refuelling of the nuclear weapons delivery aircraft. In 1963, the British Defence Ministry announced that it was developing a new nuclear stand-off missile to replace the Blue Steel, for use in the Buccaneer and in the anticipated TSR-2 aircraft [256]. (The Blue Steel was to be retired along with the V-force, although it is possible that some of the weapons will remain in service as long as any of their Vulcan delivery aircraft do.) When TSR-2 development was halted, the new weapon was redesigned for use with Britain's existing nuclear delivery aircraft [257]. In 1969, a British news report indicated that Britain was to produce

> ... a new all-British atomic bomb ... The weapon, known as the lay-down bomb, is needed for the RAFs Phantom, Buccaneer, and Vulcan bombers which are all being assigned to NATO for battlefield use in Europe. It is a small atomic bomb fitted with delayed reaction devices so that planes can drop it from low level without being destroyed by its blast [258].

The United Kingdom's aircraft types capable of delivering nuclear weapons now assigned to 'tactical' roles with NATO forces will thus clearly retain their role as delivery vehicles for nuclear weapons.

> Although the RAF no longer performs Britain's nuclear deterrent role, it still has a nuclear strike capability, for which the Vulcans in strike command are chiefly responsible. It is these aircraft which would be replaced by the Multi-role Combat Aircraft [259].
> Vulcan B-2s of Royal Air Force Strike Command now form as important a part of the strategic nuclear deterrent forces of SACEUR (Supreme Allied Commander, Europe) as they did of those of the West before missiles supplanted bombers as first-strike weapons.
> After cancellation of the American Skybolt, with which the V-bombers were to have been armed, they changed their mode of attack from high to low level and were equipped with lay-down bombs.
> This role is now performed exclusively by the Vulcan B-2s since the Victor B-2s are being converted to tanker duties; and the Vulcans assigned to SACEUR form an important part of this long-range atomic armoury with the USAF F-111s based in the UK.
> While today's Vulcan squadrons have basically the same task as that of the predecessors – the delivery of British nuclear bombs accurately upon specified targets, should the decision ever be taken to unleash the Western deterrent forces – their flight profile is now unequivocally hi-lo ...
> Nevertheless there is a severe restriction on the number of low-flying hours they can annually do, and this precaution must be seen in the context of the need for Vulcans to continue operating until MRCAs enter RAF service in about 1980 [260].
> The RAFs long-range strike force will consist by 1972 of some fifty Vulcan II V-bombers and about sixty-five Buccaneer Mark IIs. One-third of the former are to replace the Canberras that now constitute the Central Treaty Organisation's nuclear strike force; the rest are to be available to Supreme Allied Commander Europe for use against either high-explosive or nuclear land targets until their withdrawal from service in 1975 or 1976.
> Just over half the Buccaneer force will consist of planes transferred from the Royal Navy as a result of the withdrawal from service of the fleet carriers. The present intention is that all of these will be almost entirely reserved for the low-level maritime role for which they were originally designed, leaving a total of less than thirty to spearhead the deep-strike element in Central Europe.

... the 120 Phantom F-4Ms being delivered to the RAF should represent an adequate contribution to the short-range tactical strike force of SACEUR until their replacement by Jaguars between 1973 and 1975.

... The RAF is also in the process of acquiring a force of thirty Phantom F-4Ks. Originally this version was intended solely for Royal Naval use. Therefore, this force will be earmarked for maritime interception under Supreme Allied Commander Atlantic.

... Until the withdrawal from service of the carrier Ark Royal, the Royal Navy will operate thirty Phantom F-4Ks. Then these planes, too, will be transferred to the RAF; and, like the F-4Ks the RAF has already begun to receive, they will be earmarked for Saclant [261].

In the interim, the Buccaneers moved to land were to have had their range extended by a new engine, and to be fitted with nuclear weapons [262]. Published reports have described the retraining of RAF aircrews for missions requiring new delivery techniques for nuclear weapons [263]. The yields of the various British tactical nuclear air-dropped weapons described above are not known. 'Whether the revised V-force ... is considered tactical or strategic depends on the type of targets it is applied to by SACEUR' [264]. There is at least an indication that Vulcans are still carrying a megaton yield nuclear weapon 'in its new low-altitude role', perhaps as one of several different nuclear payloads that it can carry [265]. In addition, nuclear depth charges have also been available for Britain's former 76 Shackleton ASW patrol aircraft [266], and for their present successor, the 43 Nimrods [267].

The Vulcans and Buccaneers are still in service [268], but precisely how many of these British aircraft – Phantom, Buccaneer, Vulcan – are today assigned to nuclear weapon delivery apparently is not publicly known. As the Jaguar aircraft became operational with the RAF in 1974, some fraction of these aircraft also may have been assigned tactical nuclear weapon delivery roles [269]. In a list presented in 1973 in which General A. Goodpaster (SACEUR, NATO) indicated all NATO aircraft that were then nuclear capable, both those on QRA and those not on QRA, the following four British aircraft were included: Buccaneer, Harrier, Hunter, Vulcan. Some of these may be in the air-defence role. F-4 aircraft were listed also [270]. Finally, as the MRCA is acquired, it is clearly intended to carry the nuclear strike role.

As for British carrier-borne aircraft, both the Scimitar and Buccaneer S-1 and S-2 aircraft were designed with nuclear weapon capability and it is presumed that these were able to deliver nuclear weapons by LABS techniques [271]. The Scimitar was apparently operational on board UK aircraft-carriers from 1958 until 1965, when it was replaced by Phantom F-4K aircraft. Buccaneer S-1 and S-2 aircraft were operational from 1962 to the present. It appears that at least the carrier-borne Buccaneer and the Phantoms on board UK carriers were nuclear capable [272]. In the period under consideration, when these three nuclear capable British carrier-borne aircraft have been available, Britain operated three 'heavy' aircraft-carriers from 1957 to 1968, two from 1968 to 1972 and one from 1973 to the present.

Britain has had the short range (5–22 miles) Honest John ground-to-ground missile with its forces in Europe at least since 1960. These carry US nuclear warheads, under US controls [273]. They are presently being replaced by the 3–70 mile range US supplied Lance missile, also with US warheads [274]. Earlier, in the late 1950s, United Kingdom forces in Europe also had deployed the Corporal missile [275]. All the British Army's nuclear weapons and some of the RAF's are American.

IV. Tactical nuclear weapons of the USSR

Perhaps the most important message of this section is the sketchy and unsatisfactory state of knowledge concerning Soviet 'tactical' nuclear weapons, past and present. This was succinctly put by Mackintosh in 1971, 'We have, of course, no information on Soviet operational planning or nuclear capabilities' [276], and by Secretary of Defense Schlesinger in 1974 '. . . our information on Soviet tactical nuclear weapons in Europe is not very good' [277]. The phrase 'nuclear capability' when used in reference to particular Soviet delivery systems often represents a morass of ignorance and guesswork. To this author's knowledge such basic questions as whether or not the weapons of particular Soviet systems to which nuclear capability is often attributed actually carry nuclear warheads – such as fighter bombers or particular naval missiles – are still a matter of debate in classified intelligence circles. Of course there is no information on such questions in any Soviet printed source publicly known. A volume such as Douglas's recent *The Soviet Theatre Nuclear Offensive*, an analysis of Soviet open-source military literature as it applies to war in Europe, does not contain a single table or factum of information concerning the characteristics of Soviet tactical or theatre nuclear weapons [278]. A 1975 paper entitled 'A Comparison of US–Allied and Soviet Tactical Nuclear Force Capabilities and Policies' is capable of saying only the one single line – and nothing more – regarding Soviet aircraft, that 'Soviet tactical aircraft *believed to be capable* of delivering nuclear weapons are . . .' such and such types of aircraft. It adds, in general, 'The details of Soviet nuclear warheads, strategic and tactical have defied accurate definition by Western experts' [279]. The paper is also typical of the many on this question that quote each other, or a common source, for the few numbers that they use, without any indication of how that number was ever arrived at. It is not only vague Soviet sources that contribute to the continuing confusion. The following exchange took place in a 1974 US Congressional Hearing:

> *Question 19.* Does the Soviet Union have tactical nuclear weapons deployed in Eastern Europe? How many?
> *Answer.* The answer is classified and is in the Committee's files.

Question 20. What categories of Soviet nuclear weapons systems does this deployment include?

Answer. The answer is classified and is in the Committee's files.

Question 21. What is the aggregate yield of the Soviet stockpile? Are their tactical nuclear weapons bigger and 'dirtier' than ours? Are Warsaw Pact forces other than Soviet, trained and equipped to use tactical nuclear weapons?

Answer. The answer is classified and is in the Committee's files [280].

Western scholarship from ostensible authorities has also been atrocious. A 1955 statement by General Hoge that the Russian Army on the Elbe River line 'is *believed* to have atomic weapons, but we do not know to what degree' [281] is directly referenced by Hilsman and transformed into 'the announcement that Soviet troops are now being equipped with tactical atomic weapons' and '. . . they have begun to equip their forces in Eastern Germany with tactical atomic weapons, either missiles or cannon or both' [282]. According to US Department of Defense sources the USSR has not deployed nuclear 'cannon' to this day, and they did not apparently field the short range missiles with their forces before 1957 at the earliest.

Another level of dissatisfaction comes from the repeated recourse by Western analysts for their judgements to a handful of crude, superficial, impressionistic and oracular – almost pontifical – statements in an equally small number of Soviet writings or references to the subject of tactical nuclear weapons [283]. French and German doctrine has been that nuclear war would start immediately when the USSR attacks. This is taken to represent deterrence. Soviet doctrine has been that any NATO use of tactical nuclear weapons would lead inexorably to escalation and general nuclear war, and that Soviet nuclear response in Europe would be massive. This is on rare occasion taken to be deterrence, but far more often taken to stand for Soviet planning. (In any case it seems not to be very different from what actual US planning was.) It is a rare Western analyst that even considers the notion that 'Indeed, US–NATO policies and force deployments may well have encouraged the Soviets to develop forces and tactics based on a doctrine of surprise nuclear attack', assuming that it is Soviet doctrine [284]. It is universally held by Western analysts that 'More than any Western army the Soviets attempted to integrate nuclear weapons into their theatre warfare operational concepts' [285]. This judgement is made from reconnaissance of Soviet and WTO manoeuvres. It is considered an important point that Soviet tanks and armoured vehicles contain fission product air filtration systems. But both the filters and the tactics followed the introduction of large numbers of tactical nuclear weapons with NATO forces in Europe. These various problems, the paucity of hard data, the vagueness, generality, and unreality of Soviet doctrinal writings, the inferring of Soviet planning from observations of manoeuvres – and this author's general scepticism of public doctrinal statements, in particular Soviet ones – makes attempting to be definitive about Soviet tactical or theatre nuclear weapons a poor task.

One final note on terminology. The text below and the tables in the data section use NATO designations of Soviet missiles. At times, this has drawn criticism from Soviet analysts. However, it is unavoidable, as the Soviet designations for these weapons are unavailable in the West. The same holds for designations of Soviet surface vessels and submarines, though designations for Soviet military aircraft are available. Direct attempts to obtain the correct Soviet names for the weapons has proved unsuccessful. However, in the SALT negotiations in recent years, Soviet negotiators refer to their own weapons by their NATO code names.

In contrast to a very substantial amount of information on the size of the US weapons stockpile, no such information is of course available *from* the USSR and very little about the Soviet stockpile. In 1954 P. M. S. Blackett estimated the total Soviet nuclear weapon stockpile at between 75 and 300 weapons. US Senate testimony of 1963 quoted a figure of 5000 weapons, (15 per cent of the US stockpile), and Beaton estimated 10–20 per cent in 1964. Casey estimated a total Soviet stockpile of 5–10 000 Soviet warheads of all types in 1966, with no breakdown for tactical or strategic categories [286]. This last estimate may have been somewhat high, but it is very likely that the Soviet nuclear weapon stockpile has grown since in both categories.

It might be useful at the outset to indicate what the uncertainties regarding Soviet tactical nuclear weapons have been, and are:

1. How many such warheads the USSR deploys, particularly in the European theatre;

2. Whether nuclear warheads for shorter-range Soviet tactical missile systems are deployed with their launchers with Soviet troops in Eastern Europe, or whether such warheads are all or mostly kept within USSR borders;

3. Whether nuclear warheads are delivered by Soviet tactical aircraft, and if so which aircraft and how many of them;

4. Whether Soviet surface-to-surface missiles launched from naval vessels, both from submarines and surface vessels, many of which are intended as weapons against US aircraft carriers – carry nuclear warheads or not. The same question holds for the air-to-surface missiles carried by longer range Soviet maritime patrol aircraft, which are also intended as anti-aircraft carrier weapons.

Most of these questions are impossible to resolve from the information available in the open literature, and to some degree remain a matter of dispute in the intelligence community. (Nevertheless the tables in the data section of this paper follow the customary practice of indicating those Soviet aircraft considered to have nuclear weapon 'capability', though this word does not imply a known mission and role, as the same word used in connection with a Western aircraft usually does.)

One can also begin by giving a general broad description of Soviet tactical or theatre nuclear forces. Rather than building large numbers of short-range, low-yield missiles, which would be more useful for destroying discrete, well-located targets, the USSR has emphasized higher-yield mobile tactical missiles, primarily useful for terrain or blanketing fire, or for strikes against fixed logistic installations and airfields. In the description of Enthoven and Smith:

Soviet force structure raises serious doubts about their capability to fight a limited tactical nuclear war, much less one in which collateral damage and civilian casualties are to be kept to low levels. Limited nuclear wars with one side using discrete-fire techniques and the other using terrain fire are likely to be notoriously one-sided in favor of the latter. Equally important, most of the Soviets' nuclear delivery capability in Europe is based inside the Soviet Union. In short, the Soviets have neither the organization nor the force structure for a limited nuclear war fought exclusively against military targets in an engaged battle zone [287].

The difference in weapons suggests the difference in tactics. Soviet nuclear weapons are more suitable for saturating and destroying a specific geographic area. They are much less dependent on specific target acquisitions, and can therefore rely on fewer higher yield weapons. This in turn may have derived from the state of Soviet warhead technology when the weapons were being developed, or it may have been a purposeful choice.

The fact that the USSR has emphasized larger-yield and more indiscriminate nuclear terrain fire, many of the launchers for which are based within the borders of the USSR, makes it far less likely that 'limitations' could be maintained on a 'tactical nuclear war in Europe'. This seems a far more significant reason for the likelihood of such escalation than that Soviet statements say that it would occur. In fact Soviet statements probably say what they do in large degree because of the nature and location of the Soviet weapons. (It is not unlikely that the configuration of Soviet nuclear weapons that would be involved in any NATO–WTO armed interaction contributed to the introduction and increased use of the term 'theatre' nuclear weapon.) Another general overview of Soviet capabilities and planning was given in a recent US Department of Defense statement:

NATO understanding of Warsaw Pact military strategy and doctrine for a possible war in Europe is based on close study of Pact training exercises, force structure, organization, training, R&D, policy declarations and unclassified writings. Observations indicate that a major danger lies in a massive Warsaw Pact advance into Western Europe characterized by surprise, shock, and rapid air and ground exploitation.

— Surprise – Doctrine and exercises indicate that the Warsaw Pact places high value on tactical surprise with nuclear weapons. Their doctrine states that if the Warsaw Pact believes NATO is about to launch a major nuclear attack, it will seek to preempt with nuclear strikes on military targets. Moreover, there are clear indications that the Pact fully appreciates the initial advantage to be gained by a first use of theater nuclear forces in the absence of NATO indications to use nuclear weapons.

— Shock – Massive concentration of firepower on key military targets early in a conflict is a strong tenet of Pact planning. The objective is rapidly to disrupt and demoralize NATOs forces, creating opportunities for armored *blitzkrieg* attacks. Prime targets for Pact attacks are NATO nuclear delivery units, airbases, ground combat forces, command posts and support units.

— Exploitation – Warsaw Pact armored forces and their immediate support (artillery, tactical air, SAMs) are postured and trained to exploit nuclear attacks by rapid, deep, multiple thrusts to destroy remaining NATO forces and seize NATO territory. These armored forces are equipped for operations in a nuclear and chemical environment, so as to maintain movement and keep constant pressure on NATO forces.

The question is whether, in a war in Europe, the Warsaw Pact actually would follow this highly escalatory doctrine, and if so, how effective would their attacks be. National leaders are not, of course, constrained to follow the doctrine their military commanders use to guide training or exercise forces in peacetime, nor do training exercises necessarily indicate most probable tactics.

Soviet TNFs (Theatre Nuclear Forces), in addition to IRBMs, include tactical and intermediate-range aircraft, tactical rockets (Frog), surface-to-surface missiles (Scud, Scaleboard), land-based and sea-based medium and intermediate-range ballistic missiles, (SS-4s and 5s, SS-N-4s and 5s), and cruise missile configured surface ships and submarines. All these forces could be used for nuclear attacks on targets in Europe or Asia.

The Soviets continue to increase the flexibility with which they can use nuclear weapons. Older tactical aircraft are being replaced with modern dual-capable fighter and fighter-bombers such as the swing-wing Fitter C, Fencer and Flogger. Further, the quantity of delivery systems has been increasing. They are improving their theatre wide command, control and communications systems [288].

Soviet tactical nuclear weapons include surface-to-surface missiles of short and long range and both land and naval based, and presumably the air-to-surface missiles and tactical aircraft for nuclear delivery. The USSR is not known to possess nuclear surface-to-air missiles, ADMs or nuclear artillery.

The following section will look at five general questions concerning Soviet TNWs:

When they were introduced.

Their possible numbers.

Whether they are deployed outside Soviet borders, with Soviet forces in Eastern Europe.

Whether they are deployed with other WTO nations.

Their possible yields.

We will then look at the individual Soviet delivery systems for TNWs, and, finally, at some of the assessments of Soviet doctrines and tactics regarding the use of TNWs.

Soviet 'tactical', 'theatre', or 'regional' nuclear weapons were deployed sometime between 1956 and 'the early sixties'. The available sources do not agree precisely, nor do they supply much detail. However, at least for several of these early deployment years, the weapons in question were all Soviet medium bombers and MRBMs located within the USSR proper, and not in Eastern Europe. In 1977, the US Secretary of Defense stated that '. . . as early as 1956, the Soviets began deploying MRBMs and nuclear-capable light and medium bombers. Nuclear capable missiles organic to the ground forces were deployed to Soviet forces by the early 1960s' [289]. Symington stated that '. . . the Soviet Union began to deploy tactical nuclear weapons in Europe in 1957, and by the early 1960s its ground forces were fully equipped' [290]. However, Lee does not indicate the deployment of Soviet IRBM/MRBMs until 1959–1960 [291]. In addition, the first Soviet MRBMs and medium-range bombers presumably should be considered to have been targeted on SAC bomber bases and US carriers, both of whose targets were within the USSR, and it is therefore possible that the first Soviet IRBM/MRBM deployments might better be considered 'strategic' if they were intended to counter strategic systems.

The actual number of tactical nuclear warheads that the USSR deploys in total or with its forces in Eastern Europe, i.e. outside its borders, is an open question. Quoting the annual issues of *The Military Balance*, published by the IISS, nearly all publications refer to 3500 Soviet tactical nuclear warheads 'in the European area'. *The Military Balance*, in its 1970/71 issue reported the figure as 'probably about 3500 Soviet tactical warheads' [292]. However, it appears that this number is derived from the number of *delivery vehicles* that the USSR has with its WTO forces, and is *not* based on any direct evidence concerning the warheads. In 1972, Neville Brown suggested that 'half [the] figure' (of 3500) 'may be a better guess' [293]. In his 1974 publication, Record uses the number 3500 while in a 1975 one, he refers to '. . . an estimated 2250 Soviet TNW confront a NATO deployment of some 7000' [294]. Record writes that 'Although a large portion of Soviet delivery systems are deployed with Soviet forces in Eastern Europe, many, if not most nuclear warheads, may be retained inside the USSR' [295]. In fact, there are strong indications that at least until the early 1970s, all Soviet warheads *were* kept within the USSR, and were not deployed forward with USSR forces and their delivery systems outside of Soviet borders in Eastern Europe. A 1975 US Senate report is careful to say only that 'Some unclassified sources estimate that Warsaw Pact forces could be capable of using up to 3500 nuclear weapons in any attack against NATO' [296]. It says nothing about where the weapons are, or which systems might be included in that total. One reference in the open literature directly states that Soviet tactical nuclear warheads are deployed in East Europe. In a 1975 paper, Erickson stated that 'stocks of tactical nuclear weapons maintained in the air-base areas' of Soviet fighter bomber regiments in Finsterwalde, Kolberg, and Lausitz 'have recently been enlarged' [297]. There have been various Western statements that the USSR has increased its tactical nuclear weapon delivery 'capability' in the WTO area in recent years, in 1972, 1974, 1976, without any specifics as to numbers of warheads, and there has been no change in the number of '3500' that is customarily used.

Perhaps precisely because the USSR borders directly on the territory of its Eastern European WTO neighbours there is even less utility in talking about the numbers of TNWs that it may have deployed forward with its forces outside of its own borders, inaccurate as the number 3500 may be, than in stressing the particular number of TNWs that the US had deployed overseas in NATO Europe. If the number of Soviet TNWs in the Western USSR were counted, it is possible that the NATO–WTO TNW balance might look very different from its usual description. The corollary of this is that using such a broader inclusion, no one in the West probably really knows what that balance looks like. The IRBM/MRBMs and such aircraft as the Soviet Backfire bomber which in another context are referred to as 'gray area systems' – because to date they have fallen between the SALT and the MBFR negotiations – are considered a basic portion of the Soviet TNW forces. Their closest analogue is perhaps to the 'Forward Based Systems' of the US. They are not placed in a similarly labelled category since they are based within the USSR, and not outside of USSR borders. (However, it is possible that mention of these systems should have been included in the section on forward based systems, below.) It might usefully be mentioned that F.R. Germany, France and the United Kingdom have recently expressed interest in developing long-range ground-based strategic cruise missiles in a NATO context, as an expressed counter to Soviet deployment of new SS-20 MRBMs and Backfire bombers, both assumedly targeted on NATO Europe. NATO would then return to a successor to the earlier MACE system.

In May 1958, in a speech to the WTO's Political Consultative Committee meetings in Moscow, Premier Khrushchev accused NATO of planning 'to arm West Germany with atomic weapons' and declared that the Warsaw Treaty states were 'compelled by force of circumstance to consider deploying missiles in the German Democratic Republic, Poland, and Czechoslovakia' [298]. Though this has been interpreted to mean the sharing of TNWs with the named Warsaw Treaty allies, the sentence seems just as sensibly to refer to missiles deployed with Soviet forces. However, some years later during Premier Khrushchev's tenure the decision was apparently taken to provide several of the East European forces with systems that served for nuclear weapon delivery with Soviet forces, tactical missiles and advanced fighter-bombers. The first Soviet-made tactical missiles were displayed by a Soviet Warsaw Treaty ally in mid-1964, by Polish forces in Warsaw. In January 1965, at the time of WTO concern about NATO-MLF proposals, the GDR apparently requested some kind of access to, or control over, the use of nuclear weapons. The USSR apparently rejected the request [299]. The WTO forces of East European allies have been given training in nuclear warfare methods during joint military exercises, and missile systems of East European forces have participated in simulated nuclear exercises. The question is whether there are any arrangements in the WTO under which nuclear warheads for these delivery systems might be made available to the WTO allies of the USSR. The USSR has never commented on WTO procedures for controlling nuclear access, or even if any such provision has ever been agreed to by the USSR. Every indication is that there are no arrangements for transfer of Soviet TNWs to its WTO allies under procedures such as the 'double key' system in NATO. There has never been any evidence of any form of transfer of Soviet TNW warheads to WTO forces.

The USSR has now also transferred Frog and Scud missiles, and nuclear-capable aircraft, to Egypt, Iraq, Syria, North Korea and Cuba. There has been no suggestion of transfer of nuclear warheads. The allegations of the delivery and removal of Soviet nuclear warheads to Alexandria, Egypt, in October-

November 1973 during the war in the Middle East, remain unclarified to date. (There is no firm public evidence concerning the allegations, and if warheads were actually moved into the theatre, whether they were to be regarded as being for the use of Soviet forces or other forces.)

It is universally stated that the warheads of Soviet TNWs have higher yields than those of NATO. This needs at least several caveats:

— No one in the West has *any direct* knowledge of the yields of Soviet TNWs (or, for that matter, of Soviet strategic warheads). Such estimates are made at times from the payload-range capability of the Soviet missiles, or by other methods;

— It is assumed that the yields of the roughly 700 Soviet MRBM/IRBMs are relatively high. They are usually given as being in the 1 megaton range. However, this is not any different from some NATO air-dropped weapons, and not substantially different from higher Pershing missile yields.

— It is assumed that there are no very low yield Soviet TNWs, since there have apparently been relatively few Soviet nuclear tests at such yields;

— The yields of Soviet air-dropped TNWs are unknown;

— This leaves the Frog and Scud tactical nuclear missile systems. The yields of the Frog missile systems have been estimated as being in the 5–10 kt range, which would be the same as that of the US Honest John and lower than that of the Sergeant, Pershing and Lance missiles. Higher yields are attributed to the Scud missiles, but these are unlikely to be much different from Pershing yields.

Record states that 'Over two-thirds of Soviet TNW are believed to possess yields well in excess of the 13 kt Hiroshima bomb; more than 500 contain yields ranging from $\frac{1}{2}$ to 3 Mt' [300]. It is clear that the common description apparently has depended on the absence of Soviet very low yield weapons, and the assumed higher yields of the MRBM/IRBMs. However, NATO has yields equivalent to the latter groups. The description also depended on the presumed lower accuracy of Soviet TNWs.

We can now turn to some examination of the individual delivery systems for Soviet TNWs. A general summary was provided in a 1972 US Department of Defense statement, which also provided some indication of the derivation of the number 3500.

> Soviet theatre nuclear forces include about 700 ballistic missile launchers (including medium and intermediate range missiles, as well as the SS-11 dual-purpose missiles) and about 850 tactical surface-to-surface missile launchers assigned to their ground forces. In addition, their large medium bomber force of about 700 aircraft in long range aviation and 500 aircraft in naval aviation are capable of carrying nuclear weapons, as are approximately 1100 light bomber and fighter bombers in the tactical air forces. Soviet naval forces, both surface and subsurface, also carry nuclear-capable missiles. There are also small numbers of the short range SHADDOCK and SCALEBOARD missiles.

A great many aircraft in the Soviet tactical air force possess a nuclear weapon delivery capability. In addition to the multi-purpose FISHBED (MIG-21) fighters, this total includes BREWER (YAK-28) and BEAGLE (IL-28) light bombers and FITTER (SU-7) fighter/bombers, which appear to be the primary tactical air nuclear delivery vehicles.

In theater nuclear capable naval forces, the Soviets have continued a steady buildup in their three western fleets – surface and subsurface – for operations against NATO navies, NATO sea lines of communication and coastal targets. These forces include their two ASW helicopter ships, as well as other major surface combatants, submarines and small patrol boats and submarine chasers equipped with possible nuclear capable surface-to-surface and/or surface-to-air guided missiles [301].

The paragraphs contain several explicit caveats, and there are in addition others that are less evident. Major questions exist concerning the nuclear delivery roles of Soviet tactical aviation.

Of the approximately 1700 ground-launched Soviet TNW delivery systems, from about 1400 to 1500 are reportedly deployed in or targeted against Europe. The number of longer range IRBM/MRBMs was constant for many years. It was reduced slightly in 1976, and the deployment of the new SS-20 missile, which is assumed to contain MIRVs, has now been reported. US Secretary of Defense Laird's 1972 figure of 850 shorter range (FROG and SCUD) missiles (above) was the first instance in which a specific number had been supplied by any source for this group of Soviet weapons.

The Soviet Union does not deploy any nuclear artillery. The 1971–2 IISS *Military Balance* reported that Soviet artillery, the M-55 203 mm gun-howitzer, fires nuclear shells and that this artillery piece has been in service since the 'late 1950s' [302]. The same information was reported in subsequent issues of *The Military Balance*† and has been quoted and referenced by numerous other writers, including Record. However, representatives of the US Atomic Energy Commission, of the US Joint Chiefs of Staff, and General Goodpaster, SAC-EUR, have flatly denied in open testimony before US Congressional committees on several different occasions in 1973 and 1974 that the USSR has any nuclear tube artillery [303].

Ambiguities concerning the nuclear delivery role of Soviet tactical aircraft are considerable. The questions concern design, training and intention. The tables used in the data section would imply that the USSR presently maintains as many as 2400 aircraft that could be used for nuclear weapon delivery. However, Record, from whose monograph the table was extracted, writes:

> ... It is highly improbable, however, that a significant fraction of these would ever be so employed. The Soviet Union's strong preference for missiles, not aircraft, as the principal means of delivering

---

† In 1974–5, the IISS withdrew the entry, and in 1975–6, returned it with a footnote saying that 'Although shown in the table, it is uncertain whether the Soviet 203 mm artillery is nuclear capable'.

nuclear ordnance has already been noted. Moreover, Soviet tactical air doctrine, the product of a traditional land-oriented military philosophy that places a premium on massive concentrations o armor and artillery, relegates tactical air forces largely to the roles o defending the Soviet homeland and countering enemy airpower ove the battlefield. This defensive orientation contrasts sharply with US doctrine, which attaches great importance to the achievement o theater air superiority as a prerequisite to deep interdiction.

Finally, the relative design simplicity of Soviet tactical aircraf and their probable lack of sophisticated avionics packages strongly suggest a pronounced inability to deliver ordnance with the degree o accuracy necessary for battlefield deployment of TNW. The use o Soviet tactical aircraft in the nuclear mode is likely to be confined therefore, to theater targets, whose destruction in any event coulc probably be accomplished more rapidly and efficiently by the use o preferred SSM [304].

Evidence in the 1960s on Soviet training for nuclear weapons delivery witl tactical aircraft was lacking. There were also indications that the USSR did no assign a nuclear weapon delivery role to tactical aircraft for security and com mand and control reasons. The ranges of the aircraft were also relatively short There was not a single US statement in the 1960s directly stating that particula Soviet tactical strike aircraft had a nuclear weapon delivery role. Subsequen statements are hedged with caveats and the description of 'capability'. The air-to-surface missiles carried by such long- and medium-range Soviet aircraf as the TU-16 and TU-20 seem intended against US aircraft carriers. The TU-2: Blinder, an expected follow-on aircraft, was produced only in very low numbers Finally, in the 1960s, the IISS table for 'major nuclear delivery systems, aircraft printed each year in its publication *The Military Balance* did not include any Soviet tactical aircraft other than the TU-22 (Blinder). The first issue of thi publication to list Soviet tactical or 'strike' aircraft as nuclear delivery vehicle was the 1970–71 issue. It then listed the TU-22 (Blinder), IL-28 (Beagle), Yak-2{ (Brewer), SU-7 (Fitter) and MIG-23 (Foxbat).† The 1972 quote (above) from Secretary of Defense Laird again indicated several of these aircraft, and in 1974 the US Department of Defense indicated that Soviet tactical aircraft deployec after 1970 replaced tactical aircraft which had *not* been estimated to hav nuclear delivery capability.

The ability of forward-based Soviet Frontal Aviation forces tc deliver nuclear weapons has increased significantly in the past thre years with the delivery of late-model MIG-21 Fishbed aircraft anc other new fighters with a nuclear delivery capability. These havu replaced older aircraft without such a capability. We estimate th nuclear-capable Soviet aircraft in the forward area have increased b over [deleted] airframes since January 1971 [305].

---

† The MIG-23 Foxbat referred to above is now labelled the MIG-25 and is assumed to have bee: designed by the USSR as an interceptor intended against the B-70 bomber that the US was develop ing at the time.

Earlier in the same Congressional Hearing, General Goodpaster would indicate nothing more specific than that the USSR had '. . . some air-delivered weapons' for tactical use [306]. Finally, a 1977 US Department of Defense publication indicated that,

> The Soviets have also continued active training in nuclear delivery techniques with the tactical aircraft assigned to frontal aviation. The most frequently used aircraft on nuclear delivery training missions are MIG-21/Fishbed-Js, -Ks, and -Ls; SU-7 and SU-17/Fitters; MIG-23/Floggers; and various medium-range bombers. The SU-19/ Fencer is expected to be an excellent aircraft for this type of mission [307].

Despite the indications in the several above post-1970 listings of Soviet attack aircraft with TNW capability, there still seems to be substantial doubt about any TNW-assigned mission to aircraft other than the Suchoi models. In addition a contrary interpretation has been that the substantial increases in payload and range of newer Soviet tactical aircraft are not intended to supply or enhance a TNW delivery capability but to substantially increase the *conventional* ordnance capability of these aircraft against 'primary NATO airbases, nuclear storage sites, and command-and-control facilities at the onset of a European war' [308].†

Soviet surface vessels and submarines (both diesel and nuclear) carry cruise missiles of short range (up to several hundred nautical miles) capable of carrying nuclear warheads. A survey of the available US, British and Soviet [in translation] literature, indicates that there is again no absolutely unequivocal statement that these missiles carry nuclear weapons, or in fact whether their targets are US carrier strike forces, land targets, or equal or varying proportions of both. Soviet literature would lead one to believe they are intended against US carrier task forces, and this is presumed to be their primary target. US statements never discuss the function of the cruise missiles carried by heavier class Soviet surface vessels. The annual 'posture statement' by the US Secretary of Defense, throughout the tenures of Secretaries McNamara, Clifford and Laird, carried a note to its table on US and USSR strategic nuclear forces of which the following is an example.

> In addition to the Submarine-Launched Ballistic Missiles (SLBMs) on nuclear-powered submarines, the Soviets have SLBMs on diesel-powered submarines, whose primary targets the intelligence community estimates to be strategic land targets in Eurasia. The Soviets also have submarine-launched cruise missiles whose primary targets are believed to be naval and merchant vessels.

In 1974, Record indicated a total of over 500 SSM aboard about 115 Soviet naval vessels, mostly submarines.‡

---

† By January 1978, the Brookings Institution will have published a more thorough study of this subject, *Soviet Air Power in Transition*, by R. P. Berman.

‡ Data on these weapons was gathered, but has not been included in the paper.

. . . Designed primarily for antiship operations, most of these SSM could be employed directly, although with substantial loss of accuracy, against ground targets in Europe, given the proximity to the continent of most major Soviet naval deployments [309].

Very little can be said about Soviet command and control procedures for TNWs. As indicated, there is no evidence suggesting that the USSR would make TNWs available to its WTO allies even in wartime. This presumably simplifies command and control, as does the absence of very short range Soviet TNWs or any positioned close to NATO–WTO borders. There are strong indications that the control of all Soviet missiles is highly centralized, with MRBM/IRBMs under the command of the Strategic Rocket Forces, and the theatre-level Scud and Scaleboard (and probably the 'battlefield' Frog batteries) under the command of the Rocket and Artillery Forces.

The Western literature which presumes to explain Soviet doctrine for tactical nuclear warfare is quite large in contrast to the amount which treats actual Soviet TNW delivery systems. It is also an extremely contradictory body of literature. Surveying it, one finds various authors holding diametrically opposed views on the following major points:
— Whether Soviet use of TNWs would be massive or not,
— Whether Soviet use of TNWs would be early or not,
— Whether Soviet use of TNWs would be a first use, or only in response to NATO use.
— Whether the USSR would in fact use nuclear weapons at all in a conflict in the European theatre, or would instead attempt an overwhelming conventional attack.

Not the least of the problems in this area is the fact that the judgements of Western authors most often seem to be derived from a group of extremely limited and cryptic Soviet writings, from which many analysts seem to be able to obtain far more meaning and clarity than the original statements contain. Other interpretations have often stressed the situations developed in Soviet and WTO manoeuvres, while a third group has stressed Soviet force structure and the operational characteristics of Soviet weaponry.

On the development of Soviet notions on tactical nuclear warfare during Premier Krushchev's tenure, Wolfe writes as follows:

. . . By the end of the Khrushchev period, it was a generally accepted article of theater doctrine that the task of the tactical missile forces would be that of 'destroying any important enemy targets and troop formations that may survive strikes by the strategic missile force' . . .

Neither Soviet theater doctrine as it evolved during the Khrushchev decade nor the changes in organization and weapons introduced into the theater forces during that period tended to encourage the notion of imposing restraints of one kind or another upon the conduct of theater operations in Europe. A doctrine assuming that military conflict in Europe would be inseparably tied to general nuclear war and prescribing that Soviet forces must promptly seize the initiative with an all-out theater offensive did not lend itself readily to the idea of limiting theater operations once they had been set in motion. As for the theater forces themselves, their increasing nuclearization at the expense of conventional staying power, and the problems of command and control created by the more complex conditions of modern warfare, were among factors which, to some extent at least, raised questions as to whether these forces possessed the requisite capabilities for conducting carefully controlled and graduated operations. In addition to such doctrinal and technical considerations, there was also Khrushchev's own frequently stated position, adopted perhaps both out of conviction and for its deterrent effect, that limitations on warfare in the European theater, where vital interests of the great powers were closely enmeshed, would prove unfeasible . . .

If there was some shift of Soviet thinking in the Khrushchev period on the possibility of maintaining a 'firebreak' between conventional and nuclear weapons, the same can hardly be said with regard to drawing the line between tactical and strategic nuclear weapons, particularly in the case of a European conflict. With very few exceptions, both military and political analyses of this period stuck to the view that no distinction was feasible between tactical and strategic nuclear arms, and that the use of nuclear weapons in any form would mean escalation to general war in which neither side could be expected, in Khrushchev's words, 'to concede defeat before resorting to the use of all weapons, even the most devastating ones'. Or, as Marshal Malinovskii put it in November 1962, when warning that an adversary should not count on being able to employ tactical weapons without courting massive retaliation, 'No matter where a "tactical" atomic weapon might be used against us, it would trigger a crushing counter-blow' [310].

The 'nuclearization' of Soviet forces to which Wolfe refers nevertheless seems to have been substantially less than was taking place on the NATO side. Similarly, at the same time as Wolfe notes a lack of 'conventional staying power', NATO was decrying Soviet conventional superiority.

In a 1976 volume summarizing the unclassified Soviet military literature relevant to tactical nuclear warfare, Douglas presented the following nine conclusions.

1. *To the Soviets, theater nuclear weapons are a fundamental part of their warfighting capability rather than a deterrent adjunct to conventional forces.*

2. *To achieve surprise in a crisis situation, the Soviets would prefer to attack basically utilizing those forces in place at the time of the final decision, thus, providing NATO with a minimum warning of Soviet attack preparations, as little as a few hours.*

Throughout the Soviet material, the importance of surprise and the need for secrecy are major continuing themes. In fact, surprise that catches the enemy unaware, including the use of nuclear weapons, is as important as the total attacking strength. Soviet doctrine portrays NATO as the aggressor and does not lend support to the views that the Soviets would launch a 'bolt out of the blue' surprise attack in the absence of a crisis.

3. *An in-depth, massive, surprise, nuclear strike, in conjunction with an immediate, high-speed air and ground exploitation, is still the dominant Soviet concept for war against NATO.*

Although the Soviets acknowledge that war may begin conventionally, or with the limited use of nuclear weapons, they see escalation as highly likely, and intend to strike first with a massive, in-depth, nuclear strike. This strike is designed to (1) destroy NATO means of nuclear attack, (2) destroy the main groupings of combat forces and their command/control, (3) isolate the battlefield, and (4) breach the main line of defense and define the main attack corridors.

This concept calls for the landing of airborne assault troops and the beginning of a high-speed combined arms forces attack led by tank troops immediately after the nuclear strike. The ground forces are designed to be exceedingly mobile and well protected to best exploit the shock effect of the initial nuclear strike, to maximize NATOs targeting problem, to minimize their time to transit radioactive contaminated areas and, thus, presumably to minimize their losses.

4. To defeat the Soviet military strategy, it is imperative to first defeat the combined arms forces ground offensive.

Perhaps the single most important aspect regarding the Soviet nuclear threat in Europe is the political objectives that underlie their military strategy, namely, to consolidate and maintain their hold on Eastern Europe and to bring Western Europe under Soviet hegemony. Should a war in Europe break out, the Soviets intend to seize the initiative and to capture and occupy large segments of NATO territory. This can be attained only as a result of operations of the ground forces.

The Soviet nuclear strike would probably not aim at the destruction of Europe but, rather, at facilitating the accomplishment of their territorial objective. The counternuclear aspect of their nuclear strike would perform the role of preventing NATO *nuclear* forces from interfering with this exploitation. The remainder of the nuclear strike, which is considerable, is designed to hit NATO conventional forces and cripple their ability to defend against the exploitation. *Anticipation* in mounting a surprise nuclear attack is important because of the time required to get the exploitation forces coordinated and underway.

5. The Soviet ground force offensive may be particularly vulnerable to a rather moderate and easily preplanned nuclear interdiction strike directed in front of the main second echelon and reserve exploitation forces.

6. The only escalation boundary for a limited nuclear war in NATO that can be inferred from the Soviet unclassified literature is the boundary between theater nuclear war and intercontinental general nuclear war. The potential existence of this boundary appears to be growing more evident in recent Soviet writings.

The characteristics and objectives of theater and intercontinental war are so clearly different – objectives, risks, forces, strategy and tactics – that the Soviets seem to have no difficulty in envisioning a nuclear war limited to the theater – so long as the theater does not expand to include strikes on the Soviet Union itself. The more recent Soviet literature tends to project a picture of theater nuclear operations that seem to take place in an environment considerably more detached from the general war context than was the case in the early sixties. Further, the Soviets apparently view changes in the nature of the strategic forces and in the strategic balance of power, or correlation of military forces, that took place during the late 1960s and early 1970s as significantly lowering the risk of escalation from theater to intercontinental war. This, coupled with the concomitant decreasing strategic role of theater-based forces, has permitted the development of a new, detached, theater-oriented perspective.

With regard to the possibility of limiting the scope of a nuclear war in Europe to some smaller restricted area (that is, a war within a war), although one might like to identify clear affirmative trends in Soviet military literature, it is still too early to do so. Arguing against limiting the nuclear conflict to, for example, some restricted or extended battlefield area is the fundamental emphasis that the Soviets place on the need to strike throughout the entire depth of the defense, on the need to destroy all targets as soon as possible, on the decisive importance of the first strike to the total outcome of the war, on their perception of the NATO threat, and on the need to anticipate and strike first. Further, the only reference in the Soviet military literature to a limited Soviet tactical nuclear employment capability is not as a response in kind, but as a delaying action designed to mislead the enemy and provide time to switch from a conventional action to a nuclear attack which, subsequently, would be initiated with a massive, in-depth, surprise, nuclear strike. Although the Soviets may be experimenting with different forms and scales of theater nuclear warfare, this work is still referred to as 'research' in the literature and not as capability, strategy, or doctrine.

7. *The increased Soviet attention to conventional capabilities observed over the past decade does not, by itself, represent any less readiness to employ nuclear weapons in a war with NATO, nor does it represent any shift in the nature of the Soviet threat to NATO away from nuclear and toward a primarily conventional attack strategy.*

Nuclear weapons in war with NATO are still seen by the Soviets to play a very significant role in their doctrine. First, nuclear weapons will continue to occupy a dominant position for their psychological aspects or coercive threat to paralyse the NATO alliance and break off smaller states prior to or even at the beginning of hostilities should a war develop. Second, Soviet nuclear capability, which they repeatedly state would be superior to the enemy's, is seen as deterring the NATO use of nuclear weapons against Soviet forces engaged in any conventional local action. Third, the weapons provide a capability to destroy NATO nuclear means of attack and major force concentrations in a major attack on NATO and to preempt any perceived NATO nuclear strike against the Soviet forces.

8. *The manner in which the Soviets have approached nuclear warfighting may have had the side benefit of providing them with very significantly improved conventional capabilities.*

9. *At the lower end of the spectrum, the scenario where the Soviets appear now to recognize explicitly the limited use of nuclear weapons is a localized war involving a limited number of participating countries and a well-defined geographic region of military operations* [311].

It is interesting to note that many of these conclusions are based entirely on quotations from the one or two volumes by Sidorenko and Sokolovskii. Statements in these Soviet volumes which deal with the nature of tactical nuclear warfare are phrased in a tone which is a mixture of the exhortative and the predictive – such and such shall happen – which strikes this author as nothing short of bizarre and totally unrealistic. They might best be described as being dreamlike, and their most indicative characteristic is a studious disregard of any operational realities that might actually ensue under the conditions supposedly being discussed. This was almost caricatured in an unusual occasion during the 1961 Berlin crisis in which Western military attachés were taken to observe a Soviet

> ... demonstration today of how its tank-supported infantry would operate on a nuclear battlefield in time of war ... As the battalion advanced on its objective, a Soviet military commentator told the attachés that it was equipped with rockets fitted with nuclear warheads. A simulated nuclear mushroom cloud was then shown on the position [312].

A real nuclear battlefield would certainly be a good deal less simple.

Recent Western analyses continue to display a good deal of contradiction. A British military commentator indicates that 'All their (Warsaw Pact) exercises have been carried out with nuclear weapons from the word "Go" ' [313], while Erickson describing the WTO *Shield-72* exercises stressed that they were 'explicitly defensive in form', though they presumably included conditions of simulated use of TNWs [314]. Vigor and Donnelly state that 'tactical nuclear strikes are made to clear a path through the enemy defences and prevent an enemy counterattack' in the same paper in which they state that the USSR would consider 'including nuclear weapons, if this is deemed necessary to avoid defeat' [315]. In a 1975 paper, Erickson perhaps ironically pointing out that 'Soviet doctrine is a hazy affair', states that

> ... in the early 1960s, when a rudimentary form of 'flexible response' appeared on the Soviet side, conventional ammunition stocks were so low that it is unlikely that the Soviet Army could have conducted its conventional phase beyond a few hours.

but that *at present* the 'Soviet interest [is] in keeping the nuclear threshold quite *high*' [316]. (If Soviet conventional munitions stocks had been so low in the early 1960s, Soviet forces could hardly have carried out extensive conventional operations throughout the NATO theatre *following* a presumptive nuclear strike!) Erickson adds that

> ... Soviet doctrine, as ever, continues to emphasise the rapid seizure of the initiative from the 'defensive' and high speed penetration into the whole depth of the theatre: while this does imply a 'short war in a nuclear environment', one of the noticeable features over the past year or so has been increased Soviet interest in the possibility of substantial non-nuclear operations even in the *initial stage* of a major engagement, for which reason the 'attack norms' of a nuclear blitzkrieg have been scaled down to meet the conditions of a conventional phase [317].

In several papers in 1975, Lambeth carried this view even further, stressing the reliance on *non*-nuclear conventional operations by Soviet forces, and the same message seems implicit in the recent and extensive writings of Steven Canby. A 1975 article in the US publication *Army* stated that Soviet force structure was clearly directed toward conventional options [318]. A year later, Drew Middleton again wrote, however, of

> the increased strength of Soviet tactical nuclear weapons in Europe along with a new emphasis in Russian military writing on the importance of surprise and the inevitability of nuclear war once the first shot of a conventional war has been fired [319].

(For Middleton, this might be considered a not atypical bit of 'disinformation'.)

There are strong indications that NATO itself does *not* assume early Soviet use of TNWs in the case of a conflict in Europe. In a study prepared for the US Congress in 1977 concerning the planning of US general purpose forces, the following was given as the underlying assumption for US force planning in Europe and the possible circumstances of a NATO–WTO confrontation:

> It is assumed that it is possible to have a major war that·does not rapidly escalate to use of theatre or strategic nuclear weapons; further, the Warsaw Pact might begin such a major war without a preemptive nuclear strike [320].

John Morse, a former head of SHAPE's Nuclear Plans Section, also writing in 1977 bemoaned any NATO assumption that 'the Soviet will not use nuclear weapons early in any serious attack'. He claimed that 'the possibility of the use of (presumably Soviet) nuclear weapons is largely ignored, there are no plans for it, and most attention goes to other matters' [321]. Many of these views are in essence clearly contradictory and result in a confusing picture.

References to Chapter 1

1. *Military Applications of Nuclear Technology*, Part II, Hearings, Joint Committee on Atomic Energy, Congress of the United States, 93rd Congress, (22 May, 29 June 1973), p. 3.
2. *Military Applications of Nuclear Technology*, Part III, Hearings, Joint Committee on Atomic Energy, Congress of the United States, 93rd Congress, (22 May, 29 June 1973), p. 2.
3. *Military Applications of Nuclear Technology*, Part I, Hearings, Joint Committee on Atomic Energy, Congress of the United States, 93rd Congress, (22 May, 29 June 1973), pp. 15–17. Earlier, General Giller had stated, 'For my purpose, I define strategic as the Triad, the surface missiles, ocean missiles, and the SAC bombers. Other delivery systems will be included in the tactical or in the naval area, and it is merely for ease of formal rather than any particular other reason'. 93rd Congress, p. 3.
4. *Public Works for Water and Power Development and Atomic Energy Commission Appropriations for Fiscal Year 1974*, Part 4, Atomic Energy Commission, 93rd Congress, 1973, pp. 3922, 3963.

5. Casey, C. W., Col., *The Employment of Tactical Nuclear Weapons, A Search for Safe Limits*, mimeographed, April 1966, 76 pp.

6. 22 000 Tactical and 8000 Strategic; 30 000 US Nuclear Weapons, *Defense Monitor*, vol. 4, No. 2 (Center for Defense Information, Washington, DC, February 1975), 12 pp.

7. Korea and US Policy in Asia, *Defense Monitor*, Vol. 5, No. 1 (Center for Defense Information, Washington, DC, January 1976), 8 pp.

8. The Candidate on the Issues, An Interview, *Washington Post*, (21 March 1976).

9. Horton, A., *President Urged to Leave Nuclear Weapons in Korea*, (Memphis Press, 16 June 1977).

10. *Korea and the Philippines, November 1972*, A Staff Report. Senate Foreign Relations Committee, (93rd Congress, 18 February 1973), pp. 21, 23.

11. Owen Wilkes, personal communication; *Yomiuri* (Tokyo), (26 December 1974).

12. *Development, Use, and Control of Nuclear Energy for the Common Defense and Security and for Peaceful Purposes*, First Annual Report to the US Congress, Joint Committee on Atomic Energy, 94th Congress, (30 June 1975), p. 19. (A similar statement appeared in the Second Annual Report of 1976.)

13. *Development, Use and Control of Nuclear Energy for the Common Defense and Security, and for Peaceful Purposes*, Second Annual Report to the US Congress, Joint Committee on Atomic Energy, 94th Congress, (30 June 1976), p. 136. The table and an accompanying statement were provided by the Assistant for Atomic Energy to the US Secretary of Defense, (February 1976).

14. *Military Applications of Nuclear Technology*, Parts I and II, US Congress, Joint Committee on Atomic Energy, 94th Congress, (30 June 1976).

15. *Military Applications of Nuclear Technology*, Part I, US Congress, Joint Committee on Atomic Energy, 94th Congress, (30 June 1976), p. 2.

16. *Security Agreements and Commitments Abroad, Report*, Committee on Foreign Relations, US Senate, 91st Congress, (21 December 1970), p. 12.

17. Pincus, W., A new generation of weaponry: why more nukes, *New Republic Magazine*, Vol. 170, No. 6, (9 February 1974), pp. 14–17.

18. This period and the alternative recommendations of the GAC have been recently described in some detail in a short but ground-breaking volume. York, H., *The Advisors: Oppenheimer, Teller, and the Superbomb* (W. H. Freeman & Co., San Francisco, 1976), 175 pp. See also, Hewlett, R. G. and Duncan, F., *Atomic Shield, 1947/1952, Vol. II – A History of the United States Atomic Energy Commission* (University Park, Pa.: Pennsylvania State University Press, 1969); and Profiles: I. I. Rabi, Physicist, Part II, *The New Yorker*, (20 October 1975), pp. 72–80.

19. Lapp, R. E., *The New Force, The Story of Atoms and People* (Harper, New York, 1953), 240 pp.

20. Norman, L., The new look strategy, thermo-nukes in the economy package, *Combat Forces Journal*, Vol. 4, No. 7, (February 1954), pp. 115–20.

21. Wilson's Missile Order – Detailed Text, *Missiles and Rockets*, Vol. 1, No. 3, December 1956, pp. 70–1; Toftoy, H. N., Maj. Gen., Army's Role in Guided Missiles, *Missiles and Rockets*, Vol. 1, No. 2, (November 1956), pp. 36–9.

22. Pincus, W., Nukes Nobody Needs, *The New Republic*, Vol. 170, No. 16 (20 April 1974), pp. 12–4. (The Congressional Hearings in which the testimony appears is referenced in reference 38, below).

23. *The Atlantic Alliance*, Hearings, Committee on Government Operations, US Senate, 89th Congress, (May 1966), Part 2, pp. 69, 86.

24. Yefremov, A. Ye., *Europe and Nuclear Weapons* [translated], UPRS 58481, 14 March 1973.
25. Hoge Notes, No Rise in East Zone Force, *New York Times*, (16 January 1955).
26. Watson, M. S., Initial Unit Under Test with NATO, *New York Times*, (17 January 1957).
27. Les Armes Nouvelles, *Revue Militaire d'Information*, No. 284, (June 1957).
28. L'avis du Général Norstad, *Revue Militaire d'Information*, No. 284, (June 1957); Les Armes Atomiques, *Revue Militaire d'Information*, No. 283, (May 1957).
29. Roberts, C. M., US is Considering A-Arms Stockpile for NATO, Dulles Says, *Washington Post*, (16 July 1957).
30. *US Security Issues in Europe: Burden Sharing and Offset, MBFR and Nuclear Weapons, September 1973*. A Staff Report. Committee on Foreign Relations, US Senate, 93rd Congress, (2 December 1973), p. 15.
31. Hilsman, R., Strategic Doctrines for Nuclear War, in *Military Policy and National Security*, W. W. Kaufman (ed.), (Princeton University Press, Princeton, NJ, 1956).
32. Honick, M., SHAPE, A Silvered Shield, *NATO's Fifteen Nations*, Vol. 21, No. 1, (February–March 1976).
33. Finney, J. W., We Are Already Sharing the Bomb, *New York Times*, (28 November 1965); 9 Allies Planes Have Atom Arms, *New York Times*, (23 November 1965).
34. Tanner, H., 5000 A-Warheads Stored for NATO, McNamara Says, *New York Times*, (28 November 1965).
35. Horton, B., US gives NATO new nuclear plan, *San Francisco Examiner*, (16 October 1968).
36. Secretary of Defense R. S. McNamara, quoted in *Documents on Disarmament*, 1963, US Department of State. Secretary McNamara also used the phrase of tens of thousands' in an address on 14 December 1962, as did Deputy Secretary of Defense Roswell Gilpatrick in a speech in October 1962. See, Doty, R. C., McNamara Spurs NATO's Build-Up, *New York Times*, (14 December 1962).
37. *US Security Issues in Europe: Burden Sharing and Offset, MBFR and Nuclear Weapons, September 1973*. A Staff Report. Committee on Foreign Relations, US Senate, 93rd Congress, (2 December 1973), pp. 13–14.
38. *Nuclear Weapons and Foreign Policy*, Hearings, Committee on Foreign Relations, US Senate, 93rd Congress, (March–April 1974), pp. 198–9.
39. Corddry, C. W., US May Cut Stock of Europe Weapons, *Baltimore Sun*, (5 April 1974).
40. When the nukes have to go, *The Economist*, (7 June 1969).
41. *US Security Issues in Europe: Burden Sharing and Offset, MBFR and Nuclear Weapons, September 1973*. A Staff Report. Committee on Foreign Relations, US Senate, 93rd Congress, (2 December 1973), p. 19.
42. Horton, B., US gives NATO new nuclear plan, *San Francisco Examiner*, (16 October 1968).
43. *US Security Issues in Europe: Burden Sharing and Offset, MBFR and Nuclear Weapons, September 1973*. A Staff Report. Committee on Foreign Relations, US Senate, 93rd Congress, (2 December 1973), pp. 19–20. On 'Command and Control' of TNWs in NATO, see also the following comprehensive sources: Eekelen, W. F. van, Development of NATO's nuclear consultation, *NATO Letter*, Vol. 18, Nos. 7–8, July–August 1970, pp. 2–6; Wiegele, T. C., Nuclear consultation process in NATO, *Orbis*, Vol. 16, No. 2, Summer 1972, pp. 462–87; Vandervanter, E., Jr., B. Gen., Studies on NATO: an analysis of integration, RAND, Memorandum RM-5006-PR, (August 1966), pp. 25–35, Santa Monica, California; Whetten, L. L., A European view of NATO

Strategy, *Military Review*, ;September 1971), pp. 25–37; Hinterhoff, E., Maj., NATO's nuclear strategy and eastern Europe, *NATO's Fifteen Nations*, Vol. 15, No. 5, (October–November 1970), pp. 42–8.

44. *The Atlantic Alliance*, Hearings, Committee on Government Operations, US Senate, 89th Congress, (May 1966), Part 2, pp. 80–2.

45. *Nuclear Weapons and Foreign Policy*, Hearings, Committee on Foreign Relations, US Senate, 93rd Congress, (March–April 1974), pp. 187–88; see also refs. 1, 3.

46. Murray, A., Bonn defense policy stresses nuclear will, *Baltimore Sun*, (21 January 1976).

47. Tactical nuclear weapons in Europe, in *Setting National Priorities: The 1972 Budget*, Schultze, C. L. *et al.* (The Brookings Institution, Washington, DC. 1971), pp. 94–102.

48. (*a*) European security and force reductions in Europe, in *World Armaments and Disarmament, SIPRI Yearbook 1969/70*, pp. 64–91.
    (*b*) *Force Reductions in Europe, A SIPRI Monograph*, (1974), 105 pp.
    (*c*) Newhouse, J. *et al.*, *US Troops in Europe, Issues, Costs and Choices* (The Brookings Institution, Washington, DC, 1971), 177 pp.; Lawrence, R. D. and Record, J., *US Force Structure in NATO, An Alternative*, (The Brookings Institution, Washington, DC, 1974), 136 pp.

49. Address by Secretary of Defense McNamara Before the New York Economic Club, (18 November 1963), *Documents on Disarmament*, (1963), pp. 583–94.

50. Atlantic partnership and the common defense. Remarks by Assistant Secretary of Defense Paul H. Nitze, Cleveland Council on World Affairs, (2 March 1963). *DOD News Release No. 287–63*, 12 pp.

51. Enthoven, A. C., American deterrent policy in *Problems of National Strategy*, H. Kissinger (ed.), (Praeger, New York, 1965), pp. 126–7. Enthoven's paper was, however, written in 1963.

52. *Review of a Systems Analysis Evaluation of NATO vs. Warsaw Pact Conventional Forces*, Report of the Special Subcommittee on National Defense Posture of the Committee on Armed Services, US House of Representatives, 90th Congress, (4 September 1968), 15 pp.

53. Enthoven, A. C., Arms and men: the military balance in Europe, *Interplay*, Vol. 2, No. 10, (May 1969), pp. 11–4.

54. Enthoven, A. C. and Smith, K. W., What forces for NATO? And from whom?, *Foreign Affairs*, Vol. 48, No. 1, (October 1969), pp. 80–96.

55. A statement on US military forces in Europe, in *Nuclear Weapons and Foreign Policy, US Nuclear Weapons in Europe and US–USSR Strategic Doctrines and Policies*, Hearings, Committee on Foreign Relations, US Senate, 93rd Congress, 2nd Session, pp. 65–141.

56. Enthoven, A. C., US forces in Europe: how many? Doing what?, *Foreign Affairs*, Vol. 53, No. 3, (April 1975), pp. 513–32.

57. Enthoven, A. C. and Smith, K. W., How much is enough: shaping the defense program, 1961–1969, Chapter 4 in *NATO Strategy and Forces* (Harper & Row, New York, 1971), pp. 117–64.
    Relatively similar positions were taken by other authors as well, for example, the report, *A Concept for NATO*, Singer, M., Armbruster, F., Hudson Institute, HI-694-D (4 May 1966), 40 pp. stresses NATO's conventional superiority as does a second companion report, HI-672-D. See also, Schlesinger, J. R., *European Security and the Nuclear Threat Since 1945*, RAND, P-3574, (April 1967), 25 pp.

58. Getler, M., Study insists NATO can defend itself, *Washington Post*, (7 June 1973); see also Jenkins, P., Pentagon report lifts European threshold, *The Guardian*, (16 June 1973).
59. *US News & World Report*, (9 July 1973).
60. *Military Applications of Nuclear Technology*, Part II, Hearings, Joint Committee on Atomic Energy, Congress of the United States, 93rd Congress, (22 May, 29 June 1973), pp. 74–5.
61. A statement on US military forces in Europe, in *Nuclear Weapons and Foreign Policy, US Nuclear Weapons in Europe and US–USSR Strategic Doctrines and Policies*, Hearings, Committee on Foreign Relations, US Senate, 93rd Congress, 2nd Session, p. 79.
62. Kennan, G. F., *Memoirs, 1925–1950* (Bantam Books, New York, 1969), pp. 500–1.
63. Lowenstein, H. Zu and von Zuhlsdorff, V., *NATO and the Defense of the West* (Praeger, New York, 1960), p. 303.
64. Extracts from a speech, (1 February 1969), *NATO Letter*, (March 1969), pp. 24–29.
65. Finney, J. W., NATO leadership in air studied, *New York Times*, (5 May 1974). See also, White, W. D., *US Tactical Air Power, Missions, Forces and Costs* (The Brookings Institution, Washington, DC, 1974), 121 pp.
66. *NATO and the New Soviet Threat*, Report of Senator S. Nunn and Senator D. Bartlett, Committee on Armed Services, US Senate, 95th Congress, (24 January 1977), p. 5.
67. For example, NATO air forces are superior, but Soviet air power has undergone a dramatic transformation shifting away from a purely defensive role toward support of offensive action; NATO unable to withstand attack – Report, *Ottawa Journal*, (24 January 1977).
68. Luttwak, E. N., *The Missing Dimension of US Defense Policy: Forces, Perceptions and Power* (revised), ARPA-T10-76-2, ARPA, US Department of Defense, Washington, DC, (February 1976), pp. 8, 11.
69. US forces called clearly dominant, Associated Press dispatch, Paris, (24 January 1977).
70. *Annual Defense Department Report, F.Y. 1977*, Appendix E1.
71. Statement by Gen. G. S. Brown, Chairman, Joint Chiefs of Staff, Senate Armed Services Committee for FY 1976.
72. Neff., R., NATO chief accuses Soviets of violating anti-war Pact, *Christian Science Monitor*, (28 November 1973).
73. *Brief, Middle East Highlights*, No. 70, (16–30 November 1973).
74. Canby, S. L., NATO muscle: more shadow than substance, *Foreign Policy*, No. 8, (Fall 1972), pp. 38–49.
   Komer, R. W., Treating NATO's self-inflicted wound, *Foreign Policy*, No. 13, (Winter 1973–4), pp. 34–47.
   Canby, S. L., Damping nuclear counterforce incentives: correcting NATO's inferiority in conventional military strength, *Orbis*, Vol. 19, No. 1, (Spring 1975).
   Nunn and Bartlett, ref. 18 *op cit*.
   Aspin, L., Congressman, A surprise attack on NATO: refocusing the debate, *Congressional Record*, (House, 7 February 1977).
   Aspin, L., Congressman, A dubious burden: the cost of repelling a surprise Soviet attack, *Congressional Record*, (House, 14 March 1977).
   The military balance between NATO and the Warsaw Pact, Section V in *First Concurrent Resolution on the Budget*, (FY 1978), Hearings, Committee on the Budget, US Senate, 95th Congress, Vol. I, pp. 197–246.
   NATO, still strong enough to block a blitz, *Time Magazine*, (13 December 1976), pp. 46–53.

De Borchgrave, A., Nightmare for NATO, *Newsweek*, (7 February 1977), pp. 36–8. Murray, A., NATO deployment weakness found, *Baltimore Sun*, (3 June 1975); (with reference to the report to the Western European Union by General U. de Maiziere).

75. *Military Applications of Nuclear Technology*, Hearing, US Congress, Joint Committee on Atomic Energy, 94th Congress, (30 June 1976), Part I, p. 32.

76. Report on Tactical Nuclear Weapons, Members of Congress for peace through law, Congressional Record, (Senate, 20 July 1971), pp. S11626–28.

77. Hadley, A. H., *The Nation's Safety and Arms Control* (Viking Press, New York, 1961), pp. 4, 33.

78. Secret documents: nuclear bombs on Kiel, *Stern*, No. 6, (1970), pp. 170–3.
    Secret documents II: the Rhine and the Main in flames, *Stern*, No. 7, (1970).
    von Baudissin, Gen. W. Graf, NATO and Bonn make hush-hush, and nuclear bombs on Kiel?, *Stern*, No. 8, (1970).

79. Nuclear bombs on Kiel?, *Stern*, No. 8, (1970).

80. Hastings, W. H., Limited War Patterns: (I) Southeast Asia (1963), (U), Memorandum, RM-2961-1SA (The Rand Corporation, Santa Monica, Calif., July 1962), 66 pp.

81. *The Theatre Nuclear Force Posture in Europe*. A report to the US Congress, Secretary of Defense, J. R. Schlesinger, (April–May 1975), pp. 16–18.

82. *Nuclear Weapons and Foreign Policy*, Hearings, (1974), A report to the US Congress, Secretary of Defense, J. R. Schlesinger.

83. *The Atlantic Alliance*, Hearings, Part II, (1966), A report to the US Congress, Secretary of Defense, J. R. Schlesinger, pp. 79, 85.

84. NATO Nuclear Planning Group, Final Communique, NATO Press Service, Press Release-M-NPG-1(73) 17, Brussels, (16 May 1973), 2 pp.

85. *Military Applications of Nuclear Technology*, Hearings, US Congress, Joint Committee on Atomic Energy, 94th Congress, (30 June 1976), Part II, (May–June 1973), p. 100.

86. *Military Applications of Nuclear Technology*, Hearings, US Congress, Joint Committee on Atomic Energy, 94th Congress (30 June 1976), Part II, (May–June 1973), pp. 112, 116.

87. *First Use of Nuclear Weapons, Preserving Responsible Control*, Hearings, Committee on International Relations, House of Representatives, 94th Congress, (March 1976).

88. Robert McNamara, in Hearings, US House of Representatives, Committee on Appropriations, Subcommittee, Department of Defense Appropriations for 1964, 88th Congress, Part I, Washington, DC, (1963), pp. 100–2.

89. *Nuclear Weapons and Foreign Policy*, Hearings, Committee on Foreign Relations, US Senate, 93rd Congress, (March–April 1974).

90. *Military Applications of Nuclear Technology*, Hearings, US Congress, Joint Committee on Atomic Energy, 94th Congress, (30 June 1976), Part II, (May–June 1973), p. 101.

91. Hearings, Special Subcommittee on North Atlantic Treaty Organization Commitments, Committee on Armed Services, House of Representatives, 92nd Congress, (1972), pp. 13216–17, 13273.

92. Kissinger, H., *Nuclear Weapons and Foreign Policy* (Harper, New York, 1957), pp. 291–6, 308–9. See also, Speir, H., *German Rearmament and Atomic War* (Row, Peterson & Co., Evanston, Ill., 1957).

93. Weizsäcker, C. F. V. (ed.), *Kriegsfolgen und Kriegsverhütung* (Munich: Carl Hanser Verlag, 1971), 699 pp. See also, York, H., 'The nuclear "balance of terror" in Europe', *Bulletin of the Atomic Scientists*, Vol. 32, No. 5, (May 1976), pp. 8–17.

94. *Tactical Nuclear Weapons in Europe* (Brookings), 1971, *op cit.*, pp. 96, 99.

95. *Seminars, Service Chiefs on Defense Mission and Priorities, Vol. II, 11 November 1975, Army,* Task Force on Defense of the Committee on the Budget, US Senate, 94th Congress, (February 1976).

96. *United States Troop Levels in Europe, Report on Staff Survey Mission to Europe, May 1970,* Committee on Foreign Affairs, US House of Representatives, 91st Congress, (20 December 1970), p. 12.

97. Report on Tactical Nuclear Weapons, Members of Congress for peace through law, *Congressional Record, Senate* (20 July 1971).

98. Tactical Arms Limitation in Europe, 15th Pugwash Symposium, Lahti, Finland, August 1971; *Pugwash News Letter*, Vol. 9, No. 3, (January 1972), pp. 79–96.

99. *US security issues in Europe: burden sharing and offset, MBFR and nuclear weapons, September 1973.* A Staff report, (2 December 1973). Committee on Foreign Relations, US Senate, 93rd Congress, (2 December 1973). See also, Marder, M., Senate Panel study details depth of atomic presence, *Washington Post*, (3 December 1973).

100. *Nuclear Weapons and Foreign Policy*, Hearings, Committee on Foreign Relations, 93rd Congress, (March–April 1974), 316 pp.

101. *The Theatre Nuclear Force Posture in Europe, A Report to the US Congress*, Secretary of Defense, J. R. Schlesinger, (April–May 1975), unclassified version, 30 pp.

102. Record, J., *US Nuclear Weapons in Europe, Issues and Alternatives* (The Brookings Institution, Washington, DC, 1974), 70 pp.; see also, Getter, G., US urged to cut A-arms in Europe, *Washington Post*, (3 December 1974); Record, J., US tactical nuclear weapons in Europe: 7000 warheads in search of a mission, *Arms Control Today*, Vol. 4, No. 4, (April 1974), pp. 1–4.

103. Record, J., Tactical nuclear weapons in Europe: alternative postures, *Survival*, Vol. 17, No. 2, (March–April 1975), pp. 73–80
    Whitney, C. R., The nuclear overkill in western Europe, *New York Times*, (14 September 1975).
    30 000 US nuclear weapons: 22 000 tactical and 8000 strategic, *The Defense Monitor* (Washington, DC: Center for Defense Information, Washington, DC), Vol. 4, No. 2, (February 1975), 12 pp.

104. Morgan, D., Nuclear force in Europe held open to cutting, *Washington Post*, (5 April 1974).
    Corddry, C. W., US may cut stock of Europe weapons, *Baltimore Sun*, (5 April 1974).
    Corddry, C. W., Schlesinger to fly to West Germany, may study troop, nuclear arms cuts, *Baltimore Sun*, (12 April 1974).
    Finney, J. W., Schlesinger set to cut A-arms, *New York Times*, (24 April 1974). ——
    Finney, J. W., US considers reduction of atom arms in Europe, *New York Times*, (22 September 1974).

105. Secretary of Defense J. Schlesinger, Press remarks, German Ministry of Defense, (4 November 1974).
    Schlesinger says NATO must watch Warsaw Pact growth, *Defense Space Business Daily*, (6 November 1974).
    Won't reduce Europe arms, (UPI) *Chicago Tribune*, (10 December 1974).
    Russians assail US plans, *Baltimore Sun*, (6 November 1974).

106. Middleton, D., Schlesinger calls A-arms vital to guard Europe, *New York Times*, (12 December 1974).

107. Oishi, G., NATO A-arms could be negotiable, *Baltimore Sun*, (10 December 1974).
Oishi, G., US A-arms cut key topic for NATO, *Baltimore Sun*, (12 December 1974).

108. Middleton, D., NATO group cool to Kissinger plan, *New York Times*, (11 December 1975).
Lewis, P., US ready to cut N-strength in Europe, *Financial Times*, (11 June 1975).
Whitney, C. R., Europeans worried by pressure on US to reduce A-weapons, *New York Times*, (5 September 1975).
Pick, H., NATO ready for nuclear bargaining, *The Guardian*, (7 October 1975).

109. Finney, J. W., US delayed removal of warheads, *New York Times*, (24 July 1975).

110. Finney, J. W., US delaying removal of warheads, *New York Times*, (24 July 1975).

111. US to cut arsenal, (AP), Brussels, Belgium, (1976).

112. Horton, A., President is urged to leave nuclear weapons in Korea, (Scripps Howard), Memphis Press, (16 June 1977).
Beecher, W., Carter may keep N-weapons in Korea, *Boston Globe*, (10 July 1977).

113. *Security Agreements and Commitments Abroad, Report*, Committee on Foreign Relations, US Senate, 91st Congress, (21 December 1970), pp. 12–13.

114. Appendix B: announced nuclear detonations, in *The Effects of Nuclear Weapons*, (rev. ed.), US Atomic Energy Commission, 1964. See also, *SIPRI Yearbook of World Armaments and Disarmament, 1968/69*, (1969), p. 272.

115. *Major Activities of the Atomic Energy Program, July–December 1955*, US Atomic Energy Commission, (1956), p. 38.

116. Accidents of nuclear weapons and nuclear delivery systems, *SIPRI Yearbook of World Armaments and Disarmament, 1968/69*, (1969), pp. 259–70.

117. Accidents of nuclear weapons systems, *World Armanets and Disarmament, SIPRI Yearbook, 1977*, (1977), pp. 52–85.

118. Brennan, D. G., The atomic risks of spreading weapons, a historical case, *Arms Control and Disarmament*, Vol. 1, No. 1, (1968), pp. 59–60.

119. *Security Agreements and Commitments Abroad, Report*, Committee on Foreign Relations, US Senate, 91st Congress, (21 December 1970), p. 13.

120. Davis, S. R., How safe are NATO missiles? Greek A-incident surfaces, *Christian Science Monitor*, (8 December 1970).

121. Symington finds flaws in NATOs warhead security; Greek incident hinted, *New York Times*, (23 November 1970).

122. Davis, S. R., How safe are NATO missiles? Greek A-incident surfaces, *Christian Science Monitor*, (8 December 1970).

123. Pincus, W., Nukes nobody needs, *New Republic Magazine*, Vol. 170, No. 116, (20 April 1974), pp. 12–14.

124. Finney, J. W., Cyprus crisis stirred US to protect atom weapons, *New York Times*, (9 September 1974).
Gelb, L. H., US weighs status of nuclear warheads in Greece, *New York Times*, (11 September 1974).
Roberts, S. V., Greek sees worse ties if US pulls out A-arms, *New York Times*, (12 September 1974).
Cooling off the nukes, *Newsweek*, (12 August 1974).
Middleton, D., Could a US atom bomb be stolen, *New York Times*, (22 September 1974).

125. Views of Hon. C. D. Long, submitted to Accompany FY 1975 Military Construction Appropriations, (24 September 1974), House Appropriations Committee, 1974, mimeo., 5 pp. See also, Finney, J. W., Pentagon acting to guard A-arms, *New York Times*, (27 September 1974).

126. Letter from a concerned US soldier to Senator Symington, *The Defense Monitor*, (Center for Defense Information, Washington, DC), Vol. 4, No. 2, (February 1975), 12 p. 8.

127. *Development, Use, and Control of Nuclear Energy for the Common Defense and Security and for Peaceful Purposes*, First Annual Report to the US Congress, Joint Committee on Atomic Energy, 94th Congress, (30 June 1975), p. 23.

128. *Nuclear Weapons Security Primer*, US Department of Defense, (1 April 1975), quoted in First Annual Report to the US Congress, Joint Committee on Atomic Energy, 94th Congress, (30 June 1975), pp. 22–3.

129. Long, (24 September 1974), *op. cit.*

130. Oberdorfer, D., Risk seen in keeping US A-arms in South Korea, *Washington Post*, (20 September 1974).

131. La Rocque, G. R., R.Adm., in *Proliferation of Nuclear Weapons*, Hearing, Joint Committee on Atomic Energy, 93rd Congress, (10 September 1974), pp. 15–16, 22.

132. Pastore, Senator J. O., Security review of certain NATO installations, Congressional Record, US Senate, Washington, DC, (30 April 1975), pp. 7184–90.

133. *Security Agreements and Commitments Abroad*, Committee on Foreign Relations, US Senate, 91st Congress, (21 December 1970), pp. 13, 18.

134. Dunn, L. A., *Military Politics, Nuclear Proliferation, and the Nuclear Coup d'Etat*, HI-2392-P, (Hudson Institute, Croton-on-Hudson, NY, 28 February 1976), 32 pp.

135. Schlesinger, J. R., Secretary of Defense, *The Theatre Nuclear Force Posture in Europe*, A Report to the US Congress, (1 April 1975), p. 24.

136. *Development, Use, and Control of Nuclear Energy for the Common Defense and Security and for Peaceful Purposes*, First Annual Report, (30 June 1975), to the US Congress, Joint Committee on Atomic Energy, 94th Congress, p. 23.

137. *Nuclear Weapons Security Primer*, US Department of Defense, (1 April 1975), quoted in Ref. 136.

138. *Development, Use, and Control of Nuclear Energy for the Common Defense and Security and for Peaceful Purposes*, Second Annual Report, (30 June 1976), First Annual Report to the US Congress, Joint Committee on Atomic Energy, 94th Congress, (30 June 1975), p. 152. See also, Congressional Record, (30 April 1975), p. 57189.

139. See Ref. 138; Long, (24 September 1974), *op. cit.*; *Development, Use, and Control of Nuclear Energy for the Common Defense and Security and for Peaceful Purposes*, First Annual Report, (30 June 1975), *op. cit.*; Pastore, (30 April 1975), Security review of certain NATO installations, Congressional Record, US Senate, Washington, DC; Also, Floor Statement by Senator J. O. Pastore, (25 September 1974), 3 pp. mimeo.

140. For example, Rosenbaum, D., Nuclear terror, *International Security*, Vol. 1, No. 3, (Winter 1977), pp. 140–61.

141. In particular, see Pastore. *Security Review of certain NATO installations*, Congressional Record, (30 April 1975), US Senate, Washington, DC.

142. *Development, Use, and Control of Nuclear Energy for the Common Defense and Security and for Peaceful Purposes*, Second Annual Report, (30 June 1976), First Annual Report on the US Congress, Joint Committee on Atomic Energy, 94th Congress, (30 June 1975), p. 30.

143. *Development, Use, and Control of Nuclear Energy for the Common Defense and Security and for Peaceful Purposes*, First Annual Report, (30 June 1975), First Annual Report on the US Congress, Joint Committee on Atomic Energy, 94th Congress, (30 June 1975), pp. 23–24.

144. *Development, Use, and Control of Nuclear Energy for the Common Defense and Security and for Peaceful Purposes*, Second Annual Report to the US Congress, Joint Committee on Atomic Energy, 94th Congress, (30 June 1976), pp. 29–30.

145. See Ref. 144, p. 30.

146. Schlesinger, *Theatre Nuclear Force Posture in Europe*, A Report to the US Congress, (1 April 1975), pp. 24–25, 27, 29.

147. Currie, M. R., *The Department of Defense Program of Research, Development, Test and Evaluation, FY 1976*, (7 March 1975), pp. vi–25.

148. Brown, G. S., Gen., Chrm. US Joint Chiefs of Staff, *United States Military Posture for FY 1976*, Statement Before the Senate Armed Services Committee, p. 107; *To Consider NATO Matters*, Hearing, Joint Committee on Atomic Energy, US Congress, 93rd Congress, (19 February 1974), pp. 11–12.

149. Rumsfeld, D. H., Secretary of Defense, *Annual Defense Department Report, FY 1978*, p. 151.

150. Hoffman, M. R., Secretary of the Army, *The Posture of the Army*, Statement before the Committee on Armed Services, House of Representatives, (4 February 1976).

151. McElheny, V., US adding safeguards in tactical nuclear weapons, *New York Times*, (17 December 1973).

152. Beecher, W., DOD News Briefing, (26 September 1974), mimeo., 8 pp. See also, US nuclear weapons said to be 'fail safe', *Baltimore Sun*, (27 September 1974).

153. *US security issues in Europe: burden sharing and Offset, MBFR and nuclear weapons, September 1973*. A Staff Report, (2 December 1973), Committee on Foreign Relations, US Senate, 93rd Congress, (2 December 1973), p. 22.

154. *Nuclear Weapons and Foreign Policy*, Hearings, Committee on Foreign Relations, 93rd Congress, (March–April 1974), p. 209.

155. US–Soviet agreement possible on war definition, Enthoven says, *Aviation Week and Space Technology*, Vol. 78, No. 7, (18 February 1963), p. 39.

156. Scoville, H. Jr., A new weapon to think (and worry) about, *New York Times*, (12 July 1977); York, H. F. and Scoville, H., Jr., New look in nuclear solution, Letter to the Editor, *New York Times*, (16 February 1971).

157. *Nuclear Weapons and Foreign Policy*, Hearings, Committee on Foreign Relations, US Senate, 93rd Congress, (2 December 1973), p. 201.

158. Ambassador J. Martin, Final Record, Conference of the Committee on Disarmament, CCD/PV.638, (23 May 1974), mimeo., pp. 27–9.

159. Finney, J. W., US to renounce 'mini' atom arms, *New York Times*, (24 May 1974).

160. See Ref. 159.

161. Ambassador Martin, Final Record, Conference of the Committee on Disarmament, CCD/PV.638, (23 May 1974) mimeo., pp. 27–9.

162. Finney, J. W., US to renounce 'mini' atom arms, *New York Times*, (24 May 1974).

163. *US security issues in Europe: burden sharing and offset, MBFR and nuclear weapons, September 1973*. A Staff Report, (2 December 1973), Committee on Foreign Relations, US Senate, 93rd Congress, (2 December 1973), p. 22.

164. Cohen, S. T., A word from its father, something is uglier than a neutron bomb, *Washington Star*, (10 July 1977).

165. Anson, R. S., The neutron bomb, *New Times*, Vol. 9, No. 3, (5 August 1977), pp. 24–32.

166. Strauss, L. L., *Men and Decisions*, (Doubleday, Garden City, NY, 1962), pp. 418–9. See also, Bombes nucléaires 'propres' d'ici cinq ans, *Revue Militaire D'Information*, No. 286, (August 1957).

167. Gillette, R., Neutron bomb is almost 20 years old, was focus of heated controversy in 1961, *Los Angeles Times*, (13 July 1977).

168. Cohen, S. T., A word from its father, something is uglier than a neutron bomb, *Washington Star*, (10 July 1977).

169. Beecher, W., 'Clean' warheads sought for NATO, *New York Times*, (20 May 1967).

170. *Washington Post*, 1969.

171. President Nixon's Foreign Policy Report to the Congress, (25 February 1971), pp. 22–4.

172. Lawrence, R. M., On tactical nuclear war, I, *Revue Militaire Générale*, No. 1, (January 1971), pp. 46–59, and, On tactical nuclear war, II, *Revue Militaire Générale*, No. 2, (February 1971), pp. 237–61.
Possony, S. T., NATO's defense posture, *Ordnance*, Vol. 54, No. 295, (July–August 1969), pp. 41–5.
Bennett, W. S., *et al.*, A credible nuclear-emphasis defense for NATO, *Orbis*, Vol. 17, No. 2, (Summer 1973), pp. 463–79.
Gray, C. S., Mini-nukes and strategy, *International Journal*, Vol. 29, No. 2, (Spring 1974), pp. 216–41.

173. Sulzberger, C. L., Solving an ugly dilemma, *New York Times*, (16 November 1970).
Sulzberger, C. L., The new nuclear look, I, *New York Times*, (8 January 1971).
Sulzberger, C. L., The new nuclear look, II, *New York Times*, (10 January 1971).
See also a rebuttal of the Sulzberger articles, York, H. F. and Scoville, H., Jr., 'New look' in nuclear solutions, Letter to the Editor, *New York Times*, (16 February 1971).

174. Nerlich, U., *Some Comments on Modernization of Nuclear Stockpiles in Europe: A German View*, HI-1626-D (Hudson Institute, Croton-on-Hudson, NY, 24 April 1972), 11 pp.

175. Black, E. F. Brig. Gen., NATO's unmentionable option, tactical nuclear weapons, Washington Report (American Security Council), (21 December 1970), reprinted in Congressional Record, Extension E-10880-82, (30 December 1970).

176. Beecher, W., Over the threshold: 'clean' tactical nuclear weapons for Europe, *Army*, Vol. 22, No. 7, (July 1972), pp. 17–20.

177. Beecher, W., Laird says newer A-arms may be sent to Europe, *New York Times*, (16 April 1972).

178. *Military Applications of Nuclear Technology*, Hearings, Part I, (April 1973); Part II, (May–June 1973), US Congress, Joint Committee on Atomic Energy, 94th Congress.

179. Finney, J. W., US army's guns in Europe to get newer nuclear shells, Defense Chief unaware of move, *New York Times*, (11 May 1973).

180. Finney, J. W., Changes in panel could alter A-weapons program, *New York Times*, (29 May 1973).

181. *Military Applications of Nuclear Technology*, Hearings, Part I, US Congress Joint Committee on Atomic Energy, 94th Congress, (April 1973), p. 33.

182. See Ref. 178, Part I, p. 33; Part II, pp. 121–22.

183. *Department of Defense Appropriations for 1974*, Hearings, Part 7, Research, Development, Test and Evaluation, Committee on Appropriations, 93rd Congress, (1973), p. 377.

184. *Military Applications of Nuclear Technology*, Part II, US Congress, Joint Committee on Atomic Energy, 94th Congress, (May–June 1973), p. 122.

185. McElheny, V. K., US adding to safeguards in tactical nuclear arms, *New York Times*, (17 December 1973).

186. *Department of Defense Appropriations for 1974*, Hearings, Part 7, Research Development, Test and Evaluation, Committee on Appropriations, 93rd Congress, (1973), p. 377.

187. *US security issues in Europe: burden sharing and offset, MBFR and nuclear Weapons, September 1973*. A Staff Report, (2 December 1973), Committee on Foreign Relations, US Senate, 93rd Congress, pp. 2243.
Douglas-Home, C., Miniature nuclear arms developed by Pentagon for battlefield use, opposition by Europe defence chiefs likely, *The Times*, (7 May 1973).
Chalfont, Lord, Time to shoot down the Pentagon's latest bit of gee-whizzery, *The Times*, (14 May 1973).
Les Américains disposeraient dans quelques années d'armes nucléaires tactiques miniaturisées, *Le Monde*, (8 May 1973).
Noyes, H., Mini nuclear weapons 'A long-term matter', *The Times*, (11 May 1973).
Hunt, K., The alliance and Europe: Part II, defence with fewer men, *Adelphi Papers*, No. 98, (Summer 1973), pp. 17–8.
'Nuclear Puffs' might be the first atomic weapon for West Germany, [translation], *Expressen* (Stockholm), (12 June 1973).
Nuclear arms production getting easier [translation], *Dagens Nyheter* (Stockholm), (2 July 1973).

188. Mrs Myrdal, Final Record, Conference of the Committee on Disarmament, CCD/PV.620, (9 August 1973), mimeo., pp. 13–5.

189. Middleton, D., Schlesinger calls A-arms vital to guard Europe, *New York Times*, (12 December 1974).

190. *Nuclear Weapons and Foreign Policy*, Hearings, Committee on Foreign Relations, 93rd Congress, (March–April 1974); see also *To Consider NATO Matters*, Hearing, Joint Committee on Atomic Energy, (19 February 1974), 93rd Congress, printed 1975, 28 pp.

191. Finney, J. W., Small atomic arms are urged for NATO, *New York Times*, (27 January 1974).
Norman, L., The reluctant dragons: NATO's fears and the need for new nuclear weapons, *Army*, (February 1974).
Pincus, W., Why more nukes? A new generation of weaponry, *New Republic 170*, Vol. 6, (9 February 1974), pp. 14–7.
Pincus, W., Nukes nobody needs, *New Republic 170*, Vol. 16, (20 April 1974), pp. 12–4.
Rhodes, R., Los Alamos revisited, *Harpers Magazine*, (March 1974).

192. Getler, M., US eyes miniature warhead, *Washington Post*, (24 January 1977).

193. *Comparison of the Pershing II Program With the Acquisition Plan Recommended by the Commission on Government Procurement*, Report to the Congress, US General Accounting Office, PSAD-77-51, (24 January 1977), p. 6.

194. *International Defense Review*, Vol. 6(2), (April 1973).

195. *Nuclear Weapons and Foreign Policy*, Hearings, (March–April 1974), Committee on Foreign Relations, 93rd Congress, p. 209.
196. *The Theatre Nuclear Force Posture in Europe*, A Report to the US Congress, Secretary of Defense, J. R. Schlesinger, (April–May 1975).
197. NATO ministers issue Monterey nuclear communique, *Defense Space Business Daily*, (20 June 1975).
198. Finney, J. W., US assigns more missile submarines to the defense of NATO, *New York Times*, (18 June 1975).
199. *US security issues in Europe: burden sharing and offset, MBFR and nuclear weapons, September 1973*. A Staff Report, (2 December 1973), Committee on Foreign Relations, US Senate, 93rd Congress.
200. *Public Works for Water, Power Development and Energy Research, Appropriations Bill, 1977, Hearings, Part 7, Energy Research and Development Administration*, Subcommittee, Committee on Appropriations, House of Representatives, 94th Congress, (1976), pp. 917–1124.
201. Statement, Hon. M. R. Hoffman, Secretary of the Army, Committee on Armed Services, House of Representatives, 94th Congress, (4 February 1976).
202. Boffey, P. M., Arms control: impact statements called a 'farce' and a 'mockery', *Science*, Vol. 194, (1 October 1976), pp. 36–7; Schneider, B. R., Stonewalling on the arms control impact statements, *Bulletin of the Atomic Scientists*, Vol. 33(1), (January 1977), p. 5.
203. *Analysis of Arms Control Impact Statements Submitted in Connection with the FY 1978 Budget Request*, Prepared for the Committee on Foreign Relations, US Senate and the Committee on International Relations, US House of Representatives, by the Congressional Research Service, Library of Congress, 95th Congress, April 1977, 414 pages.
204. See Ref. 203, pp. 153–168.
205. See Ref. 203, pp. 338–348.
206. *Public Works for Water and Power Development and Energy Research Appropriation Bill, 1978, Hearings, Part 6, US Energy Research and Development Administration*, Subcommittee of the Committee on Appropriations, House of Representatives, 95th Congress, 1977, pp. 1011–1216.
207. *Congressional Record*, Senate, (1 July 1977), pp. S-11427–42. (For the record of that part of the day's debate that was held in open session.)
208. Transcript of President Carter's news conference, *New York Times*, (13 July 1977).
209. *Congressional Record*, Senate, (13 July 1977), pp. S-11741–S-11789.
210. Wren, C., Soviet criticizes US proposal for small NATO nuclear arms, *New York Times*, (5 February 1974).
211. Tucker, A., US navy irradiates and burns monkeys to death, *The Guardian*, (29 May 1977).
212. Soviet calls rights drive pretext for arms buildup, *New York Times*, (20 June 1977).
213. Klose, K., Soviets denounce US neutron project, *Washington Post*, (10 July 1977). Browne, M. W., US neutron bomb criticized in Soviet: Tass says radiation weapon runs counter to human rights cause, *New York Times*, (10 July 1977).
see also:
Soviets say US stance on bomb harms relations, *Washington Post*, (18 July 1977).
Soviet press opposes new bomb, *Baltimore Sun*, (28 July 1977).
214. Isnard, J. and Lavallard, J.-L., Essais nucléaires et force de dissuasion, *Le Monde*, (10 July 1973).
215. La panoplie atomique française, *Le Monde*, (24 July 1973).

216. Isnard, J. and Lavallard, J.-L., Essais nucléaires et force de dissuasion, *Le Monde*, (10 July 1973).

217. France develops parachute drop of atomic bombs, *New York Times*, (15 February 1967).

218. Isnard, J., Pluton, ou les vertus de l'ambiguité, *Le Monde*, (25 July 1975).

219. France, *Aviation Advisory*, NL-79/4/68, pp. 8–10.

220. French plan five-year hardware program, *Aviation Week and Space Technology*, Vol. 94(10), (8 March 1971), p. 36. (French strategic air-dropped weapons are delivered by a force of Mirage IV bombers, with the aid of twelve KC-135 aircraft for inflight refuelling purposes, and between 1968 and 1972, were variously described as being 50 kt, 'about 50 kt', 60 kt, 80 kt, or 'about 100 kt'.)

221. Farrell, R. E., French reach for technology in nuclear mach 2 transport programs, *Aviation Week and Space Technology*, Vol. 78(10), (11 March 1963), p. 281.

222. France's maritime Jaguars, *Flight International*, Vol. 97(3185), (26 March 1970). See also *Aviation Week and Space Technology*, Vol. 94(10), (8 March 1971), p. 36.

223. France promises nuclear response to any aggression, *Aerospace Daily*, (Washington, DC), (29 June 1977). See also, Isnard, J. and Lavallard, J.-L., Essais nucléaires et force de dissuasion, *Le Monde*, (10 July 1973).

224. Selon un rapport de la Commission de la Défense Nationale, l'armement nucléaire Français représente 20 megatonnes, *Le Monde*, (7 November 1973).

225. France's aerospace industry, *Flight International*, Vol. 104(3376), (22 November 1973), p. 865.

226. *Aviation Week and Space Technology*, Vol. 94(10), (8 March 1971), p. 36 and Isnard J. and Lavallard, J.-L., Essais nucléaires et force de dissuasion. *Le Monde* (10 July 1973).

227. Isnard, J., Pluton, ou les vertus de l'ambiguité, *Le Monde*, (25 July 1975).

228. France's aerospace industry, *Flight International*, Vol. 104(3376), (22 November 1973), p. 865.

229. France will not station arms outside country, *New York Times*, (24 June 1977). (This source gives the range for Pluton as 'up to 93 miles'.)
    Mauthner, R., Bonn–Paris A-missiles plan, *Financial Times*, (24 June 1975).
    Isnard, J., Le Pluton: un sujet de discorde entre Bonn et Paris, *Le Monde*, (January 1973).

230. *Aviation Week and Space Technology*, Vol. 78(10), p. 36.
    Farrell, R. E., French nuclear, space work moves toward major goals, *Aviation Week and Space Technology*, Vol. 80(11), (16 March 1964), pp. 270–2.
    Beecher, W., France plans larger A-arsenal, including small nuclear arms, *New York Times*, (13 April 1967).

231. France, *Aviation Advisory*, NL-79/4/68, pp. 8–10.

232. Fourquet, A. M., Gen., The role of the forces, *Revue de Défense Nationale*, (May 1969), reprinted in *Survival*, Vol. 11(7), (July 1969), pp. 206–11.

233. Debré, M., La France et sa défense, *Revue de Défense Nationale*, Vol. 28(1), (January 1972), pp. 5–21.

234. The purpose of Pluton, Address by the French Prime Minister, M. Chirac, (10 February 1975), in *Défense Nationale*, May 1975, reprinted in *Survival*, Vol. 17(5), (September–October 1975), pp. 241–3.

235. Général Guy Mery, Comments to the Institut des Hautes Etudes de Défense Nationale, *Défense Nationale*, (June 1976), reprinted in *Survival*, Vol. 18(5), (September–October, 1976), pp. 226–8.

236. France's Defense Policy, Address by V. Giscard d'Estaing, President of the French Republic. At the Institut des Hautes Etudes de Défense Nationale, Paris, (1 June 1976), mimeo., 14 pp. Reprinted in *Survival*, Vol. 18(5), (September–October 1976), pp. 228–230. See also, another subsequent address by General Guy Mery again dealing in good part with French tactical nuclear weapons in *Défense*, Bulletin de L'Institut des Hautes Etudes de Défense Nationale, (May 1977), pp. 19–23.

237. Hargrove, C., NATO and the French: A question of deciding when to press the button, *The Times*, (18 July 1977).

238. La Bombe, ou Dé A à H, extracts from the complete text transcript, Radiodiffusion Télévision Française, 1970.

239. Isnard, J., Pluton, ou les vertus de l'ambiguité, *Le Monde*, (25 July 1975).

240. *Armed Forces Management*, Vol. 14, No. 2, (November 1967).

241. Browning, J., French communists modify nuclear weapons policy, *Christian Science Monitor*, (2 June 1977).

242. Wilson, A., French wooing NATO planners, *Observer*, (18 May 1969).

243. Brenner, M. J., France's new defense strategy and the Atlantic puzzle, *Bulletin of the Atomic Scientists*, Vol. 25, No. 9, (November 1969), p. 5.

244. Nanes, A. S., Arms and man, *Current History*, Vol. 33(194), (October 1957), p. 204.

245. Middleton, D., RAF to disperse bombers abroad, *New York Times*, (25 October 1960).

246. Brown, N., *British Arms and Strategy, 1970–1980*, (Royal United Services Institution, May 1969), p. 43.

247. Brown, N., *Nuclear War: The Impending Strategic Deadlock*, (Pall Mall Press, London, 1964), p. 129.

248. Beaton, L., Nuclear stocktaking, *The Guardian*, (4 April 1964).

249. Brown, N., *British Arms and Strategy, 1970–1980*, (Royal United Services Institution, May 1969), p. 43.

250. Parliamentary Debates (Hansard), (HMSO, London, 8 April 1970), p. 529.

251. Beaton, L., Nuclear stocktaking, *The Guardian*, (4 April 1964).

252. Donaldson, E. M., Navy takes on nuclear defence, *Daily Telegraph*, (2 July 1969). (However, in 1966, one knowledgeable source wrote that 'The British deterrent force consists of some 180 medium-range bombers of the Vulcan and Victor classes, which can carry at least one and possibly more thermonuclear bombs of 10-megaton capacity'. Burns, E. L. M. Gen., *Megamurder*, (Pantheon, New York, 1966), p. 150.

253. First RAF bomber unit equipped with Blue Steel now operational, *Aviation Week and Space Technology*, Vol. 78(9), (4 March 1963), pp. 78–9.

254. V-bombers train for low-level strikes, *Aviation Week and Space Technology*, Vol. 80(8), (24 February 1964).

255. Gretton, P., Sir V. Adm., The Defence White Paper, *Royal United Services Institution Journal*, Vol. 3(642), (May 1966), p. 123.

256. Coleman, H. J., British planning Blue Steel replacement, *Aviation Week and Space Technology*, Vol. 78(8), (25 February 1963), p. 29.

257. Brown, N., *British Arms and Strategy, 1970–1980*, (Royal United Services Institution, May 1969), p. 43.

258. Pincher, C., The new deterrent: Britain ready to decide on NATO bomb, *Daily Express*, (21 June 1969).

259. Getting MRCA airborne, *Flight International*, Vol. 96(3162), (16 October 1969).

260. Low level RAF deterrence, *Flight International*, Vol. 104(3327), (29 November 1973).

261. Brown, N., *British Arms and Strategy, 1970–1980*, (Royal United Services Institution, May 1969), pp. 32–4. See also, World's air forces, Great Britain, *Flight International*,

Vol. 99(3250), (24 June 1971), p. 929; and Ellis, P., From ark to omega, *Flight International*, Vol. 100(3270), (11 November 1971), pp. 766–72.

262. RAF Buccaneers, *Flight International*, Vol. 99(3231), (11 February 1971), pp. 202–7.

263. The Royal Air Force at fifty: Bomber Command, *Flying Review International*, Vol. 23(5), (May 1968), pp. 250–3.

264. *Op. cit.*

265. *NATO's Fifteen Nations*, Vol. 15(4), (August–September 1970), inside front cover; see also, Taylor, J. W. R. and Mondey, D. *Spies in the Sky*, (Ian Allan, London, 1972), p. 89.

266. Brown, N., *British Arms and Strategy, 1970–1980*, (Royal United Services Institution, May 1969).

267. Fairhall, D., Nimrod – the mighty hunter, *The Guardian*, (5 October 1974).

268. *The Military Balance, 1976–1977*, (International Institute for Strategic Studies, London), pp. 19–20.

269. *The Military Balance, 1974–1975* and *1975–1976*, (International Institute for Strategic Studies, London). See annual table entitled: I. Nuclear Delivery Vehicles: Comparative Strengths and Characteristics, (B) Other NATO and Warsaw Pact Countries, (ii) Aircraft.

270. *Military Applications of Nuclear Technology*, Hearings, Joint Committee on Atomic Energy, US Congress, 93rd Congress, (May–June 1973), Part II, pp. 98–9.

271. *Flight International*, Vol. 76(2634), (4 September 1959), pp. 116, 135.

272. Smart, I., *Future Conditional. The Prospect for Anglo–French Nuclear Cooperation*, Adelphi Paper No. 78, (International Institute for Strategic Studies, August 1971), p. 4.

273. *Nuclear Weapon Programme*, Twelfth Report from the Expenditure Committee, (19 July 1973), (HMSO, London), pp. XVI.

274. Richardson, D., Lance into battle, *Flight International*, Vol. 111(3556), (30 April 1977), pp. 1192–5.

275. Heymont, I., The NATO nuclear bilateral forces, *Orbis*, Vol. 9(4), (Winter 1966), p. 1033, which quotes the *New York Times* (11 April 1958), p. 47.

276. Mackintosh, M., Soviet arms and capabilities in Europe, *Royal United Service Institution Journal*, (March 1971), pp. 22–30.

277. *Nuclear Weapons and Foreign Policy*, Hearings Committee on Foreign Relations, 93rd Congress, (March–April 1974), p. 159.

278. Douglas, J. D. Jr., *The Soviet Theatre Nuclear Offensive*, (US Air Force, Washington, DC, 1976), 127 pp.

279. Cohen, S. T. and Lyons, W. C., A comparison of US–Allied and Soviet tactical nuclear force capabilities and policies, *Orbis*, Vol. 19(1), (Spring 1975), pp. 72–92.

280. *Nuclear Weapons and Foreign Policy*, Hearings, Committee on Foreign Relations, 93rd Congress, (March–April 1974), p. 200.

281. Hoge sees no rise in East Zone force, *New York Times*, (16 January 1955).

282. Hilsman, R., Strategic doctrines for nuclear war, in W. W. Kaufman (ed.), *Military Policy and National Security*, (1956), pp. 51, 66.

283. Foremost amongst the Soviet sources quoted are:
Sokolovsky, V. D. Marshal, *et al.*, *Military Strategy*, (1st, 2nd, and 3rd editions), [in translation], Foreign Technology Division, (Wright Patterson AFB, Ohio, 1968).
Savkiin, Y. Y., Col, *The Basic Principles of Operational Art and Tactics*, Moscow, 1972, [in translation], (USAF, Washington, DC, 1974), and *Characteristics of Modern Warfare*, translated and reprinted in *Strategic Review*, (Fall, 1974).
Sidorenko, A. A., *The Offensive*, Moscow, 1970, [in translation], (USAF, 1974),

particularly Chapter 4; The employment of nuclear weapons and destruction of the enemy by fire.

284. Cohen, S. T., A comparison of US–Allied and Soviet tactical nuclear force capabilities and policies, *Orbis*, Vol. 19(1), (Spring 1975), pp. 72–92, p. 92.

285. Karber, P. A., The tactical revolution in Soviet military doctrine, (The BDM Corporation, McLean Virginia, 2 March 1977), p. 1.

286. Casey, G. W., (Col.) The employment of tactical nuclear weapons, a search for safe limits, (1 April 1966), mimeographed.

287. Enthoven, A. C. and Smith, W. K., *How Much is Enough? Shaping The Defense Program, 1961–1969*, (Harper & Row, New York, 1971), p. 127.

288. Rumsfeld, D. H., *Report of Secretary of Defense to the Congress*, On the FY 1977 Budget and the FY 1977–1981 Defense Programs, (27 January 1976), pp. 80–1.

289. Rumsfeld, D. H., Report of the Secretary of Defense to the Congress on the FY 1978 Budget and the BY 1978–1982 Defense Program, (17 January 1977), p. 81.

290. *Nuclear Weapons and Foreign Policy*, Hearings, Committee on Foreign Relations, 93rd Congress, (March–April 1974), p. 159.

291. Lee, W. T., *Influence of NATO/Europe on Soviet Military Policy*, SSC-ISR-TN-2 (Stanford Research Institute, Stanford, Cal., 30 April 1970). p. 9.

292. *The Military Balance, 1970–71* (London: IISS), p. 95.

293. Brown, N., *European Security, 1972–1980* (London: Royal United Services Institute for Defense Studies, 1972), p. 67.

294. Record, J., *Sizing Up the Soviet Army* (The Brookings Institution, Washington, DC, 1975), p. 40.

295. Record, J., *US Nuclear Weapons in Europe, Issues and Alternatives* (The Brookings Institution, Washington, DC, 1974), p. 37.

296. *Development, Use, and Control of Nuclear Energy for the Common Defense and Security and for Peaceful Purposes*, First Annual Report to the US Congress, Joint Committee on Atomic Energy, 94th Congress, 1975), p. 20.

297. Erickson, J., Soviet military capabilities in Europe, *Journal of the Royal United Services Institute for Defense Studies*, Vol. 120, No. 3, (March 1975), pp.

298. Wolfe, T. W., *Soviet Power and Europe, 1945–1970* (The Johns Hopkins Press, Baltimore, Md., 1970), p. 151.

299. See Ref. 298, pp. 487–8.

300. Record, *Sizing Up the Soviet Army* (The Brookings Institution, Washington, DC, 1975), p. 40.

301. Statement of Secretary of Defense M. R. Laird in the FY 1973 Defense Budget and FY 1973–74 Program, (17 February 1972), (House Armed Services Committee, Washington, DC), mimeo., pp. 45–6.

302. *The Military Balance, 1971–2* (London: IISS), p. 59.

303. See, for example, *Military Applications of Nuclear Technology*, Part I. Joint Committee on Atomic Energy, US Congress, 93rd Congress, (May 1973), p. 11.

304. Record, J., *US Nuclear Weapons in Europe, Issues and Alternatives*, (The Brookings Institution, Washington, DC, 1974), p. 40.

305. *To Consider NATO Matters*, Ref. 148, p. 27.

306. *To Consider NATO Matters*, Ref. 148, p. 19.

307. How DOD assesses the balance of US, USSR, and PRC general purpose forces, *Commanders Digest*. Vol. 20, No. 4, (17 February 1977), p. 21.

308. The Defense Budget, in *Setting National Priorities, The 1978 Budget*, J. Pechman (ed.), (The Brookings Institution, Washington, DC, 1977), p. 93.

309. Record, J., *US Nuclear Weapons in Europe, Issues and Alternatives*, (The Brookings Institution, Washington, DC, 1974), p. 41.
310. Wolfe, T. W., *Soviet Power and Europe, 1945–1970*, (The Johns Hopkins Press, Baltimore, Md., 1970), pp. 208–10.
311. Douglas, J. D. Jr., *The Soviet Theatre Nuclear Offensive*, (US Air Force, Washington, DC, 1976), pp. 3–7.
312. Topping, S., Russia exhibits atomic infantry, *New York Times*, (18 August 1961).
313. Stanier, J. W., and Bairsto, P. C., *A British View of a Conventional Strategy for the Central Front in NATO, Part II*, Report of a Seminar Held at Royal United Services Institute for Defence Studies, London, 23 October 1974, (26 March 1975), mimeo.
314. Erickson, J., Shield-72, Warsaw Pact military exercises, *Journal of the Royal United Services Institute for Defence Studies*, Vol. 4, No. 117, (December 1972), pp. 32–3.
315. Vigor, P. H. and Donnelly, C. N., The Soviet threat to Europe, *Journal of the Royal United Services Institute for Defence Studies*, Vol. 120, No. 3, (March 1975).
316. Erickson, J., Soviet military capabilities in Europe, *Journal of the Royal United Services Institute for Defence Studies*, Vol. 120, No. 3, (March 1975).
317. See Ref. 316.
318. Barber, R. E., Col., The myth of Soviet nuclear war strategy: evidence shows emphasis on conventional, *Army*, Vol. 25, No. 6, (June 1975), pp. 10–17.
319. Middleton, D., Anxieties about NATO, *New York Times*, (10 December 1973).
320. *Planning US General Purpose Forces: Overview*, Budget Issue Paper, (January 1977), Congressional Budget Office, US Congress, Washington, DC, p. 13,
321. Morse, J. H., Capt. (USN), Questionable NATO assumptions, *Strategic Review*, Vol. 5, No. 1, (Winter 1977), pp. 21–8.

Additional references concerning the June–July 1977 US debate on new tactical nuclear weapon designs

1. Pincus, W., Neutron killer warhead buried in ERDA budget, *Washington Post*, (6 June 1977).
2. US said to build neutron warhead, *Baltimore Sun*, (6 June 1977).
3. Neutron bomb, *Atlanta Constitution*, (6 June 1977).
4. Pincus, W., Carter weighing radiation warhead, *Washington Post*, (7 June 1977).
5. A new warhead we don't need, *Washington Post*, (8 June 1977).
6. Pincus, W., Hatfield seeking to bar funds for neutron killer warhead, *Washington Post*, (9 June 1977).
7. The true precision bomb, *Boston Globe*, (9 June 1977).
8. Antevil, J., N-bomb precision – blessing and curse, *New York Daily News*, (10 June 1977).
9. The new killer bomb, *Los Angeles Times*, (16 June 1977).
10. Horton, A., Warhead designed to kill troops, spare property, near congress OK, *Pittsburgh Press*, (20 June 1977).
11. Nine nuclear weapon devices schedule for US production, *Aviation Week and Space Technology*, Vol. 20, (June 1977).

12. Pincus, W., Monkeys get radiation in neutron bomb tests, *Washington Post*, (22 June 1977).
13. Neutron bomb funds allowed to stand by Senate committee, *Washington Star*, (23 June 1977).
14. No people – bomb ban, *Baltimore Sun*, (23 June 1977).
15. Pincus, W., Panel votes 'killer' warhead funds, *Washington Post*, (23 June 1977).
16. Pincus, W., Pentagon pushes neutron shell for artillery forces, *Washington Post*, (24 June 1977).
17. Corddry, C., Soviet defense against neutron warhead would lag up to 20 years, US aide says, *Baltimore Sun*, (25 June 1977).
18. Pincus, W., Pentagon wanted secrecy on neutron bomb production, *Washington Post*, (25 June 1977).
19. No neutron warheads, *Washington Post*, (26 June 1977).
20. Pincus, W., ACDA to weigh in on killer warhead, *Washington Post*, (28 June 1977).
21. Pincus, W., New bomb advances in Senate, *Washington Post*, (2 July 1977).
22. Tolchin, M., Neutron bomb fund debated by Senate in a secret session, *New York Times*, (2 July 1977).
23. Senator Hatfield on neutron weaponry, *Washington Star*, (4 July 1977).
24. Battle over the N-bomb, *Newsweek*, (4 July 1977).
25. Pincus, W., Neutron arms deemed danger to SALT, *Washington Post*, (6 July 1977).
26. McGrory, M., Whistling through that neutron bomb, *Washington Post*, (7 July 1977).
27. Neutron bomb is reported already tested, *Washington Star*, (7 July 1977).
28. Will, G., Those blasts against neutron weapons, *Washington Post*, (7 July 1977).
29. Decision by Carter on neutron arms indicated in August, *New York Times*, (7 July 1977).
30. Weinraub, B., Pentagon hopes to deploy neutron warheads by 1979, *New York Times*, (8 July 1977).
31. Pentagon protest of neutron bomb brings 4 arrests, *Washington Post*, (8 July 1977).
32. Pincus, W., Neutron warhead wouldn't be deployed until '79, Hill told, *Washington Post*, (8 July 1977).
33. The reduced-blast warhead, *Wall Street Journal*, (8 July 1977).
34. Corddry, C., Neutron tests called successful, *Baltimore Sun*, (8 July 1977).
35. The neutron warhead, *Washington Star*, (8 July 1977).
36. No neutron bombs now (Editorial), *Chicago Sun Times*, (8 July 1977).
37. Towell, P., Neutron bomb poses dilemma for Congress, *Congressional Quarterly*, (9 July 1977), pp. 1403-7.
38. Upsetting the nuclear balance, (Editorial), *Boston Globe*, (9 July 1977).
39. The real flaws in neutron weaponry, (Editorial), *Milwaukee Journal*, (9 July 1977).
40. Technology of neutron arms is not new, *Washington Star*, (9 July 1977).
41. Weapons, clean and dirty, (Editorial), *Baltimore Sun*, (10 July 1977).
42. Cool the neutron arms debate, (Editorial), *Philadelphia Inquirer*, (10 July 1977).
43. Those 'neutron bombs'; a choice that blurs? *Philadelphia Bulletin*, (10 July 1977).
44. A cleaner nuclear weapon, (Editorial), *Chicago Tribune*, (10 July 1977).
45. New bomb, (Editorial), *Atlanta Constitution*, (11 July 1977).
46. Dr Strangelove is not loose, (Editorial), *Los Angeles Times*, (11 July 1977).
47. Pincus, W., Production of neutron arms backed, *Washington Post*, (12 July 1977).
48. The neutron weapon, (Editorial), *New York Times*, (12 July 1977).
49. Pincus, W., Senate to vote soon on starting neutron warhead production, *Washington Post*, (12 July 1977).

50. Scoville, A. and Teller, E., A new weapon to think (and worry) about, *New York Times*, (12 July 1977).
51. Weinraub, B., Carter says he backs production of neutron arms, *New York Times*, (13 July 1977).
52. Excerpts from transcript, President Carter's news conference, *New York Times*, (13 June 1977).
53. The neutron bomb (Editorial), *Washington Post*, (13 July 1977).
54. Heinz, J. Sen., The neutron bomb: let's get the facts, *The Philadelphia Inquirer*, (13 July 1977).
55. Wills, G., And for small jobs, madam, try no-mess neutron bomb, *Baltimore Sun*, (13 July 1977).
56. Corddry, C. W., Carter leans to radiation bomb option, *Baltimore Sun*, (13 July 1977).
57. Haig urges neutron bomb for allied arsenal, *Baltimore Sun*, (13 July 1977).
58. Gillette, R., Neutron bomb is almost 20 years old, was focus of heated controversy in 1961, *Los Angeles Times*, (13 July 1977).
59. Taylor, W., Chances slim of bottling up neutron bomb, *Washington Post*, (13 July 1977).
60. Pincus, W., Senate refuses, 58–38, to kill neutron bomb, *Washington Post*, (14 July 1977).
61. Drozdiak, W., Neutron weapons could deter aggression, NATO leaders say, *Washington Post*, (14 July 1977).
62. Corddry, C. W., Neutron arms win Senate vote, *Baltimore Sun*, (14 July 1977).
63. Taylor, E., A new starter in the death race, *Philadelphia Inquirer*, (14 July 1977).
64. The neutron bomb, (Editorial), *Christian Science Monitor*, (15 July 1977).
65. Weinraub, B., What role for neutron bomb, *New York Times*, (17 July 1977).
66. Frye, A., The high risks of neutron weapons, *Washington Post*, (17 July 1977).
67. Getler, M., Bonn party chief says US bomb a 'perversion', *Washington Post*, (18 July 1977).
68. Kissinger in dark on neutron weapon decision, *Washington Post*, (19 July 1977).
69. Nunn, S. Sen. and Heinz, H. J. Jr. Sen., Controversy over the neutron bomb, *Baltimore Sun*, (23 July 1977).
70. Getler, M., Debate on bomb grows in Bonn, *Washington Post*, (24 July 1977).
71. Lentz, E., West Germany aroused over neutron bomb prospect, *New York Times*, (24 July 1977).
72. Germany eyes deployment of US neutron bomb, (AP), (27 July 1977).

Appendix 1

Nuclear weapon delivery systems distribution in NATO

No systematic data have been gathered on the numbers or distribution of nuclear weapon delivery capable aircraft within NATO, or on the training of national military units of NATO member nations in the handling of nuclear weapons. Tables 1A.1 and 1A.2, concerning missiles, are taken from Heymount [1]. The data in Tables 1A.1 and 1A.2 are incomplete in that they stop at 1966 and there is no mention of Matador, Mace or Pershing. In addition, Honest John missiles were in service with UK forces. Lance is to be procured by the UK, F.R. Germany, Italy, Belgium and apparently France, and the Netherlands. When a successor missile system is introduced, various NATO nations apparently retain the older missile – or at least there seems no available information to the contrary.

Tables 1A.3, 1A.4, 1A.5 and 1A.6 were prepared by Sally Anderson. Table 1A.7 is from *Arms Control Today*, (April 1974).

Two references have been found to weapon systems of non-nuclear NATO nations which seem to have been nationally designed to be nuclear-capable. It is difficult to understand the rationale for the first, and no corrob ration can be found for the second:

(*a*) The Italian cruiser *Giuseppe Garibaldi* was fitted with four Polaris tubes in its stern. (*Jane's Fighting Ships*, 1965–66 to 1969–70.) The ship test-fired Polaris rockets in the Caribbean in 1962. However, at some time in the early 1970s, the ship was refitted, and the missile tubes were replaced. A second ship in the class, the *Vittorio Veneto* was at one point also reported as being intended to be fitted with Polaris tubes [2], but apparently was never built in that way. (These ships were also fitted for Terrier and Asroc missiles.)

(*b*) The second report is more abstruse. In 1963, a new West German submarine ordered by F.R. Germany, Denmark, Norway and Turkey was 'reported to have tubes for eight torpedoes which can be atom tipped' [3].

Table 1A.1. Nuclear-capable weapons systems in the possession of NATO nations †

	8 in howitzers	155 mm howitzers	Honest John	Corporal	Nike Hercules	Sergeant
Canada			×			
Belgium	×	×	×		×	
Denmark			×		×	
France		×	×		×	
Greece			×		×	
Italy	×	×	×		×	
Netherlands			×		×	
Norway		×			×	
Turkey	×		×		×	
United Kingdom				×		
F.R. Germany			×		×	×

Table 1A.2. Deliveries of nuclear-capable weapons systems under various US military assistance programmes†

	Cumulative FY 1950–65	Estimated July 1965 to July 1966
Nike missiles	2 927	261
Corporal missiles	113	—
Honest John missiles	3 673	185
8 in howitzers	123	1

† Tables 1A.1 and 1A.2 from: Heymont, I., The NATO nuclear bilateral forces, Orbis, Vol. 9, No. 4, Winter 1966, pp. 1025–41. *Source:* Department of Defense, *Military Assistance Facts*, (15 February 1965), p. 19.

Table 1A.3. US guided missiles, given and sold to foreign governments, 1950–1973, that are nuclear capable with US forces

Type and designation	Assistance	Sales	Total
Surface-to-surface:			
Corporal	113	—	113
Jupiter	63	—	63
Pershing	—	133	113
Sergeant	38	195	233
Honest John†	3 566	2 688	6 254
SLBM:			
Polaris (A-3)	—	102	102
Air-to-surface:			
Bullpup	3 768	1 603	5 371
Walleye	—	104	104
Surface-to-air:			
Nike	3 460	1 766	5 226
Terrier	114	172	286
Total missiles:	**11 122**	**6 763**	**17 885**

† 1950–1972

110

Table 1A.4. Description of nuclear-capable guided missiles transferred abroad

Designation	Range miles	Yield kt
Bullpup	7–10	20
Corporal	75	20
Jupiter	1 500	5 000
Nike-Hercules	84	1
Pershing	96–450	60–400
Polaris (A-3)	2 800	
Sergeant	27–84	60
Terrier	21	1
Walleye	12	100
Walleye II	35	100
Lance	3–70	1–100
Honest John	5–22	20

Table 1A.5. Description of nuclear missiles deployed abroad

	Range miles	Yield kt
USA:		
Honest John	5–22	20
Sergeant	27–84	60
Nike-Hercules	84	1
Lance (to replace Sergeant and Honest John)	3–70	1–100
USSR:		
Frog	10–45	kt range
Scud A	50	kt range
Scud B	185	kt range

Table 1A.6. Foreign deployment: as of 1973

	Number	Deployed in
USA:		NATO countries, Spain,
Honest John	136	Taiwan, South Korea
Sergeant	19	
Nike-Hercules	108	
Total missiles:	**263**	
USSR:		WTO countries, Egypt,
Frog	261	Iraq, Syria, N. Korea,
Scud	195	Cuba
Total missiles:	**456**	

Table 1A.7. US and NATO tactical nuclear weapons in Europe

Type of tactical nuclear weapon	Number of weapons/ delivery vehicles	Number of weapons warheads or bombs†	Maximum range of weapons *miles*
Atomic Demolition Munitions (ADM), atomic mines	300	300	Not applicable
Artillery, M-109 155 mm and M-110 203 mm	1 010	3 030	10–18
Surface-to-surface missiles Honest John and Sergeant (being replaced by Lance); Pershing	175 108	175 324‡	25–85 variable, 60–450
Tactical aircraft§ US aircraft, F-4 and F-111 Other NATO aircraft	316 612	1 020 1 224	500–1 000 combat radius 420–600 combat radius
Surface-to-air missiles (SAM)	720	720	60
Land-based antisubmarine warfare (ASW) aircraft	380	380	1 800 combat radius
Totals	**3 621**	**7 173‖**	

† All weapons are in low-kiloton/high-kiloton range, except for some air-deliverable bombs in the low megaton range.
‡ The US maintains approximately 216 of these in reserve.
§ These do not include 98 nuclear-capable aircraft on the two US aircraft carriers regularly stationed in the Mediterranean Sea.
‖ Total number is based upon the assumption that the maximum capability of each weapon will be utilized. Some weapons such as aircraft have flexibility to carry a range of payloads. Of the total number of tactical nuclear weapons approximately 150 are British and over 7000 are American. France also maintains about 100 tactical nuclear weapons in Europe in the form of aircraft and Pluton surface-to-surface missiles.

Sources: US Congress, Joint Committee on Atomic Energy, Subcommittee on Military Applications, *Military Applications on Nuclear Technology:* Parts I and II, 93rd Congress, 1st Session, (16 April, 22 May and 29 June 1973); Trevor Cliffe, Military technology and the European balance, *Adelphi Papers,* No. 89 (London: International Institute for Strategic Studies, August 1972; Robert P. Berman, unpublished research; White Paper 1971–72 on the Security of the Federal Republic of Germany and on the State of the German Federal Armed Forces, English version.
Table taken from *Arms Control Today* (Arms Control Association, Washington D.C.), Vol. 4, No. 4, (April 1974), p. 4.

Table 1A.8. Yields of some US tactical nuclear weapons

Weapon	Yields 'DOD Unclassified'†	'Other Unclassified'†	CDI, Defense Monitor§	Casey‖	Other
Atomic Demolition Munitions, (ADMs)	—	—	—	sub and low to high kt	
155 mm artillery shell	subkiloton	2 kt	—	1–2 kt	
8 in howitzer	subkiloton to kiloton	1 kt	—	1–2 kt	
Honest John	kiloton range, several yields	low kt	up to 100 kt	5–10 kt	
Sergeant	kiloton range, several yields	low kt	up to 100 kt	10–60 kt	
Lance	kiloton range, several yields	1–100 kt	up to 50 kt	—	
Pershing	kiloton range, several yields	60–400 kt	up to 400 kt	up to 1 Mt	
Air dropped bombs	several types and yields, from subkiloton to megaton	from 0.2 kt to over 1 Mt	from 5 kt to more than 1 Mt	kt–Mt	
Walleye ASM	—	—	5–10 kt	—	

† *Nuclear Weapons and Foreign Policy*, Hearings, Committee on Foreign Relations, US Senate, 93rd Congress, (March–April 1974), p. 201.
§ 30 000 US nuclear weapons, *The Defense Monitor*, (Center for Defense Information, Washington, DC), Vol. 4, No. 2 (February 1975), p. 5.
‖ Casey, G. W., Col., The employment of tactical nuclear weapons, a search for safe limits, 1 April 1966, mimeographed.

Ground-to-air-missiles

The following nuclear missiles have been deployed with US and NATO forces in Europe:

Jupiter	Honest John
Thor	Corporal
Mace	Sergeant
Matador	Pershing
Davy Crockett	Lance
Little John	

The first three of these explicitly functioned as strategic missiles. A complete survey of these weapons would indicate:

The yields of the warheads which they carry or carried.

Their range.

The years in which they were deployed; year of initial operational capability, and system lifetime.

The numbers deployed.

The number of nations in which they were deployed.

Systematic material on all these missiles and variables has not been gathered here, although some of this information does appear in the material below.

Topics omitted

This survey of materials omits any substantive discussion of several categories of tactical nuclear weapons such as:

Nuclear-tipped surface-to-air missiles.

Nuclear-tipped torpedoes.

Nuclear-tipped air-to-air missiles.

Nuclear depth charges and ASW rocket-torpedoes.

Nuclear air-to-surface missiles.

The names, ranges and estimates of yields for several of these are given below.

Nike-Hercules surface-to-air missile; range, 80 miles, yield, up to 5 kt.

Talos surface-to-air missile; range, 70 miles, yield, 5 kt.

Terrier surface-to-air missile; range, 25 miles, yield, about 1 kt.

ASROC, ASW Rocket; range, 6 miles, yield, 1 kt.

SUBROC, ASW Rocket; range, 30 miles, yield 1 kt.

Mark 57 and Mark 101 nuclear depth bombs, carried by US P-3 and S-3 aircraft, and by ASW helicopters, estimated yield of 5–10 kt.

The naval surface-to-air nuclear missiles, Talos and Terrier, have to some degree already been phased out, and are to be replaced by the Aegis missile system, and the Nike-Hercules will be replaced by the Standard missile system.

The survey also therefore omits any discussion of the question of the possibility of a tactical nuclear engagement at sea, and whether such an occurrence could remain isolated and not escalate. See Reference 4.

Polaris submarine capability assigned to NATO

On 17 May 1961, President J. F. Kennedy announced that the US would 'commit to the NATO command five – and subsequently still more – Polaris atomic missile submarines, which are defensive weapons, subject to any agreed NATO guidelines on their control and use and responsive to the needs of all members but still credible in an emergency' [5]. The first US Polaris submarines went on patrol in the Atlantic on 18 November 1960. Between 1961 and 1964, the US Polaris boats contained 16 missiles with one warhead each with a yield of approximately 0.8 Mt. Between 1964 and 1970, these missiles were MRVed and there were three 200 kt warheads in each (designated for a single target). (The IISS stated that a larger number of US Polaris submarines were committed to SACLANT.) Beginning in 1970, the US converted its Polaris missiles to MIRVed Poseidon missiles. Each submarine then contained approximately 160 individually targeted MIRVed warheads (ten per missile) of 50–60 kt each. It is not clear what number of Polaris submarines were in fact actually assigned, or if the US continued to assign five SLBM submarines to NATO under these circumstances, since this would have effectively meant an increase in NATO assigned targets from 80 (5 × 16) to 800 (5 × 160). In 1966 (before MIRVing took place) Vandevanter specifically refers to 'the three Polaris submarines assigned to SACEUR', and a later report in 1974 (after MIRVing) refers to the assignment of 'about three' US Polaris–Poseidon submarines to NATO.

As a result of the 1962 agreement by which the US supplied Polaris missiles without warheads to the UK, the UK subsequently committed its Polaris submarines 'to the planning control of SACEUR',

> 'This is taken to mean that they accept NATO targeting and would only be used independently where, in the words of the Nassau Agreement, "HM Government may decide that supreme national interests are at stake", or, according to the interpretation preferred by the Labour Government, if NATO were dissolved . . .
> The use of British nuclear weapons, whether tactical or strategic, would, as stated by Mr Healey, depend in the last resort on the decision of the British Prime Minister . . .
> Assignment to NATO means that in normal circumstances they will not be used independently of NATO plans and decisions. It does not mean that NATO or SACEUR can use them independently of the British Government and without its consent' [6].

The four British submarines carry 16 A-3 missiles, each containing three British manufactured MRV warheads. Two boats are on patrol most of the time, at least one boat at all times. The first UK Polaris vessels went on patrol in 1968.

In 1975, the US announced that it was assigning additional Poseidon missiles 'to the defense of Western Europe as part of a plan to strengthen the North Atlantic Treaty Organization's nuclear forces' [7]. It was stated that the Poseidon's MIRVed warheads 'had sufficient accuracy to be used in a tactical role', and for fixed targets this is certainly so. It was indicated that 'warheads could be used for attacking military bases and supply lines near a battlefront. These targets until now were largely assigned to fighter bombers'. This means that the additional Poseidon have been used to replace some number of QRA aircraft. That would serve to relieve some of the question of vulnerability of QRA aircraft and improve the survivability of the NATO 'tactical' nuclear force. There are indications that NATO was willing to consider the reduction of the number of QRA aircraft at least since 1973 due to the question of vulnerability and preemption. At the same time, it releases the equivalent number of aircraft for ground support with conventional munitions.

References

1. Heymount, I., the NATO nuclear bilateral forces. *Orbis*, Vol. 9, No. 4 (Winter 1966).
2. *Technology Week*, Vol. 20, No. 7 (13 February 1967).
3. *Undersea Technology*, Vol. 4, No. 5 (May 1963).
4. *FAS Public Interest Report*, Vol. 28, No. 10 (December 1974), pp. 146, 8. Special issue on navy general-purpose forces.
5. American Foreign Policy, current documents, Department of State Publication 7808, (1965), p. 486, Washington, DC.
6. Burrows, B. and Irwin, C., *The Security of Western Europe; Towards a Common Defense Policy*, (C. Knight, London, 1970), pp. 61–2.
7. Finney, J. W., US assigns more missile submarines to the defense of NATO, *New York Times*, (18 June 1975).

Forward based systems

The phrase 'forward based systems' is used to describe those longer range US 'strike' or 'attack' aircraft which the US bases in Europe or on aircraft carriers, which are specifically designated as nuclear weapon delivery systems, and whose range permits them to reach targets within the USSR. This is the narrower of the two possible usages which are found in the literature, as will be indicated below. These systems have come under discussion both in the SALT negotiations between the US and the USSR and in the MBFR negotiations. There is an excellent and concise study of the systems available, and very little about them will therefore be said here [1].

Nerlich lists the three elements which presently make up the 'forward based systems' as:

1. US F-111 aircraft based in the United Kingdom.
2. F-4 aircraft in Turkey, northern Italy, and in the Federal Republic of Germany.
3. Aircraft carrier based A-6s and A-7s in the Mediterranean.

The US also has F-4s based in other NATO countries. Some portion of these aircraft are part of those that stand on Quick Reaction Alert (QRA).

F-104 aircraft, which have been outfitted solely for nuclear weapon delivery, and whose flight range also permits them to reach targets within the USSR are not included in this definition since they are only operational with the air forces of NATO nations other than the US.

Pershing missiles are also not included in the category of 'forward based systems', although their range permits them to reach targets just within the Soviet borders from locations in the Federal Republic of Germany and their target assignments probably overlap to some degree with those of FBS and other QRA aircraft. Used in this narrower way, the designation is used only to apply to US attack aircraft.

A broader usage for the phrase 'forward based aircraft' also occurs, as is apparent in the following excerpt from a 1973 Hearing of the US Joint Committee of Atomic Energy.

> ### Forward-Based Aircraft in NATO
>
> *Senator Symington:* Will you advise us of the number, type and range of forward-based aircraft in NATO and how many of these are quick reaction alert? You might supply that for the record.
>
> *General Goodpaster:* The total figure I would like to give you. The quick reaction alert figure is of the order of [deleted] that we maintain on quick reaction alert.
>
> *Senator Symington:* Would you identify by nation in both categories (QRA and non-QRA) how many are United States, and how many are not United States?
>
> *General Goodpaster:* I had better provide that for the record, Mr Chairman.
> [The information supplied follows:]
> There is a total of [deleted] forward-based aircraft in NATO. Of this

total, there are [deleted] dual capable nuclear/conventional aircraft, of which [deleted] are quick reaction alert (QRA), ([deleted] percent) of the [deleted] aircraft are US. Of the [deleted] on QRA, [deleted] (or nearly [deleted] percent) are US.

Approximate aircraft ranges

Type of aircraft		Range (*statute miles at altitude*)
A-6	F-102	
A-7	F-104	
Buccaneer	F-111	
F-4	G-91	(all aircraft ranges
F-5	Harrier	were deleted)
F-35	Hunter	
F-84	LIG	
F-86	Mirage	
F-100	Vulcan	[2]

In this case it is clear that what is being referred to by the phrase both in the question and in the data provided in reply by the US Department of Defense, are all NATO nuclear-capable aircraft, land-based and carrier-based. The 1973 Senate Foreign Relations Committee Report which states 'In all, there are in NATO over 2000 US and other NATO forward based nuclear-capable aircraft' was also apparently using the phrase in this broader way [3].

References

1. Nerlich, U., *The Alliance and Europe: Part V, Nuclear Weapons and East–West Negotiations*, Adelphi Papers, No. 120, (Winter 1975–1976), pp. 35.
2. *Military Applications of Nuclear Technology*, Part II, Hearings, Joint Committee on Atomic Energy, US Congress, 93rd Congress, (May–June 1973), pp. 98–99.
3. *US security issues in Europe: burden sharing and offset, MBFR and nuclear weapons, September 1973*, A Staff Report. Committee on Foreign Relations, US Senate, (2 December 1973), p. 14.

Aircraft capable of nuclear weapon delivery

Most of the points regarding nuclear capability of tactical 'strike' aircraft, or fighter-bombers, have already been covered in other sections. They will only be drawn together here in outline form, and references for them will not be repeated.

— US aircraft, some portion of which are on Quick Reaction Alert (QRA) include FB-111, F-111, F-4, A-6, A-7, F-100, and F-104.
— Aircraft of NATO allies also are on QRA, and a list of these was presented earlier in the text (p. 118).
— In 1965 it was announced that aircraft of nine NATO allied nations (UK, Belgium, Canada, France, Greece, Italy, Netherlands, Federal Republic of Germany and Turkey) were fitted for delivery of US nuclear weapons in Europe. France and Canada later withdrew on separate occasions from this association, leaving seven NATO allies in this role.
— The F-104s that were provided to NATO allies were by and large solely intended to function for nuclear weapon delivery.
— As of 1973 the number of different US aircraft that could deliver tactical nuclear weapons ranged between 17 and 20. This is far more than the number people ordinarily consider in this role.
— As of the end of 1973, there were over 2000 US and NATO aircraft with nuclear weapon delivery capability.
— US aircraft carriers supplied a substantial nuclear weapon delivery capability, and this subject is covered in the next section.
— Air-delivered nuclear weapons have been available to NATO commanders since the early 1950s.
— The yields of air-dropped tactical nuclear weapons are quite considerable at the high ends of their range. These range from 2.5 kt to 1.4 Mt. An F-4 aircraft can carry – or did so in the past – several types of air-dropped nuclear weapons, and the following are estimates of their yields.

   — The B-28 bomb, the oldest in the inventory and probably now phased out; over 1 Mt
   — The B-43 bomb; Megaton range
   — The B-57 bomb; 10–20 kt
   — The B-61 bomb; 100–500 kt
   — The Walleye optically guided glide bomb

One F-4 can obviously carry over 1 Mt in weapons, and these were often based quite close to the USSR, in Turkey, for example. In 1964 Neville Brown referred to a 'standard NATO store' (payload) of 1 Mt, in apparent reference to air-dropped weapons. At the end of the 1960s the International Institute for Strategic Studies indicated that the 'average' explosive yield of the various kinds of NATO air-dropped weapons was in the range of 100 kt.

— In the early years of the US nuclear targeting plans the Tactical Air Command had the role of delivering a substantial portion of the nuclear yield targeted on the USSR.
— France and the UK have their own air-dropped tactical nuclear weapons.
— For many years, evidence as to the nuclear weapon delivery capability, or mission, of Soviet fighter bombers was very ambiguous. Since 1970 it seems more likely that these aircraft, as well as larger Soviet medium range aircraft, also have nuclear weapon delivery roles.

Aircraft carrier borne nuclear strike forces

Several of the sections in this data portion of the paper were included to stress aspects within the subject area of tactical nuclear weapons that receive relatively less attention elsewhere. This holds for this section, as well as the subsequent one on the two early US cruise missiles in Europe, the Matador and Mace.

A category of delivery system that has by and large escaped general attention in this regard is aircraft-carrier-based strike aircraft. At least until the mid-1960s, the nuclear strike mission of the US aircraft carrier was substantial. In 1960, US Chief of Naval Operations Adm. Arleigh Burke stated that 'there were more nuclear bomb carrying planes aboard five Navy carriers in the Mediterranean and Far East than in Russia's entire heavy bomber fleet' [1]. In the following year, Secretary of Defense McNamara supplied further details: 'From the decks of a single carrier of the Forrestal class, fifty attack aircraft can be launched armed with megaton nuclear weapons. Six carriers of this class, as well as nine other attack carriers, are deployed throughout the world's oceans, and two other attack carriers are currently in maintenance' [2]. The 'nine other attack carriers' carried a slightly reduced nuclear complement. In 1964, when a Polaris submarine squadron was deployed to North Pacific waters, the third carrier of the Seventh Fleet 'was released from deterrent duty' in the area [3]. Although there were subsequent statements of changes in the mission priorities of the attack carrier force, there is little evidence of their reduced reliance on nuclear weapons. Certainly there have been no indications of actual reduction in nuclear weapons aboard these vessels, despite ostensible changes in mission [4]. As indicated these aircraft are usually classed within the forward-based systems, discussed above. As also indicated in the text by the quotation from Sen. Symington on page 4–5, there is the question as to whether these aircraft, and the nuclear weapons that they deliver, should be considered 'strategic', or 'tactical'.

AJ-1 aircraft were apparently on routine deployment in the Mediterranean on board aircraft carriers at least as early as 1953 [5]. US aircraft carrier nuclear armed strike forces were also regularly deployed in the Norwegian sea, the same area into which the first Polaris submarines were subsequently deployed. Soviet military literature of the period stresses the need to neutralize these forces, which prompted the development of Soviet short range surface launched surface-to-surface missiles, carried both by Soviet submarines and surface vessels. Large sections of the Soviet Naval Fleet were largely designed as platforms for these anti-carrier missiles. The following two tables indicate national aircraft carrier capability from 1946 to 1970, and the types of carrier born aircraft that were capable of delivering nuclear weapons and that were assigned to that role. However, the tables can only be considered applicable to those years in which the US, UK or France respectively deployed nuclear capable aircraft on board such carriers. These years appear to be:

— For the US, after 1953.
— For the UK; unclear. The two early UK carrier-borne nuclear delivery capable aircraft, the Scimitar and Buccaneer, were first deployed respectively in 1958–59, and 1962–63. However there is no specific information on when these may have actually carried nuclear weapons (see Section III, on Tactical Nuclear Weapons of the UK).
— For France; unclear. Etendards were deployed quite early on board French aircraft carriers but it is not likely that they could have had tactical nuclear weapons before these had been developed and deployed for France's land based tactical delivery aircraft, which was not until late 1972, and it would appear that if they have been given to the French naval air arm at all, no date for this is known, and it apparently was not before 1974 (see Section II, on Tactical Nuclear Weapons of France).

References

1. Burke disputes air force, *Missiles and Rockets*, Vol. 6, No. 11, (14 March 1960).
2. McNamara, R., address in Atlanta, Georgia, (11 November 1961).
3. Hayes, J. D., Rear Adm., Sea Power, July 1964–June 1965, A Commentary, *Naval Review*, (1966), p. 244.
4. Admiral Moorer explains requirements for attack aircraft carriers, *Navy Magazine*, Vol. 12, No. 12, (December 1969).
5. *Honeywell Flight Lines*, Vol. 4, No. 3, (May 1953).

Table 1A.9. Designation for US aircraft carriers appearing in Jane's

1946–50	1950–3 (?)	(?) 1953–7	1957–
Large fleet carrier	Large aircraft carrier (CVE)	CVA	Attack aircraft carrier, CVA
Fleet aircraft carrier	Aircraft carrier, (CV)	CVA + CVS = attack and support	Support aircraft carrier, CVS
Light fleet carrier	Small aircraft carrier	CVS	Small aircraft carrier, CVL
Escort aircraft carrier	Escort aircraft carrier	CVL CVE	Escort and helicopter escort aircraft, CVE and CVHE

Table 1A.10. Aircraft carriers – Jane's

	USA			UK		USSR		Australia	France	Others†	World total
	CVA	CVAN	CVS etc.	Heavy	Light	Heavy	Light	Light	Light	Light	
1946–47	27		70	6	8	1		0	2	3	**117**
1947–48	27		75	6	7		1	0	2	2	**120**
1948–49											
1949–50	28		75	6	7		1	2	2	2	**123**
1950–51	28		75	6	6			2	2	2	**121**
1951–52	27		75	7	9			2	3	2	**125**
1952–53	26		72	8	9			2	3	2	**122**
1953–54	17		84	6	10			3	3	3	**126**
1954–55	18		83	6	11			2	4	3	**127**
1955–56	20		83	6	10			2	4	3	**128**
1956–57	21		83	5	11			2	4	3	**129**
1957–58	20		86	3	10			2	4	4	**129**
1958–59	20		86	3	4			2	5	5	**125**
1959–60	17		63	3	6			2	5	5	**101**
1960–61	16	1	42	3	6			2	4	5	**79**
1961–62	16	1	41	3	4			2	3	5	**75**
1962–63	14	1	43	3	4			2	4	5	**76**
1963–64	14	1	43	3	4			2	4	5	**76**
1964–65	15	1	42	3	4			2	4	5	**76**
1965–66	15	1	42	3	4			2	4	5	**76**
1966–67	16	1	39	3	4			2	4	6	**75**
1967–68	16		39	3	4		1	2	4	6	**75**
1968–69	16		39	2	4		2 ?	2	4	6	**76**
1969–70	15		40	2	4		2	2	4	5	**75**

† Brazil, Canada, Argentina, India in 1969–70.

Note – It is not as simple as one might imagine to derive these figures; one cannot just pick up a copy of Jane's Fighting Ships, turn to the US, UK or USSR and find a number for 'aircraft carriers total'. The problem is that there are several numbers or several ways to derive such numbers. For example the US entries have had since 1945 a synopsis of the number of ships of each type entitled 'Strength of Fleet'. For the UK, such a synopsis exists for some years but less so in recent years. Starting 1956–57 such a synopsis appeared also in the USSR section. In addition, from 1959–60 a matrix appears which lists the number of naval vessels by type and by nation. One can always calculate the total number of a particular vessel type by counting every vessel of a given type: for example aircraft carriers, for a nation for that year. Upon doing these things one discovers that the matrix categories for aircraft carriers are not always the same as the categories within the national section and that the numbers derived by these three methods often do not agree with each other.

Table 1A.11. Types of carrier-borne aircraft capable of delivering nuclear weapons

Year	USA						UK		France
1945–46									
1946–47									
1947–48									
1948–49	(P2V)								
1949–50	(P2V)	(AJ-1) →							
1950–51	(P2V)								
1951–52									
1952–53									
1953–54									
1954–55			A3D						
1955–56			A3D						
1956–57			A3D-2						
1957–58			A3D-2	A4D-1					
1958–59		(AJ)	A3D-2	A4D-1			Scimitar		
1959–60			A3D-2	A4D-1			Scimitar		
1960–61			A3D-2	A4D-2N	A3J-1		Scimitar		
1961–62			A3D-2	A4D-2N	A3J-1		Scimitar		
1962–63		(A6A)	A3D-2	{ A4D-2N / A4D-5	{ A3J-1 / A3J-3		Scimitar	Buccaneer S-1	{ Dassault Etendard 4M
1963–64			A3B	A4C	{ A5A / A5C		Scimitar	Buccaneer S-1 and 2	
1964–65		(A6A) →	A-3D	A4E	A-5A		Scimitar	Buccaneer S-1 and 2	
1965–66			A-3B	A4E	A-5A		Scimitar	Buccaneer S-1 and 2	
1966–67			A-3B	A-4E	A-5A		(F4K)(c)	Buccaneer S-1 and 2	
1967–68			A-3B	A-4E / A-4F	A5A	(A-7)	(F4K)	Buccaneer S-1 and 2	
1968–69			(a)	(a, b)	(a)	(A-7)	(F4K)	Buccaneer S-1 and 2	
1969–70			(a)	(a, b)	(a)	(A-7)	(F4K)	Buccaneer S-1 and 2	↓

(a) Jane's lists the same aircraft in its 1968–69 and 1969–70 volumes but omits the mention of nuclear armament. Jane's would supply no further information as to whether this was a deliberate alteration in the notation. Other sources indicate that carrier borne RA-5C reconnaissance aircraft nevertheless retain their underwing pylons for attachment of nuclear stores.

(b) Though the A7A (Corsair) became operational in late 1961 and is a replacement for the A4 (Skyhawk), Jane's does not list nuclear armament for it, nor is it so listed by the International Institute for Strategic Studies.

(c) Though Jane's lists the F4K as carrier-borne, and this aircraft is capable of carrying nuclear weapons, Jane's does not list nuclear armament for it.

Matador

Matador was the first of the US Air Force missiles to reach operational status. It was 'initially operational' in October 1951 and was placed with units in Germany on 9 March 1954. By 1959 Matador units were also in Taiwan and in South Korea. Numbers were given by the US as 'total divisions – three, divisional strength – about 50', implying some 150 missiles.

Deployment varied over the years as follows:

1954–1957	— Federal Republic of Germany.
1958	— 'units' in Federal Republic of Germany, one 'unit' in Taiwan.
1959	— Federal Republic of Germany, Taiwan, Korea and Libya.
1960	— Federal Republic of Germany, Taiwan, Korea, Libya; 'Also in hands of NATO allies ... Matadors being replaced by more advanced Maces. Matadors in Federal Republic of Germany are being turned over to German troops.' 'Mideast' listed in place of Libya.
1961	— same as 1960.
1962	— Federal Republic of Germany, Taiwan, Korea. Being phased out of US arsenal. 2 squadrons in Germany turned over to German troops.
1963	— same as 1962, with additional statement of 'Being turned over to Federal Republic of Germany and Nationalist Chinese'.
1964	— phased out of US forces; in use by Federal Republic of Germany and Nationalist China with nuclear warheads maintained under US control.
1965	— same as 1964.
1966	— same as 1964.
1967	— same as 1964.

References to Matador disappear after 1967 from the journals listed below. Statements could not be found which specified that Matador (or Mace) were no longer deployed in Western Europe either with US or with German forces as of such and such a date. Matador was replaced in the European theatre by Mace, which was in turn replaced by the Pershing missile.

Matador: References

1. *Missiles and Rockets*, World Missile/Space Encyclopedia issues, 1957 to 1967.
2. *Air Force and Space Digest*, Annual Gallery of USAF Weapons; Missiles, 1958 to 1962.
3. McGuire, F. G.: Mace B Bases Readied on Okinawa. *Missiles and Rockets*, Vol. 8, No. 1, p. 18, (13 March 1961).

Mace

Mace was designed as a replacement for the most advanced (C) version of Matador. The Mace A version became operational with US troops in the Federal Republic of Germany in the spring of 1959. In 1960 there were at least 2 Divisions of the Mace A version with 50 missiles per division. The Mace B version was deployed on Okinawa by 1962 or 1963. Mace A was radar guided, and Mace B inertially guided. The missile had a range of 1200 miles or greater.

By 1962 there were three squadrons in the Federal Republic of Germany and fixed site coffin-type concrete and steel hardened facilities were built. In 1962 Mace A and B versions were both deployed in Europe and Mace B in Okinawa. Mace A was fired from a mobile launcher and Mace B from the hardened facilities. At peak deployment there were 5 Mace A squadrons and 1 Mace B squadron in hardened sites in Europe, 2 Mace B squadrons in Okinawa. There were 20–50 missiles per squadron, and a total of 1000 Mace missiles were procured. The entries concerning the Mace in *Air Force and Space Digest* in 1967, 1968 and 1969 all carry the exact same wording: 'One Mace B squadron remaining in Germany is being phased out in favour of (the) Army's Pershing missile'. It is therefore not clear precisely when Mace B ended its deployment in Germany. In 1970 Holst indicated that the Mace missile was still deployed in Europe. Through this period Mace B had remained as part of the US forces. Mace carried a warhead in the very high kilotons or of 1 Mt. One source gave a yield as high as 5 Mt in 1962 [8]. From its range, yield, and deployment locations, Mace would seem to have been as much a strategic missile as were the Thor and Jupiter missiles that had been deployed to Europe, but it was called a tactical missile.

Mace: References

1. *Missiles and Rockets*, World Missile/Space Encyclopedia, issues, 1958–1967.
2. McGuire, F. G.: Mace B Bases Readied on Okinawa. *Missiles and Rockets*, Vol. 8, No. 1, p. 18, (13 March 1961).
3. *Aviation Week and Space Technology*, Annual inventory of aerospace power issues; 1958 to 1969.
4. *Air Force and Space Digest*, Annual gallery of USAF weapons, missiles and gallery of missiles and space weapons; (1958–1968).
5. *Military Export Reporter*, (11 January 1968), p. 14, 'Japan; Government says nuclear missiles must remain for defense'.
6. Japanese magazine article concerning disorders on Okinawa; pp. 44–57, 'Kagena air base and at Naka'; (1969).
7. US to inactivate unit of missiles on Okinawa, *International Herald Tribune*, (19 December 1969).
8. The balance of nuclear terror, *Interavia*, Vol. 17, No. 11, (November 1962).

Table 1A.12. Characteristics of Matador and Mace

Name; missile	Designation	Range	Speed	Launch weight *lb*	Altitude	Guidance	Cost ($) per missile	Number procured	Warhead	Period in service	Sited at
Matador	TM 61 A; B; C; D; By 1960, C only. In 1963 MGM-IC	700 miles. C model has greater range	(max.) mach 0.9	10 000 12 000 13 800	40 000+ feet	Radar + Loran controlled	100 000	150 (?)	Nuclear; 3 000 lb 'tactical missile'	Oct. 1951 or March 1954 to 1963 or present depending on functional interpretation of 'control'	West Germany 1954 to ? Korea: 1959 to 1963. Taiwan: 1959 to ?
Mace	TM 76 A B. In 1963, CGM-13 B	A: >650 naut. miles B:1 500 miles	650(+) mph mach 0.9 (+). Warhead delivered supersonic	13 800 15 500 A: 14 000 B: 18 000	40 000+ feet	A; Atran map matching radar, low-level penetration B; Inertial	250 000	1 000	Nuclear; 'large yield' tactical; 'hundreds of kilotons'	Spring 1959 to the present	West Germany 1959 to 1968(?) Okinawa 1962–1969

Atomic Demolition Munitions

In 1960 the US Department of Defense announced that, 'Atomic demolition munitions in various sizes with tactical low yields are available and more are being developed' [1]. There is a possibility that similar devices were prepared for use by the US Navy [2].

The US ADMs are presently stockpiled in US custody in Europe. They are designed to destroy particular points of access. In time of impending war they could be buried in the ground or attached to bridges along the frontiers where they could be detonated by electronic command in the path of an advancing army. Problems associated with these weapons include when to emplace them, when to detonate them, and whether Presidential authority over their firing should be delegated in view of the fact that available time might be at a minimum. NATO has had continual problems with the question of the use of ADMs. At one time both Turkey and the Federal Republic of Germany seemed interested in making greater use of them. After separately studying them in the Nuclear Planning Group (NPG) both countries recommended further study of the problems involved. The question appears to have been under discussion in 1965 [3], and again when Turkey made a case at the September 1967 Ankara meeting of the NATO NPG for use of nuclear land mines on the Turkish–USSR border [4]. After the October 1970 Ottawa meeting of the NPG it was announced that the NPG had

> ... reached agreement on political Guidelines to cover the possible use of nuclear land mines in Europe. The mines, if they ever were used, would help seal off invasion routes that could be used by Warsaw Pact armoured columns, should war erupt in Europe ... Defense Department officials in Washington say the new guidelines will give NATO's military commanders the authority to plan for movement of the mines from rear storage points closer to the place where they would be used in an impending crisis. ...
> The United States has maintained stocks of what are believed to be several hundred of these atomic demolition devices in Europe for several years. They are said to be stored primarily in West Germany, Turkey, Greece, and Italy. ... [5].

The report ended with the suggestion that there was concern in Washington over the problems the mines could create for a US President if they were actually installed. Since they were far forward along NATO's border, in the case of a conflict there might be the choice of having the mines overrun by enemy forces, or if detonated, causing a very early escalation from conventional to nuclear war. However, several recent US Congressional Hearings have been explicit in stressing that it is the Federal Republic of Germany that does not permit the prior preparation of the emplacement holes for the ADMs (which NATO's SACEUR has apparently always been anxious to do) and which is reluctant to agree to their use, at least within the Federal Republic of Germany [6]. The US has been trying to get around the problem of 'prechambering', an optimum pre-dug hole depth, by the development of a technological adaptation, earth penetrator nuclear warheads that could be dropped from aircraft.

The 1973 Senate Committee on Foreign Relations Report summarized the situation on ADMs as follows:

> Atomic mines – Atomic Demolition Munitions, to use the proper full name, or ADMs as they are colloquially known – are a particularly sensitive question in Europe and especially in Germany [deleted] because their use would require an early decision to escalate to nuclear weapons and because they would presumably be used in friendly areas before the arrival of the enemy. The controversy over ADMs, which began about a decade ago, has involved two questions: the prior preparation of emplacements for the munitions, 'prechambering' as it is called; and the pre-delegation of authority to emplace and fire. There are general guidelines within NATO concerning the use of ADMs. They state *inter alia*, that [deleted].
> [Deleted]
> There are [deleted] American ADM teams in Germany. About [deleted] of these teams are to support US forces and the remainder support allied (mostly German) forces. Another [deleted] ADM teams are in Italy. Each team has five or six men. They are continually tested on an unannounced schedule and deployed by air during exercises. There are no ADM teams in [deleted].
> The West German Government will not allow any prechambering for ADMs. At SHAPE we were told that no chambers have been specifically constructed in Europe for ADMs, although there are chambers for conventional explosives. These chambers are not considered to be as satisfactory as chambers prepared specifically for ADMs, some because they are shallower and would result in greater fall out. At 7th Corps Headquarters we were told that [deleted]. Although these chambers are controlled by the German territorial army [deleted].

References

1. Army demolition men get atomic munitions, *New York Times*, (5 February 1960).
2. Navy: design, development, prototype and production of tactical atomic demolition munitions trainers, Device 18 B-4, (Hayes International Corporation (company catalogue). Birmingham, Alabama, 1966).
3. Middleton, D., Atom minefields weighed by NATO, *New York Times*, (11 January 1965).
4. Smith, T., Atomic land mine study ordered for Turkish defense, *New York Times*, (30 September 1967).
5. Getler, M., 8 NATO nations in accord on guidelines for A-mine use, *International Herald Tribune*, (2 November 1970).
6. *To Consider NATO Matters*, Hearing, Joint Committee on Atomic Energy, US Congress 93rd Congress, (19 February 1974), pp. 13–14.
   — *US–USSR Strategic Policies*, Hearing, Committee on Foreign Relations, US Senate, 93rd Congress, (4 March 1974), p. 54.
7. *US security issues in Europe: burden sharing and offset, MBFR and nuclear weapons, September 1973*, A Staff Report. Committee on Foreign Relations, US Senate, 93rd Congress, (2 December 1973), p. 15.

Permissive Action Links (PAL) controls, and double key arrangements

Weapons-handling procedures for US nuclear weapons have been designed so that

> Positive measures have been taken to prevent deliberate unauthorized arming, launching, firing, or releasing of nuclear warheads and missiles. Switches or controls are either locked or sealed in a safe position to prevent rash actions or to prevent someone from turning a knob in confusion or panic. Still another measure is the common use of configurations that require two or more persons to take independent actions to close the required number of switches – after the seals or locks on the switches are opened [1].

In addition to administrative and mechanical systems the USA began to install a system of electronic locks on certain of its nuclear weapons both in the USA and based overseas after July 1962. These were called permissive action links, (PAL), and were added as an additional safeguard against accidental or unauthorized use, and because US nuclear warheads were deployed for use by foreign forces. Apparently most or all of the kinds of nuclear weapons usually considered tactical nuclear weapons had these devices added to them. (Not all nuclear weapons have PAL controls, and since specific information is not available on precisely which do or do not, it is an assumption that all TNWs do.) The PAL system makes it impossible to fire a nuclear warhead which has such controls until a coded electronic message from SHAPE releases the trigger mechanism within the weapon.

With miniaturization of electronic components it was possible to make more complex PAL systems, providing greater assurance against unauthorized use. PALs can now have six different codes for locking and unlocking of designated weapons, and PAL design has apparently reached the stage where the fifth and sixth design variations are now under development for installation into particular warheads.

Reference

1. Heymont, I., The NATO nuclear bilateral forces, *Orbis*, Vol. 9(4), (Winter 1966), pp. 1038–9.

Appendix 2

General tables – USA and USSR

Table 1A.13. US Navy: nuclear-capable delivery systems deployed as of 1976

System	Delivery vehicle	Range, *km*	IOC date[a]
SAM			
Terrier	DDG, CG, CGN, CV	> 15	1962
Talos	CG	>120	1959
ASW			
ASROC	FF, FFG, DD, DDG, CG, CGN	—	1961
ASTOR	SS, SSN, SSBN	—	1963
	Helo)	S-3: 5 000 (ferry range)	
	P-3 A/C	P-3: 9 000 (ferry range)	
	Allied Maritime Patrol		
SUBROC	SSN	—	1964
Strike			
Tactical bombs	CV (A-6, A-7, S-3 A/C)	A-6: 5 190 (ferry range) A-7: 1 150 (radius of action)	1959–present
Strategic			
Polaris	SSBN	4 630	1963
Poseidon	SSBN	4 630	1964

[a] Date of initial operational capability (IOC).

Non-US Navy: nuclear-capable delivery systems currently deployed and under development[a]

[a] US NATO countries have been provided with numerous weapon systems which could have a nuclear capability but for which there are no plans to provide nuclear weapon support.

Table from: *Development, Use and Control of Nuclear Energy for the Common Defense and Security and for Peaceful Purposes*, Second Annual Report, Joint Committee on Atomic Energy, 94th Congress (US Government Printing Office, Washington, 30 June 1976), pp. 141–42.

Table 1A.14. US Army: nuclear-capable delivery systems deployed as of 1976

System	Range, *km*	IOC date[a]
Honest John	39	1958
Sergeant	140	1962
Lance	> that of Honest John	1973
Pershing	725	1963
Nike·Hercules	140	1958
Sprint	—	1974
Spartan	Several hundred	1974
MADM[b]	n.a.	1965
SADM[c]	n.a.	1964
8-in howitzer	14.5	1956
155-mm howitzer	14.5	1963

[a] Date of initial operational capability (IOC).
[b] Medium Atomic Demolition Munition (MADM).
[c] Special Atomic Demolition Munition (SADM).
Table from: *Development, Use and Control of Nuclear Energy for the Common Defense and Security and for Peaceful Purposes*, Second Annual Report, Joint Committee on Atomic Energy, 94th Congress (US Government Printing Office, Washington, 30 June 1976), pp. 141–42.

Non-US Army nuclear capable delivery systems currently deployed and under development

System	Non-US forces
Honest John	Belgium, Denmark, France, FR Germany, Greece, Netherlands, UK and South Korea
Sergeant	FR Germany
Lance[a]	Belgium,[b] FR Germany,[b] Italy, Netherlands[b] and UK[b]
Pershing	FR Germany
Nike Hercules	Belgium, Denmark, France, FR Germany, Greece, Italy, Japan, Netherlands, Norway, South Korea, Taiwan and Turkey.
8-in howitzer	
155-mm howitzer	

[a] Non-nuclear Lance has been provided to Israel.
[b] Planned but not yet operational.
[c] Most NATO countries having sizeable armed forces possess both the 8-in and 155-mm howitzers.

Table 1A.15. US Air Force: nuclear-capable delivery systems deployed as of 1976

System	Range[a], *nautical miles*	IOC date[b]
Minuteman II	6 300	1965
Minuteman III	6 300	1970
Titan II	6 300	1963
B-52	6 000–9 000	1956
FB-111	2 900	1959
SRAM	120	1972
F-4	1 600	1963
F-111	2 500	1967
F-101	1 000	1957
F-106	1 500	1959

[a] Ranges may not reflect operational conditions and may vary considerably depending on loading, configuration, mission, flight profile, inflight refuelling, etc. Aircraft range is basic ferry range. In all cases it was indicated that the possible ranges exceeded those given above.
[b] Date of initial operational capability (IOC).

Non-US Air Force nuclear-capable delivery systems currently deployed and under development

System	Range[a], *nautical miles*	Non-US forces[b]
F-100	1 600	Turkey
F-101	1 000	Canada
F-104	1 450	Belgium, Denmark, Greece, FR Germany, Italy, Netherlands, Norway and Turkey
F-4	1 600	Greece, FR Germany, Iran, Israel, Japan, South Korea, Spain, Taiwan, Turkey and UK
F-16	2 200	To be determined

[a] See note *a* above.
[b] For many NATO countries possessing these aircraft, there are no plans to provide nuclear weapon support.

Table from: *Development, Use and Control of Nuclear Energy for the Common Defense and Security and for Peaceful Purposes*, Second Annual Report, Joint Committee on Atomic Energy, 94th Congress (US Government Printing Office, Washington, 30 June 1976), pp. 141–42.

Table 1A.16. Ground-launched TNW deployed with US forces in Europe

Name of system	Number deployed[a]	Range (nautical miles)	Yield (kilotons)	Dual-capable	Initial operational capability
Honest John	36	4.5– 22.0	20.0	Yes	1953
Sergeant	36	2.4– 84.0	Low	Yes	1962
Lance	36	2.6– 70.0	1.0–100.0	No	1973
Pershing	108	96.0–390.0	60.0–400.0	No	1962
Nike Hercules	144	1.0– 20.0	1.0	Yes	1958
M109 155 mm howitzer	326[b]	9.0	Low	Yes	1962
M-110 8 in howitzer	360	8.0	Low	Yes	1964
ADM	Unknown	1.0– 3.0	Low	No	1950s

[a] Nominal estimates, based on the number of units deployed.
[b] Combined US and allied deployments.

Sources: Estimates based on data appearing in International Institute for Strategic Studies, *The Military Balance, 1972–73*, and *The Military Balance, 1973–74* (London: IISS, 1972, 1973); T. N. Dupuy and Wendell Blanchard, *The Almanac of World Military Power* (2nd ed., R. R. Bowker Company, 1972); Trevor Cliffe, *Military Technology and the European Balance*, Adelphi Papers, 89 (London: International Institute for Strategic Studies, 1972); R. T. Pretty and D. H. R. Archer (eds.), *Jane's Weapon Systems, 1971–72* (McGraw-Hill, 1972); and Richard Rhodes, 'Los Alamos Revisited,' *Harper's*, vol. 248 (March 1974).

Table 1A.17. US nuclear-capable tactical aircraft available for European contingencies

Source	Type	Number now deployed in Europe	Deployable in Europe by $M + 30^a$	Strike radius (nautical miles)	Maximum speed[b]	Initial operational capability
USAF	F-4	528	888	700	2.4	1962
	F-111	72	288	1 500	2.2	1967
	F-105D/B[c]	0	158	1 000	2.2	1960
	F-100C/D[c]	0	352	600	0.9	1955
USN	A-6	24	48[d]	900	0.9	1956
	A-7A/B/E	60	120[d]	900	0.9	1966
Total		**684**	**1 854**			

[a] Cumulative total 30 days following the decision to mobilize.
[b] Mach number.
[c] Drawn from USAF Reserve and Air National Guard.
[d] Represents the reinforcement of the Sixth Fleet by two additional carriers.

Sources: Author's estimates based on data appearing in International Institute for Strategic Studies, *The Military Balance, 1972–73* and *The Military Balance, 1973–74*; Dupuy and Blanchard, *The Almanac of World Military Power* (1972); Pretty and Archer, *Jane's Weapon Systems, 1971–72*; and John W. R. Taylor (ed.), *Jane's All the World's Aircraft, 1973* (McGraw-Hill, 1972).
Table taken from Record, J., *US Tactical Nuclear Weapons in Europe, Issues and Alternatives*, (The Brooking Institution, 1974), pp. 22–4.

Table 1A.18. Distribution of Soviet ground-launched and sea-based tactical nuclear weapons by estimated yield†.

Less than 20 kilotons		20–500 kilotons		500–1 000 kilotons		1 000–3 000 kilotons	
		SCUD-B	(75)[b]	SKEAN	(100)	SANDAL	(500)
SCUD-A	(50)[a]	FROG 6-7	(75)				
SCUD-B	(75)[c]	SHADDOCK	(560)				
FROG 2-5	(400)[d]	SCALEBOARD	(100)				
		SARK/SERB	(85)				
Subtotals	**(525)**		**(1 020)**		**(100)**		**(500)**

[a] Numbers in parentheses are estimated numbers of systems.
[b] Assumes SCUD inventory to be two-thirds of SCUD-B inventory.
[c] Assumes one-half SCUD-B inventory has yields selectable from low to high kiloton range.
[d] Assumes one-third of FROG inventory to be FROG 6-7 models. Older models are assumed to have fixed yield, newer FROGs to have selectable yields.

Table 1A.19. Soviet ground-launched tactical nuclear weapons[a]

NATO code name	Number deployed	Probable location	Yield	Dual capable	Range (nautical miles)
FROG 2-7[b]	600	Eastern Europe	Multikiloton	Yes	4–50
SCUD-A	} 200	Eastern Europe	Multikiloton Kiloton-	Yes	48–60
SCUD-B		Eastern Europe	megaton[c]	Yes	48–185
SHADDOCK[d]	100	USSR only	Multikiloton	No	60–300
SCALEBOARD	100	USSR[f]	Multikiloton[c]	?	4–480
SANDAL	} 600[e]	USSR only	Megaton Kiloton to	No	900–1 000
SKEAN		USSR only	megaton	No	1 750–2 000

[a] Represents entire estimated inventory, the bulk of which is deployed in or targeted against Europe.
[b] A family of SSM representing modifications of a single basic design.
[c] Selectable yields.
[d] Principally a naval weapon, although some are clearly configured for use by ground forces against ground targets.
[e] The SHADDOCK, SANDAL, and SKEAN deployments in the USSR represent the 700 Soviet MR/IRBM targeted on Europe that form such a prominent topic of many contemporary arms control proposals.
[f] Unconfirmed reports indicate that some SCALEBOARD may be deployed in Eastern Europe.

Sources: Estimates based on data appearing in International Institute for Strategic Studies, *The Military Balance, 1972–1973* and *The Military Balance, 1973–1974* (London: IISS, 1972, 1973); T. N. Dupuy and Wendell Blanchard, *The Almanac of World Military Power* (2nd ed., R. R. Bowker Company, 1972); Trevor Strategic Studies, 1972); and R. T. Pretty and D. H. R. Archer (eds.), *Jane's Weapon Systems, 1971–1972* (McGraw Hill, 1972).

Table 1A.20. Soviet nuclear-capable tactical aircraft[a]

Type	NATO code name	Number deployed	Strike radius (nautical miles)	Maximum speed[b]	Initial operational capability
MiG-21	FISHBED J	} 1 350	350	2.2	1970
MiG-23	FLOGGER		600	2.4	1971
SU-7B	FITTER		600	1.7	1959
YAK-28	BREWER		500	1.1	1962
IL-28	BEAGLE		1 500	0.8	1950
TU-22	BLINDER[c]	260	700	1.5	1962
TU-16	BADGER A/B/C[d]	800	1 700	0.8	1955

[a] Represents entire estimated inventory.
[b] Mach number.
[c] Includes units in naval aviation forces.
[d] Includes about 300 units in naval aviation forces.

Sources: Author's estimates based on data appearing in International Institute for Strategic Studies, *The Military Balance, 1972–1973* and *The Military Balance, 1973–1974*; Dupuy and Blanchard, *The Almanac of World Military Power* (1972); Pretty and Archer, *Jane's Weapon Systems, 1971–1972*; and John W. R. Taylor (ed.), *Jane's All the World's Aircraft, 1973* (McGraw-Hill, 1973).
Table taken from Record, J., *US Tactical Nuclear Weapons in Europe, Issues and Alternatives*, (The Brookings Institution, 1974), pp. 38, 39, 41.

Table 1A.21. USA versus USSR nuclear ground support systems: 1966

Organizational level	United States			USSR			Remarks
	Weapons system	Range	Yield	Weapons system	Range	Yield	
Battalion	Davy Crockett	2–4 km	0.25–0.5 kt	none	none		
Division	8 in howitzer	17 km	1–2 kt	FROG 1	64 km	5–10 kt	It is not believed that
	155 mm howitzer	18 km	1–2 kt	FROG 2	27 km	10 kt	FROGs are assigned
	Honest John	39 km	5–10 kt	FROG 3	40 km	5–10 kt	to divisions but
				FROG 4	40 km	5–10 kt	rather fire in support of divisions
Corps-army or Soviet front	175 mm howitzer	32 km	low kt	SCUD	160 km	high kt	
	Sergeant	220 km	10–60 kt				
	Pershing	640 km	up to 1 Mt	SHADDOCK	350 km	high kt	
Other	Nike Hercules	128 km	low kt	none	none		air defence
		150 000 ft		none	none		
	Atomic demolitions		low to high kt	atomic demolitions		low to high kt	available on request
	Tactical air		kt–Mt	tactical air		kt–Mt	available on request

This table and the following two are taken from G. W. Casey, *The Employment of Tactical Nuclear Weapons, A Search for Safe Limits*, 1 April 1966, mimeographed.

Table 1A.22. US nuclear delivery systems for support of ground combat: 1966

Organizational level	Weapon	Range, kilometres	Estimated yield kt
Battalion	Davy Crockett	4	.25
Division	8 in howitzer	17	2
	155 mm howitzer	18	2
	Honest John	40	5–10
Corps	175 mm howitzer	32	low kt
	Sergeant	220	10–60
Army	Pershing	640	high kt
	Nike Hercules	130 slant range 150 000 ft ceiling	low kt
Available to all levels on request	Atomic Demolitions		Sub kt to kt
	Tactical Air Support		kt to Mt

Note: 1. 'Tens of thousands' of nuclear warheads are available for these delivery systems.

Table 1A.23. Soviet nuclear delivery systems for support of ground combat: 1966

Weapon	Range kilometres	Estimated yield, kilotons
FROG 1	64	5–10
FROG 2	27	10
FROG 3	40	5–10
FROG 4	40	5–10
SCUD	190	Several kt
SHADDOCK	560	Several kt
Atomic demolitions	Available on request	Small to high kt
Tactical aircraft	Available on request	Small to high kt

Notes: 1. FROG abbreviation for *Free Rocket Over Ground.*

2. It is estimated that the Soviets have between 5–10 000 warheads for all their nuclear systems (as compared to 50 000 US warheads for tactical nuclear systems above). How many of the 5–10 000 Soviet warheads are for the above systems is not known.

3. As explained in the text the forces manning the above systems are called Operational-Tactical Rocket Forces and are separate, but in support of the regular ground forces. It is believed that the FROG rockets are deployed in support of Soviet Divisions and the SCUD and SHADDOCK support Soviet Fronts.

2. Tactical nuclear weapons in Europe

O. Šuković

I. Introduction

While most public discussion and writing on nuclear warfare in recent years has been focused on 'strategic' weapons, and a great deal of information about these weapons is widely accessible, much less attention has been given to another kind of warfare – 'tactical' nuclear warfare – for which a large array of nuclear weapons has been built. These weapons are less known to the public than strategic weapons, which is surprising in view of the increasing attention being focused on them.

There are several reasons why it is appropriate to examine the role of tactical nuclear weapons (TNWs) in the strategic concepts of NATO and the WTO and to reappraise the relationship between the three major elements of the military posture in both these alliances: strategic forces, theatre nuclear forces and conventional forces. First of all, recent official statements on US nuclear strategy have considerably increased the danger to mankind inherent in the existence of nuclear arsenals. For instance, in January 1974 former Defense Secretary J. Schlesinger announced that the USA intends to adopt what was termed a counterforce strategy as a strategic nuclear option and, to this end, to improve the accuracy of delivery of its nuclear weapon systems. Counterforce strategy does not replace deterrence; rather it supplements it with the additional capability to strike, either pre-emptively or in response to an attack, at the opponent's military targets, including hardened missile silos. Such a strategy requires a large number of accurate powerful nuclear warheads targeted not against cities and industrial and transportation centres but against military installations. In addition, a counterforce strategy implies the capability of fighting a nuclear war if deterrence fails to prevent its outbreak.

The statement of former US Secretary of Defense Schlesinger on 1 July 1975 that 'we will make use of nuclear weapons should we be faced with obvious aggression likely to result in defeat in an area of great importance to the US in terms of foreign policy' caused considerable concern. Further, the capabilities and objectives of the US nuclear arsenal as a whole are being re-examined because of the effects of parity with the USSR and the SALT negotiations. Along with this, there is considerable pressure to supply NATO forces with 'mini-nukes', a new family of precision-guided miniaturized weapons. The NATO proposal of December 1975 put forward in Vienna at the Mutual Force Reduction (MFR) negotiations for withdrawal from western European territory of 1000 nuclear warheads (together with 54 nuclear-capable F-4 aircraft

137

and 36 Pershing SSM) and 29 000 men, in exchange for the withdrawal of 69 000 Soviet ground troops and 1700 medium tanks, may have significant consequences for the nuclear position in Europe.

Bearing in mind the need for re-examining the posture and policy behind the enormous stockpile of TNWs in Europe in the light of these new developments in policy and technology, the following questions will be examined in this paper: the problem of defining TNWs; the strategic doctrines of NATO and the WTO; a comparison of NATO and WTO TNWs in Europe; issues and proposals and finally conclusions. The paper will present factual and general information on tactical nuclear warfare without going deeply into the many complex problems connected with this subject and some of which are dealt with in papers concerning specific questions.

II. Defining TNWs

Before dealing with the problems connected with strategic doctrines, it is necessary to make a few remarks about the terminology used in this chapter. It is common to talk about 'strategic' and 'tactical' nuclear weapons, but it is not always clear which weapons belong to the first and which to the second category. The difficulties in distinguishing between these two types of weapons have led to many definitions, none of which is generally accepted or completely satisfactory. According to some writers, the most important criterion for classification is the range of the delivery systems, therefore TNWs are considered all to be weapons that are employable in a theatre of operations (for example, non-Soviet Europe) but which could not or would not be used against the USA or the USSR. Thus 'the main difference lies in the distance they are able to travel. The range of the TNW is not sufficient to cause any serious damage to the Russian mainland' [1]. Other writers make the basic factor for defining TNWs the intention of the parties in the conflict. Thus, according to one writer, TNWs are simply 'nuclear weapons designed to support land forces' [2], while another considers them to be those weapons 'designed to influence the land or air defense battle directly, rather than indirectly (for example by interdicting an enemy's lines of communication or by carrying a threat to his homeland)' [3]. Tactical nuclear weapons have been defined as 'the use of nuclear weapons for limited tactical military purposes' [4].

After pointing out that there is no clear-cut distinction between tactical and strategic nuclear weapons, one writer draws the distinction according to the intended use: 'Theater-based weapons are intended to be used on the battlefield against purely military targets and avoiding collateral damage as far as possible'. He continues that 'another possible distinction between tactical and strategic nuclear weapons is the level of command at which they are deployed. Such weapons in the hands of local field commanders are obviously intended for tactical use' [5].

Another group of writers does not make distinction between strategic and tactical nuclear weapons, but speaks of tactical or strategic use of nuclear weapons. One writer from this group sees the distinction in the fact that '"tactical" use is . . . taken to mean use within the battlefield area or directly connected with the manoeuvre, movement or supply of combatant forces' [6]. Another writer elaborates this further by pointing out that

> A military action is usually called 'tactical' if it is directed primarily against the military forces employed by the enemy, while a 'strategic' action tries to destroy his military resources'. Since most weapons can be used in both 'tactical' and 'strategic' functions, nuclear weapons cannot be defined by their technical qualities, like yield or range. Therefore NATO nowadays prefers to use the phrase 'tactical use of nuclear weapons' rather than 'use of tactical nuclear weapons'. This refers to an employment of nuclear weapons in the context of direct defence or deliberate escalation, but not general nuclear response.

With this in mind, he uses the term TNW 'similarly to the US "theater nuclear weapons", that is including all US nuclear weapon systems in Europe as well as the French and British systems "for tactical use"' [7].

While defining TNWs some writers combined the intended use and the zone of employment. Thus officially tactical nuclear war is designated as

> A conflict between the land forces and associated air and naval forces of two or more nations in which nuclear weapons are limited to the defeat of opposing forces in a theatre of operations. Implicit in this definition is the condition that a strategic nuclear exchange on the belligerents' homeland does not occur [8].

> The term 'TNW' . . . denotes, in its widest sense, weapons assigned to the general purpose forces for use against targets within the battlefield area or at sea or against targets which are directly connected with the employment of forces in combat zones [9].

Other writers base their definition on yield. According to some of them, all warheads below one megaton should be considered tactical, while more powerful weapons are designed as strategic; others believe that the limit should be set much lower – at the 500 kiloton (kt), 100 kt, or even the sub-kiloton level. Implicit in this definition is an attempt to define as tactical those weapons that generate the least amount of collateral damage.

After pointing out that there are two basic approaches to defining TNWs (one based on yield and the other on range) and discussing the disadvantages of these approaches and existing definitions, Record concludes that 'it is thus clear that the term TNW is very ambiguous. Much more satisfying is the concept of tactical versus strategic use of nuclear weapons, which focuses on the types of designated targets rather than inherent capabilities of the weapon system themselves'. He defines TNWs as

> all nuclear weapon systems other than strategic bombers (and the bombs they carry), nuclear cruise missiles and land- and sea-based intercontinental ballistic missiles. Although, conceivably, any of these

strategic weapons could be employed for tactical purposes, both their design and the doctrines governing their use make such usage unlikely [10].

Bearing in mind the ambiguity of the existing definitions and the difficulties of drawing a line between strategic and tactical nuclear weapons, the TNWs of the USA and the USSR will be understood in this chapter as all nuclear weapon systems other than strategic bombers, strategic submarines, SLBMs and ICBMs (listed in the SIPRI Yearbook 1976, Appendix 1A) [11], while all British and French nuclear weapons will be included. Finally, it should be mentioned that a weapon that might be tactical for one country may inevitably be strategic for another especially where the nuclear weapon systems of other European countries and their territories are concerned.

III. Strategic doctrines of NATO and the WTO†

Although much has been written on strategic doctrines (especially on those of the USA and NATO) a brief account of the past and current strategic doctrines of the two alliances is necessary for a better understanding of present US and Soviet tactical nuclear postures in Europe.

The strategic doctrine of NATO

Soon after the end of World War II a state of apparent understanding that prevailed between the USA and the USSR had been replaced by a policy of 'containment'. This policy was backed up primarily by the US monopoly of the atomic bomb.

The signing of the NATO treaty in 1949 marked the completion of the second major component of US strategy. The treaty was a recognition of the fact that the western European countries could not match the Soviet Union's conventional strength and of the desire of these powers to use the US nuclear forces for the defence of Europe. This initial concept of the alliance – a clear statement of US intentions backed up by actual nuclear forces and potential conventional forces, the whole constituting a credible deterrent – survived, with minor revisions, until 1967 as the official policy of NATO [12].

NATO strategy did not long remain unchallenged. The nuclear superiority of the USA was threatened in 1949 by the testing of an atomic bomb by the Soviet Union. The beginning of the Korean War was also a challenge to the deterrent credibility of nuclear weapons. Largely as a result of these events, the NATO Council approved at its Lisbon meeting in 1952 force goals of 96 NATO divisions by 1954.

The Eisenhower administration, having studied the cost involved in carrying out the Lisbon goals, quickly abandoned the notion of matching the Soviet Union locally and announced its 'New Look' policy based on the

---

† This is an abridged and updated version of the strategic doctrines elaborated in: SIPRI, *Force Reductions in Europe*, (Almqvist & Wiksell, Stockholm, 1974), pp. 23–49.

strategy of 'massive retaliation'. The new strategy maintained that each and every conflict above the level of a simple border incident or brush fire should lead to intervention with nuclear weapons [13]. Soon after its adoption, criticism of the New Look produced changes. Secretary of State Dulles suggested a modified doctrine on 29 September 1954:

> Now you may ask does this mean that any local war would be automatically turned into a general war with atomic bombs being dropped all over the map? The answer is no. The essential thing is that we and our allies should have the means and the will to assure that a potential aggressor would lose from his aggression more than he could gain. This does not mean that the aggressor has to be totally destroyed. It does mean a capacity to inflict punishing damage [14].

The Radford Plan, which called for only 30 divisions, was submitted by the USA to NATO in 1957 and marked the formal abandonment of the Lisbon goals of 96 divisions. The plan was based on the assumption of a permanent inferiority on the part of NATO in conventional forces. It was believed that the proposed 30 divisions of NATO troops could not hope to offer significant resistance without access to nuclear weapons. As part of the implementation of this strategy a large number of TNWs were shipped into Europe. These were placed under the double veto system which required the consent of both the USA and the host country before they could be used.

In 1957 the Soviet Union test-fired an intercontinental ballistic missile (ICBM) and soon it was believed that the Soviet Union had already surpassed the US in the development of long-range missiles. According to Kissinger, it was in order to close the supposed 'missile gap' [15] that the USA pressed its allies to permit the installation on their territories of IRBMs. The presence of these missiles in Europe tended to establish an inextricable link between the defence of Europe and that of the USA. An attack on Europe would damage the strategic balance and immediately threaten the survival of the USA. Thus strategic deployment rather than a decision taken at the moment of attack would determine the US response [16].

The strategy of massive retaliation met with criticism from the date of its inception. As early as 1946, Brodie had suggested that the important thing about nuclear weapons was not to use them [17]. By the latter half of the 1950s he had been joined by other strategists who suggested that massive retaliation was hardly credible as a doctrine [18].

As crisis succeeded crisis, it became apparent that the USA was unwilling to invoke massive retaliation. The doctrine was not a credible response except in the most extreme circumstances. The Kennedy administration sought, therefore, to replace a strategy calling for a single all-out response to any attack with something which offered more options. The new policy was summed up in the phrase 'flexible response'.

The new strategy was accompanied by powerful rearmament. The concern of the administration that the USA should be able to meet the enemy at every level with forces so superior as to deter any aggression, or at least offer a major chance of success if fighting broke out, motivated rearmament over the entire strategic field. Alongside it the strategy of counterforce was developed. McNamara saw counterforce as meaning that the USA would develop weapons of retaliation to correspond to the weapons deployed by the USSR on a total or selective basis according to the strategic situation. The strategy also offered an 'assured destruction capability', by which was meant a guaranteed capability to cause the enemy unacceptable damage. The strategy ensured that the first-strike capability of the enemy could never eliminate the defender's second-strike capability. How the new strategy was to be put into operation is unclear and official explanations are inconsistent.

The realization that massive retaliation had shortcomings and that conventional options were expensive caused increased interest in the possibilities displayed by TNWs, especially among NATO military commanders. The case for TNWs rested on several assumptions: that TNWs could be substituted for manpower; and that the yield of weapons could be limited, thereby limiting damage and civilian casualties and preventing escalation of the conflict. These assumptions proved to be incorrect. Later studies showed that more manpower would be required to fight a tactical nuclear war; that the prospects for limiting collateral damage were poor due to weapon inaccuracy; and that the risk of escalation was increased by the absence of a natural firebreak between tactical and strategic use of nuclear warfare.

The reasons for retaining the tactical nuclear options were: (a) the presence of several thousand TNWs in Europe, including bombs, missiles and artillery shells, (b) the contribution these weapons made to the deterrence of conventional as well as nuclear aggression; (c) the requirement to deter a first-use of tactical nuclear weapons by the Soviet Union and (d) the hedge they represented against the possibility of failure in other parts of the NATO posture.

The new strategic doctrine was not officially adopted by NATO until May 1967, after much debate. Previously the European members of NATO had accepted US hegemony in defence questions as a necessary condition of an automatic nuclear response by the US in the event of aggression in Europe. The doctrine of flexible response removed the automatic nuclear response. A nuclear reaction would now take place only after deliberation on the part of the USA and in stages. The European members were concerned to commit the USA to their defence. They saw NATO as a means of ensuring US protection, by which they meant US nuclear protection. The US nuclear umbrella might be unpleasant but it was necessary and therefore tolerable. With the development of the Soviet nuclear capability, the question arose as to whether the USA was prepared to sacrifice North American cities in the event of a Soviet invasion of Europe. The 'credibility of guarantees' became a central question in the alliance.

The problem was resolved not by a reassessment of the role of nuclear weapons but by searching for a device that would reduce Allied misgivings about their dependence on the USA. Britain and France found their answer in creating their own independent nuclear forces. In response, the USA proposed the creation of a 'multilateral force' (MLF). At first glance MLF appeared to be a satisfactory solution, but it gradually became clear that the proposal raised impossible problems regarding political sovereignty and military command.

The strategy of flexible response, unlike that of massive retaliation, met with strong resistance among certain European states, notably France. Its major defect lay in the limited number of options which were available, in what is termed 'truncated flexibility'. It was further pointed out that any imbalance in conventional, tactical nuclear or strategic nuclear capabilities would tend to invalidate the theory of a strategy of flexible response and face the deficient side with practical decisions which allowed of no alternative options. NATO attempted to solve the problem of its deficiency in conventional forces by making a distinction between tactical and strategic nuclear weapons with regard to deterrent value, and suggesting that while strategic nuclear weapons may not represent a credible alternative to conventional forces, tactical nuclear forces are a reasonable substitute. The problems posed by the use of tactical nuclear forces, which have already been discussed, made this situation only a a temporary stopgap.

In the initial years of the strategy of flexible response the European members of NATO failed to implement it, probably as a result of both political reservations and economic pressures. However, in recent years there have been signs of greater willingness to accept burdens implicit in the new strategy.

Confronted with economic and political problems, the Nixon administration re-examined the doctrine of flexible response and attempted to adapt the strategy to the changing circumstances. As a result a strategy of 'realistic deterrence' was accepted which was based on the concept that strategic forces retain their central role as a deterrent against nuclear attack and the need for the USA and its allies to maintain strong conventional capabilities.

The new strategy has done nothing to change US global policy, because 'realistic deterrence' relies on a policy of strength expressed in the term 'assured destruction'. The enemy is still to be deterred by the nuclear balance. To this extent realistic deterrence does not differ from the strategy of flexible response.

The need for a further change in doctrine appeared in President Nixon's Foreign Policy Message to the Congress in 1970:

> Should a President, in the event of a nuclear attack, be left with the single option of ordering the mass destruction of enemy civilians, in the face of the certainty that it would be followed by the mass slaughter of Americans? Should the concept of assured destruction be narrowly defined and should it be the only measure of our ability to deter the variety of threats we may face? [19].

The next year the President expanded on his 'Doctrine of Sufficiency'

> I must not be – and my successors must not be – limited to the indiscriminate mass destruction of enemy civilians as the sole possible response to challenges. This is especially so when that response involves the likelihood of triggering nuclear attack on our own population. It would be inconsistent with the political meaning of sufficiency to base our force planning solely on some finite – and theoretical –

capability to inflict casualties presumed to be unacceptable to the other side [20].

In 1972–74, a number of articles and statements appeared in the USA calling for a revision in US strategic weapons policy [21]. The debate focused on four major deterrent issues. A few writers, most notably Brennan [22], challenged the fundamental basis of US deterrent strategy. They criticized dependence on MAD as enshrined in the 1972 ABM treaty and proposed a 'defence-oriented' strategy to protect the civilian population. They argued that active and passive defence was not only desirable but, over time, feasible and that it should be the primary objective of defence planners. On the other hand, supporters of a 'deterrent' strategy responded that the two great powers' hostage relationship was a 'matter of physical fact', arising from the destructiveness of the offensive nuclear forces on both sides. They admitted, however, that the mutual threat to destroy millions of people was not a preferred strategy but was the only viable one.

Given the mutual hostage relationship, a second issue raised in the debate was the desirability of deterring war by the threat of killing millions of people. Iklé proposed an alternative strategy of threatening destruction of 'military, industrial and transportation assets – the sinews and muscles of the regime initiating war' [23]. Russett also challenged the idea that deterrence must depend on destroying enemy cities and suggested a 'countercombatant strategy' which was 'militarily feasible' and 'preferable on moral grounds'. Aggression would be deterred by a threat to destroy enemy nuclear strike forces, internal security forces, military bases and military-related industries [24].

The third issue debated was the wisdom of the USA continuing, as a matter of declared policy, to abstain from improving the counterforce capabilities of its missile systems. May [25] argued that in order to ensure military victory, the US deterrent strategy should be based on the threat of counterattack, so that US military forces remaining after a nuclear war would be superior to those of the enemy. The final issue was how to deter the Soviet Union from launching a pre-emptive counterforce attack once she had the capability to hold sufficient weapons in reserve to threaten American cities. Senator Jackson's solution was to ensure that the US had the capability to respond flexibly at any level of Soviet aggression. Panofsky admitted that the controlled use of nuclear weapons might potentially minimize the risk of escalation but warned that it would remove an important factor which presently contributed to deterring the outbreak of large-scale war [26].

The debate was heightened by the statement of US Secretary of Defense J. Schlesinger in January 1974 that the USA intended to adopt a counterforce strategy as a strategic nuclear option and, to this end, was improving the accuracy of delivery of its nuclear weapon systems. In order to 'shore up deterrence across the entire spectrum of risk' and to 'limit the chance of uncontrolled escalation' if war occurred, Schlesinger announced that the US would introduce flexibility into her nuclear deployment planning. He argued that 'what we need is a series of measured responses to aggression which bear some relation to the provocation, have prospects of terminating hostilities before general nuclear war breaks out, and leave some possibility for restoring deterrence' [27]. To justify the development of a counterforce system, the Department of Defense draew on three related arguments: (*a*) the USSR was developing the same capability, (*b*) the US needed a 'flexible' nuclear strategy and (*c*) the new system would be a 'credible' deterrent to nuclear war. In the course of studying the desirability of flexibility, the US government had in effect re-evaluated the utility of nuclear weapons in a world in which both great powers had invulnerable second-strike forces. They concluded, at least implicitly, that nuclear weapons continued to be useful for deterring nuclear attacks (and perhaps for helping deter conventional conflicts) over a wide range of contingencies and for re-establishing deterrence if war occurred.

Commentators reported a change in the US nuclear war targeting strategy and raised the possibility that the new doctrine would undermine deterrence by making the use of nuclear weapons more likely, antagonize the USSR, provoke anxieties in western Europe and generate pressures for a large US defence budget.

On the question of whether to introduce flexibility, critics of the new doctrine argued that plans for limited nuclear attacks undermined deterrence by making nuclear war more likely. Flexibility would tend to reduce inhibitions against the use of nuclear weapons by making nuclear war more thinkable, acceptable and respectable. Serious concern was also expressed that the new doctrine would generate requirements for new weapons and improvements in US counterforce capabilities against hardened targets such as missile silos. Such improvements would be provocative and destabilizing, for it would appear to the USSR that the USA was acquiring a first-strike capability. According to W. Panofsky, procurement of highly accurate nuclear warheads would be 'destabilizing by being physically indistinguishable from weapons designed for a pre-emptive attack against the opponent's retaliatory forces'; this might prompt the USSR to strike first in a crisis, or to adopt a launch-on-warning strategy. Moreover, additional US forces might fuel an arms race and thereby potentially undermine US–Soviet relations and future arms-control negotiations [28].

On the other hand, Schlesinger argued that the doctrine did not necessarily require new weapons or major improvements in existing systems: 'The evolution in targeting doctrine is quite separate from, and need not affect the sizing of the strategic forces'. Nevertheless, he went on to announce that the USA intended to maintain, as a 'principal feature' of her strategic posture, 'forces to execute a wide range of options in response to potential actions by the enemy'. However, he suggested that the US might want to change the mix of the current forces, improve command and control procedures, and deploy smaller warheads. Some concern was also expressed that the new doctrine on strategy raised the prospect of a nuclear war confined to Europe, leaving the territories of the two great powers as sanctuaries.

Thus the new US emphasis on counterforce weapons, the introduction of long-range cruise missiles, the discussion on the replacement of existing TNWs by precision-guided low-yield nuclear weapons and the reinstatement of the first use policy are the main components of recent developments in nuclear doctrine and could have a far-reaching effect on the future of Europe and mankind.

The strategic doctrine of the WTO

In the early postwar years Soviet strategy had two main aims: to maintain large conventional forces as a deterrent against the Western nations and to break the US monopoly of nuclear weapons. Consequently the USSR gave fresh attention to training and equipping its theatre forces for campaigns in Europe, giving priority to those already stationed in Eastern Europe. The priority given to conventional land and air forces did not, of course, mean that military strategists were indifferent to the military–technical revolution and well in advance of the expectations of the Western countries and much to their surprise, the USSR exploded its first atomic bomb in August 1949. Four years later it tested its first thermonuclear bomb.

In addition, efforts were being made to develop long-range bombers and to develop and improve Soviet strategic delivery capabilities, which were still oriented towards medium-range operations in Eurasia rather than towards intercontinental missions.

Some steps were also taken to rebuild the armed forces of the Eastern European countries. Bilateral defence treaties were concluded to this end in 1948. This did not entail an alliance of the NATO type, but the armed forces of the East European countries were modified to conform to the Soviet military organizational pattern and they began to receive sizeable quantities of Soviet arms and equipment.

Military strategy after 1953 was concerned with directing resources into supplying the armed forces first with strategic and subsequently with tactical nuclear weapons. The years up to 1961 were dominated by the aspiration of reaching a balance in strategic nuclear weapons with the USA, while Soviet strategy continued to be based on the possession of large conventional forces. During the latter part of this period, which extended some years beyond the end of Khrushchev's term, each side had sufficient strategic nuclear weapons to deter the other and Soviet strategy was primarily based on nuclear weapons.

At the same time that Khrushchev became leader of the party and of the state, nuclear weapons became a subject of serious theoretical and political debate. The debate resulted in the view that nuclear war would postpone the development of socialism and that 'war is no longer a fatal inevitability'. This change in political doctrine took place in the same year that the USSR introduced its first operational long-range bombers and could, for the first time, strike directly at the territory of the US. Khrushchev was concerned to adapt Soviet military thinking to the revolution in the technology of warfare.

Resistance to some of Khrushchev's reforms by military leaders who advocated traditional military concepts, together with the pressure of world events (the Berlin crisis of 1961), brought about various modifications in the military policy outlined by Khrushchev in January 1960. Speaking at the XXIInd Party Congress in October 1961, Marshal Malinovski, Minister of Defence, gave a less optimistic picture of the Soviet defence outlook. He claimed that 'in realistically appraising the situation' one must hold that the West was making serious preparations for a surprise nuclear attack. While he shared Khrushchev's view that a future war would 'inevitably' be a nuclear-missile war and that the use of such weapons in the early stages would have a decisive influence on the war's outcome, he also brought in a traditional notion that final victory could be assured only by 'combined action of all arms and services' [29].

The central issues in Soviet military theory and planning concerned the relationship between conventional theatre warfare in Europe and strategic operations on a global scale. The essential question was whether a future war would be

> a land war with the employment of nuclear weapons as a means of supporting the operations of the ground forces [or] . . . a fundamentally new kind of war in which the main means of solving strategic tasks will be missiles and nuclear weapons [30].

The other controversial question was whether Soviet military preparations should be aimed primarily at deterrence or at improving the Soviet capacity to fight a war if deterrence should fail. Two schools of military thought were engaged in this debate: the modernists, who stressed the need to discard the old doctrinal view and to exploit modern technologies of war; and the traditionalists, who argued for a moderate rate of military innovation and cautioned against extremes.

In spite of this debate it can be said that, beginning in the early 1960s and continuing throughout the remainder of this period and well into the next, a military strategy primarily based on nuclear weapons has dominated. This is demonstrated by, among other things, the massive deployment of nuclear weapons.

The role of nuclear weapons is stressed:

> In modern warfare, nuclear weapons can be employed for various missions: strategic, operational and tactical . . . It permits the execution of military missions in a considerably shorter time than was possible in past wars [31].

The revised edition of Sokolovski's *Military Strategy* acknowledged the possibilities of TNWs being used by both sides in a limited war, stating: 'in the course of a local war it may happen that the belligerents will employ TNW, without resorting to strategic nuclear weapons' [32]. The introduction of TNWs, however, marked the limit at which the authors were prepared to set the escalation threshold. At this point, they reverted to the now standard argument, that use of nuclear weapons in any form would mean escalation to world war: 'However, the war would hardly be waged very long with the use of TNWs only. Once matters reach the point where nuclear weapons are used, then the belligerents will be forced to launch all their nuclear power. Local war will be transformed into global war' [33].

During the same period the theatre forces of eastern Europe were gradually consolidated and improved. The year 1955 saw the signing of the Warsaw Treaty, but until the end of the decade the contribution of the military forces of the other members of the WTO appear to have carried little weight in Soviet planning. In the early 1960s the importance of the WTO in the common defence of the socialist countries began to be emphasized, while in 1964 the USSR began to furnish the other member countries with potential nuclear delivery systems in the form of tactical missiles with a range of about 150 miles. The nuclear warheads remained in the hands of the USSR but the acquisition of delivery systems and participation in simulated nuclear exercises marked a significant step towards nuclear sharing at some future time. Although the strategy primarily based on nuclear weapons was still valid in the mid-1960s, the new Brezhnev–Kosygin leadership gradually introduced what was called 'flexibility with caution'. Under them, programmes in the strategic field have fallen largely into two categories: those aimed at a build-up of strategic delivery forces and those of ABM deployment. In the summer of 1966 an accelerated programme of ICBM development was set in motion. The USSR also accorded special importance to the development of mobile land-based missiles. High priority was given to missile-launching submarines.

By 1965 it had become evident that Soviet strategy was growing less categorical in advancing the theory of nuclear-missile war. Since 1967 this has become even more apparent since some well-known military theorists and leaders have criticized the doctrine that gave strategic nuclear weapons such overriding importance. The importance of preparing Soviet forces for a wide range of operations below the level of general nuclear war began to be recognized more explicitly. General S. M. Shtemenko, Deputy Chief of General Staff of the Soviet military forces, stated early in 1965 that Soviet military doctrine did not 'exclude' the possibility of non-nuclear warfare nor of warfare restricted to the use of TNWs 'within the framework of so-called "local" wars' [34]. General Lomov even envisaged the possibility of a local war limited to conventional means or to TNWs in Europe, although he warned that 'the probability of escalation into a nuclear world war is always great and might under certain circumstances become inevitable'. He insisted that Soviet forces should be prepared for operations 'with conventional arms alone' or with 'limited employment of nuclear weapons' [35].

It is claimed that the development of Soviet military power since 1966–67 has been marked by the 'harmonious and even development of all types of forces necessary for preparation of any kind of war'. This development has been based on the judgement that the military balance has essentially lessened the possibility of direct military confrontation between the USA and the USSR; that military–technical progress has led to a balance in strategic nuclear weapons and this in turn to the SALT agreements; and that, on the other hand, the importance of local war and of interventions with conventional forces has grown and that such a war is also possible in Europe. Manoeuvres and exercises both by the USSR and the WTO have confirmed, according to official statements, the capabilities of all kinds of armed forces to carry out combined offensive operations using conventional weapons. Consonant with this last consideration there has been a continued strengthening of Soviet conventional forces. At the same time there has been a steady deployment of new strategic weapons. While the USSR, like the US, has recognized the importance of negotiations and agreements on the limitation of strategic armaments, it has, like the US, not ceased to improve its offensive and defensive systems as allowed under the SALT agreements.

During the past few years the strategy calling for 'harmonious development of all types of armed forces' appears to have gained complete acceptance. Soviet strategists have paid particular attention to the study of local wars and to the two questions of how they can be prevented from escalating and how a quick victory can be achieved. The opinion that every local war, particularly a local war in Europe, must lead to world war has been replaced by the view that this will be the case only where a direct confrontation of the nuclear powers occurs. Military theory is increasingly taking the view that the use of nuclear weapons in war is unlikely; both sides would aim at avoiding their use.

Major progress has been made in strengthening the military potential of the joint WTO forces. During this period a main feature of Soviet policy towards the alliance has been the promotion of joint field exercises. The USSR has also continued the programmes instituted under Khrushchev for re-equipping and modernizing the armed forces of the Eastern European countries.

During the past few years it appears that there may have been differences of opinion among WTO members on various issues. It is supposed that at least two issues were raised by the Soviet nuclear monopoly within the WTO: that of access to nuclear weapons or nuclear sharing and that of participation in nuclear planning and strategy. The USSR was also faced with a broader set of problems relating to command and decision-making within the WTO.

The Soviet Union distinguishes between three levels at which military operations can be conducted: strategic, operational and tactical. The dividing lines between these levels cannot be drawn accurately since they overlap. The operational level is usually associated with the front. When talking about TNWs the USSR refers to nuclear weapons systems organic to a front or its subordinate formations. The Soviet distinction does not automatically imply a restriction to small-calibre weapons, although they clearly recognize the value of small-yield weapons.

It can be concluded that the WTO does not regard conventional and nuclear war as separate entities. Despite a recent trend to improve its conventional forces and to recognize that a conventional war in Europe need not escalate to nuclear war, the WTO strategy, doctrine and forces are still strongly oriented towards nuclear operation. The USSR apparently sees an escalation of war in Europe to nuclear war as likely.

IV. Comparison of USA–NATO and Soviet–WTO tactical nuclear weapons†

US–NATO capabilities

The United States maintains about 7200 tactical nuclear warheads in the European area. Although the exact number is classified, former Secretary of Defense R. S. McNamara stated in 1967 that there were then about 7000 TNWs in the European area [36]; the following year his successor, Clark M. Clifford, put the figure at 7200 [37]. It is believed that this number has been increased meanwhile, but again it was officially stated that on 5 August 1975 the USA had *about* 7000 nuclear warheads deployed in Europe [38].

NATO tactical nuclear forces in Europe consist of SSMs, SAMs, artillery, tactical aircraft, ADMs (Atomic demolition munitions) and SLBMs. US tactical surface-to-surface missiles (SSMs) in Europe consist primarily of Pershing, Lance and Honest John, with Lance currently being deployed to replace the older Sergeant missile and Honest John rocket. Only the Pershing may be classified exclusively as a long-range theatre nuclear weapon (it is also the only SSM maintained on Quick Reaction Alert). All four SSMs are mounted on self-propelled carriers. Launch preparation time ranges from less than an hour for the Lance to about two hours for the Pershing. Warhead yields vary from about 1 kt to 400 kt.

US forces also possess about 150 Nike Hercules surface-to-air missiles (SAMs). Nike Hercules is a dual capacity SAM system deployed in NATO Europe which can counter extremely high-altitude/high-speed WTO aircraft. Nuclear warheads for Nike Hercules are designed for air defence, but it has rather limited range and is relatively inaccurate against ground targets.‡

Deployed *nuclear capacity artillery* consists of M-110 203 mm (8 in), M-109 155 mm and M-115 203 mm (8 in) howitzers. The first two systems are self-propelled, dual-capacity and intended for battlefield use only. Projectile yield ranges from as low as 100 tons to several kilotons. The third system is towed. Studies are being undertaken to develop an improved 8 in nuclear projectile and to improve the current 155 mm nuclear projectile.

US forces in Europe have about 2800 *nuclear-capacity tactical aircraft*. Land-based aircraft consist of F-105D, F-4, F-111 A/E, F-15A and A-7D. All five types are dual-capacity. A certain number of these aircraft are kept on peacetime quick reaction alert, launchable within a short period of time.

---

† This comparison is based on: J. R. Schlesinger, *The Theater Nuclear Force Posture in Europe*, A Report to the US Congress in Compliance with Public Law 93-365, (1975), pp. 16–18; *The Military Balance* (1976–7), an IISS publication (1976), pp. 73–77 and J. Record [10], pp. 19–28 and 37–41.

‡ Nike Hercules surface-to-air missiles are not listed in *The Military Balance* 1976–7.

Nuclear bombs carried by these aircraft range from about 0·1 kt to over 1 Mt. Besides these, US tactical aircraft include the carrier-based strike aircraft A-4, A-6A, A-7 A/B/E and F-4.

Atomic demolition munitions (ADMs)

ADMs are nuclear demolition devices that are manually emplaced and detonated by timer or command. They are purely defensive devices and must be placed underground before detonation. It is generally believed that no ADMs are presently prechambered in NATO Europe, although it is supposed that they are stored on the continent.

Other NATO countries possess a large number of tactical delivery systems, most of them of US design. However, the warheads remain under the physical control of US forces. Only France and the United Kingdom possess their own delivery systems and nuclear warheads.

It is estimated that the following missile systems are operated by other NATO countries: IRBMs (SSBS-S2), SRBMs (Sergeant, Pershing, and Honest John). In addition, France has begun deployment of its own dual-capacity Pluton SSM (with a range of 75 miles and a 15 kt yield). Besides these, Great Britain and France have a certain number of SLBMs (UGM-27C Polaris A3, MSBS M-1 and M-2, with ranges of from 1900 to 2880 m and yields up to 500 kt). Deployed nuclear-capacity artillery in NATO countries consists of M-110 203 mm, M-101 155-mm and M-115 203 mm howitzers.

As far as tactical nuclear aircraft are concerned, a certain number of US design are assigned to an operational squadron of NATO forces. These include the F-4 and F-104. There are also planes of British and French design such as Vulcan B2, Jaguar, Buccaneer S2 and Mirage IVA.

Soviet–WTO capabilities

The Soviet Union is believed to maintain about 3500 tactical nuclear warheads in the European area. Although a large number of Soviet delivery systems are deployed with Soviet forces in East Europe, many if not most nuclear warheads may be retained inside the USSR.

Soviet tactical nuclear systems consist of IRBMs (SS-5 Skean), MRBMs (SS-4 Sandal), SRBMs (SS-1b Scud A, SS-1c Scud B, SS-12 Scaleboard and FROG 1-7) and LRCMs (SS-N-3 Shaddock). There are also sea-based missiles, such as SLBMs (SS-N-4 Sark and SS-N-5 Serb) and SLBMs (SS-N-3 Shaddock). As shown in Table 2.3, yields of Soviet tactical weapons range from several kilotons to several megatons, which makes them more destructive than US weapons. Ranges of Soviet SSMs vary from 10 to 45 miles for the FROG series to 2300 miles for the Skean IRBM. Ranges of Soviet sea-based systems vary from 350 miles for Sark to 750 miles for Serb. It is also believed that the Soviet Union has M-55 203 mm gun/howitzers. IISS has estimated that the USSR has about 2500 nuclear capacity aircraft with a maximum range of between 900 and 2500 miles. Soviet forces are not known to possess either tactical nuclear surface-to-air missiles or atomic demolition munitions.

It should be noted that all tactical nuclear weapons in East Europe are Soviet-designed. Other WTO countries have some dual-capacity SRBMs, such as SS-1b Scud A, SS-1C Scud B and FROG 1-7. The ranges of these missiles are between 10 and 185 miles, while the yields are in the kiloton range. In addition, they possess some tactical nuclear capacity aircraft such as IL-28 Beagle, SU-7 Fitter and SU-20 Fitter. The range of these aircraft is between 900 and 2500 miles. The warheads for all these systems are held in Soviet custody.

V. Issues and proposals

As mentioned above, one of the reasons for the deployment of TNWs in Europe was the belief that NATO could not match the Soviet conventional forces which were still put at 175 divisions. From the perspective of the US military doctrine of the 1950s, the decision to send nuclear weapons to Europe was a natural one, reflecting the fact that it was assumed that these weapons would be used when necessary and the fact that, at that time, nuclear weapons were viewed as simply another piece of 'conventional' artillery. Moreover, there was widespread belief that nuclear weapons could somehow compensate for the fact that the enemy was believed to have a much larger ground army. In short, it was believed that the USA could gain a distinct advantage from the use of nuclear weapons and that it should therefore be willing to be the first to use them on the military battlefield. In this context, the USA approached its European allies to request permission to store US nuclear weapons on European soil, particularly on the territory of the Federal Republic of Germany.

The willingness of other NATO countries to accept these weapons on their territories stemmed mainly from two reasons: (a) to enhance deterrence of major conventional aggression and Soviet first use of nuclear weapons and (b) TNWs were believed to augment the credibility of US willingness to employ its strategic deterrence in defence of Europe. The decision to rely on nuclear weapons rather than on more costly conventional weapons for defence was also stimulated in no small part by the experience of the USA in Korea and by the failure of western European countries to meet the conventional force goal of 96 divisions agreed upon at the Lisbon conference in 1952. The belief that TNWs could redress the conventional force imbalance in Europe is still offered as a rationale for the deployment of US TNWs in Europe.

The first question which is crucial for this argument is that of conventional balance in Europe. Supporters of nuclear defence argue that the present balance of conventional military power in Europe – particularly in the Central Region – is hopelessly unfavourable to NATO and that the prospect of unilateral US troop reductions and a declining western European investment in conventional arms can only worsen the balance in the future. As a result, proponents of nuclear defence remain convinced that 'the West cannot field the quantitative and qualitative conventional capabilities necessary to halt, let alone defeat, a conventional attack . . . by the Warsaw Pact' [39]. Many writers imply that Europe's conventional indefensibility is so apparent as to make suggestions to the contrary unworthy of analysis. Ph. Karber, for example, often refers to 'NATO's obvious inability to fight a conventionally superior Warsaw Pact ', whereas Possony asserts that 'the notion that West Europe could be defended conventionally has always been a myth' [40].

On the other hand, a number of analysts strongly suggest the opposite. Enthoven and Smith [41], by criticizing traditional but meaningless division counts and concentrating on more important structural and qualitative indicators of military capabilities, concluded already in 1968 that Europe could be defended not only without resort to nuclear weapons, but also within the framework of existing conventional forces at the disposal of NATO. Moreover, Enthoven considers that the essential facts remain as valid today as they were in mid-1968: NATO has more total active-duty military manpower than the WTO; NATO and WTO land forces manpower in the critical central region are about equal (total manpower in division forces as in the central region of mid-1968 was 677 000 for NATO and 619 000 for WTO) [42]; and NATO tactical aircraft are about equal in numbers and distinctly superior in quality to those of the WTO.

Using the broader, but equally relevant, concept of total manpower in land forces (i.e. including training bases, supply overheads, and so on), Secretary of Defense J. Schlesinger reported to the Congress in March 1974 that the manpower in the ground forces 'which the Pact could launch against the Center with very little warning . . . amounts to about 925 000 men', while 'to counter this immediate threat, NATO has in the central region of Europe about $29\frac{1}{3}$ divisions . . . manpower in ground forces amounts to about 890 000 if France is included'.

One important category in which NATO is at a large numerical disadvantage is tanks. However, this numerical disadvantage is to some extent offset by NATO's modern antitank defences and the superior quality of NATO's tanks, whose effectiveness was so well demonstrated by both sides during the Middle East War of 1973. In tactical air forces, NATO has a distinct advantage because of the notably higher quality of tactical aircraft and armour at the disposal of the Atlantic Alliance. The total inventories on the two sides are about equal.

If these data are well established facts, Enthoven is asking how it is possible to exaggerate the WTO and depict NATO as hopelessly outnumbered. He considers that the main vehicles of exaggeration are identifiable and easily refutable. The most common method has been to count divisions instead of counting soldiers, guns, tanks or vehicles. But the size and content of the divisions vary so much as to make their number meaningless for the purpose of force comparison. Not only are there more men in a NATO division than in a WTO

153

division, but NATO generally has more men outside its divisions. NATO has over 30 000 soldiers in central Europe for each of its divisions there, while the WTO has only about 16 000 soldiers for each division [43]. Another method of exaggerating the threat is to assume that all WTO divisions are at the same readiness and equipment status as the Soviet divisions in the GDR. Still another method of distorting the balance is to include on the NATO side only the forces that the NATO countries commit to the alliance while counting all the WTO units. The forces that are most often excluded in force balance comparisons are the French.

A similar approach is contained in the Annual Defense Department Report, FY 1975, which projects 58 WTO divisions as the 'immediate' threat and from 80 to 90 divisions as the 'mobilized threat' to West Europe. A more recent series of studies undertaken by the Department of Defense reached a similar judgement projecting 85 WTO divisions as the 'designated (ground) threat' to western Europe. These numbers, as we have seen, are not so impressive in reality because Soviet and East European divisions possess substantially less firepower and but little more than one-half the manpower field by opposing US and allied divisions.

A recent study by Lawrence and Record [44] estimates the build-up of manpower available for combat in the Central Region after mobilization (M-day) as follows:

	M-day	M + 30	M + 60 (in thousands)
Total NATO	660	1 045	1 105
Total WTO	576	1 076	1 241

'It is apparent' the authors consider, 'that the present (M-day) balance of forces in the center is generally not unfavourable to NATO (including France) given the widely held presumption that the Pact, as the attacker, would need to muster a minimum superiority over NATO of two to one – and many believe three to one – in all of the (above) categories if it is to have a reasonable chance of successfully invading and occupying western Europe. With few variations the same holds true following 30 and even 60 days of simultaneous or nearly simultaneous mobilization, unless a number of highly questionable assumptions are made about the USSR's willingness to strip its Far Eastern defences, the combat readiness of most Russian divisions, and the degree of East European support the Soviet could count on for an offensive against western Europe' [45].

According to recent studies, the weakness of NATO's conventional defence posture does not appear attributable to an insufficiency of forces but rather to the way in which those forces are organized and disposed. Nevertheless, given the assumption of the WTO as aggressor and the high probability of substantial 'political' warning time before a major invasion, a successful conventional defence of Europe seems not only feasible but also well within the grasp of the forces now deployed on the continent. In sum, it is not at all 'obvious' that western Europe is conventionally undefended and casual claims to that effect confront a weighty body of contrary evidence.

Another assumption was that TNWs could be substituted for manpower imbalance and therefore that the development of TNWs was the key to low-cost defence. This line of reasoning was reflected in a speech delivered in 1954 by General A. M. Gruenther, Supreme Allied Commander in Europe: 'If seventy divisions, for example, are needed to establish a conventional line of defense between the Alps and the Baltic, then seventy minus X divisions equipped with atomic weapons would be needed' [46]. Careful studies later suggested the opposite. More manpower would probably be needed to fight a tactical nuclear war than a non-nuclear war. The reason was quite simple. When both sides have TNWs and when these weapons are used in large enough numbers or large enough yields, the engaged front-line divisions are rapidly destroyed. Since NATO had only meagre reserves, it would be unable to form a new front against Soviet reserve forces. Without a front to make the Soviets concentrate, these would not present good nuclear targets for NATO. In short, while the advocates of the tactical nuclear option argued that NATO must resort to nuclear weapons because of the weight of Soviet manpower, the chief result seemed to be that the battlefield was dominated by Soviet reserves, rather than front-line divisions.

The very character of a tactical nuclear war, it is argued, would not only dictate a much larger battlefield (and thus negate the potential economies of dispersion) but also generate substantially higher rates of attrition. These implications were discerned as early as 1956 by General Matthew B. Ridgway:

> There are a number of sound and logical reasons why a field army of the atomic age may have to be bigger than its predecessors of the past. The complex new weapons themselves – the atomic cannon, rockets and guided missiles – require far more men to serve and maintain them than did the simpler field pieces of World War Second and Korea. The prospect of sudden and enormous casualties, inflicted by the enemy with his own new weapons, makes necessary the training of replacements in great numbers of the dead, and a medical establishment larger than ever to care for the sick and wounded. In the main, though, the changing shape of the battlefield itself sets the requirements for more men. The battle zones of World War Second within which actual ground combat took place were rarely as much as twenty-five miles in depth. Penetrations of armored and airborne forces in the battle areas of the future may well extend two hundred miles or even more in depth, and only by great dispersion, in the wars of the future, will ground elements be able to survive [47].

Ridgway's contention that the security, handling and maintenance of TNWs would require substantial manpower that might otherwise be employed in a conventional role has been confirmed by numerous later studies. It is interesting to point out that in recent studies it is claimed that miniaturization of existing TNWs would spare manpower and cost now needed for a credible non-nuclear deterrent.

Another assumption underlying the limited nuclear war concept was the feasibility of adequate limitation on yields, targets and number of weapons to keep collateral damage and civilian casualties down and thus to prevent the war from escalating to a general strategic level. Although theoretically the controlled use of small-yield nuclear weapons against strictly military targets could keep collateral damage low, the prospects for these limitations working out in an actual war appear to be very low because it would be extremely difficult for either side to determine whether the restraints were being maintained, and the first side to violate them would have an overwhelming advantage.

It is perfectly true that the primary use of TNWs is against purely military targets, but in Europe there are many targets which are extremely important on the military side and are at the same time large cities. Collateral damage has always been a major inhibiting factor in the acceptance and use of TNWs in Europe. The problem stems partly from the heavy population of prospective battlefields in Europe and the historical and social values of urban areas close to the most likely targets. As a result even a small-scale exchange of TNWs will result in a large number of civilians killed and an immense amount of destruction.

These negative effects of the use of TNWs appeared already to have been confirmed in two tactical nuclear war games conducted in 1955. Although both exercises entailed only unilateral and, by contemporary standards, limited employment of TNWs on the part of NATO, their result strongly indicated that reliance on tactical nuclear defence would lead to devastation of the very Western Europe that NATO is designed to protect (see chapter 3, page 173) [48, 49]. Further war games and studies undertaken by the Department of Defense in the 1960s continued to reaffirm these conclusions:

> When the defense of Europe is seen to entail its nuclear destruction, the European incentive to permit the use of TNWs on its soil diminishes rapidly [50].

But even a limited tactical nuclear war carries in itself the danger of escalation to a general nuclear war. A well defined 'firebreak' exists between conventional and nuclear war. Because of differences in the effects of conventional and nuclear explosives, violations of the firebreak are readily detectable; thus there is no ambiguity about the kind of war being fought. This firebreak is the most obvious discontinuity in the spectrum of modern warfare. The concept is widely recognized and has been observed for 30 years. The possibility of establishing a similar firebreak for an engaged tactical nuclear battle abounds with uncertainties. On the other hand, the USSR has emphasized higher-yield, mobile tactical missiles as being primarily useful for terrain or blanketing fires, or for strikes against fixed logistics installations and airfields. Indeed, the Soviet force structure raises serious doubts about its capability to fight a limited tactical nuclear war, much less one in which collateral damage and civilian casualties are to be kept to low levels. In short, the USSR has neither the organization nor the force structure for a limited nuclear war fought exclusively against military targets in an engaged battle zone.

156

The danger of escalation has long been recognized. As early as 1959, Bernard Brodie maintained that:

> The use of any kind of nuclear weapons probably increases markedly the difficulties in the way of maintaining limitations on war. For one thing it is much easier to distinguish between use and non-use of nuclear weapons than between the use of nuclear weapons below some arbitrary limit of size and use well above that limit. Their discontinuity in effects and in identification coincides with a moral feeling of the subject [51].

Three years later, Blackett maintained that 'it can never be quite certain that a tactical nuclear war would, in fact, turn into a strategic one; but the likelihood is sufficiently high to make it absolutely necessary to plan for the probability that it would' [52]. McNamara himself, while not specifically rejecting tactical nuclear weapons, reached a similar conclusion concerning their limitability:

> Nuclear weapons, even in the lower kiloton ranges, are extremely destructive devices and hardly the preferred weapons to defend such heavily populated areas as Europe. Furthermore, while it does not necessarily follow that the use of tactical nuclear weapons must inevitably escalate into global nuclear war, it does present a very definite threshold beyond which we enter a vast unknown [53].

While the subsequent advent of strategic parity between the two great powers has admittedly reduced the prospect of a tactical nuclear war evolving into a strategic exchange, the danger that it might do so remains, given the irrationality and uncertainty inherent in armed conflict and, more importantly, the absence of convincing recognition on the part of the Soviet Union of a distinction between tactical and strategic nuclear war [54].

As has already been pointed out, TNWs are looked upon as providing a credible linkage between the NATO forces deployed in Europe and the strategic forces of the USA. It is important to remember that until the last years of the Eisenhower administration, the use of TNWs was not contemplated in isolation from strategic nuclear war; under the doctrine of massive retaliation, which called for the immediate use of the US strategic deterrent in response to any form of major Soviet aggression, TNWs were to be 'used in a mop-up action after a strategic exchange'. But as the reasons for deploying TNWs became of increasingly doubtful validity the Kennedy administration attempted to put forward a new justification for them – that TNWs serve to enhance the credibility of the US strategic deterrent by providing a link between the possible failure of conventional defence of Europe and US willingness to employ its ultimate weapons. With the coming to the office of the Kennedy administration, McNamara began a dialogue with the European political leaders. He attempted to persuade them that the NATO alliance should put less reliance on the use of nuclear weapons and should seek to develop a conventional capability for direct defence against limited or even massive Soviet attacks. The Europeans resisted this approach, not because they had any faith in the efficacy of a limited nuclear war strategy in Europe but rather because they had come to regard the US TNWs as a bridge between conventional forces and US strategic nuclear power which they had come increasingly to see as the backbone of the deterrence. By the 1960s Europeans no longer believed that the probability of US involvement was sufficient to deter a Soviet attack. On the contrary, they came to feel that deterrence depended on the Soviet Union believing that there was a substantial probability in the event of a conflict, of US strategic nuclear weapons being used against Soviet territory.

Nevertheless, the significance of TNWs even as a link to the strategic deterrent has appreciably diminished in an era of secure US and Soviet strategic retaliatory capabilities. To Europeans and to many Americans, parity has served to reduce the credibility of the strategic deterrent and, by extension, of ideas that the use of TNWs might pave the way for its employment. The logic underlying this conclusion appears unassailable, unless one assumes a disposition on the part of the US to provoke its own possible destruction under circumstances other than direct attack upon its own territory. It was De Gaulle's rejection of precisely this assumption that led to France's withdrawal from NATO's integrated military command and its decision to build a *force de frappe* free of foreign constraints. Later, the waning credibility of a US strategic nuclear response to an invasion of Europe has in fact been officially admitted [55].

For nearly a decade discussion has been going on as to whether the USA should retain or fundamentally change the stockpile of 7200 relatively large and 'dirty' nuclear bombs. Along with the opponents to the use of TNWs, even the supporters of these weapons, according to a recent analysis [56], have come to the conclusion that the current US tactical nuclear stockpile is ill suited to meet European political demands for limiting collateral damage, because the fission content of the weapons is high, their yields are too big, and their accuracy is not satisfactory. Faced with these problems, the supporters of TNWs are trying to find a solution in reducing the yields of existing weapons and ultimately in the development and deployment of a new family of TNWs whose effects would more resemble large conventional munitions than existing nuclear ordnance. It is claimed that the limited destructiveness of miniaturized TNWs confers upon NATO a true nuclear war fighting capability that would at the same time be 'clearly not escalatory' [57].

In this context a statement of former US Defense Secretary Laird is significant. He said that in the light of the SALT negotiations, both TNWs and the conventional land, air and naval forces assigned to defend western Europe would assume greater importance. That is why, Laird said, the Nixon administration was considering the deployment of smaller and cleaner TNWs in western Europe and calling for a new doctrone to strictly control their use [58].

Throughout the Kennedy and Johnson administrations, officials resisted even the development of markedly 'cleaner' weapons for fear that they might be considered more usable if war broke out. The use of even a few TNWs, many officials during that period argued, could easily get out of control and lead to an all-out strategic nuclear exchange between the USSR and the USA.

Although the technology of miniaturization is still largely in the research and development stage, for the past several years pressures within the US military establishment for procurement of mini-nukes have been strong enough to suggest that miniaturization was under earnest consideration by the Nixon administration and continues to be under the Ford administration.

In May 1974 Martin, the US ambassador to the CCD, publicly stated 'categorically that the US government has no intention whatsoever to treat such tactical systems as interchangeable with conventional arms'. However, he went on to say that this policy would not preclude the US from undertaking 'qualitative' improvements in present types of TNWs, including reducing yields and improving the accuracy of delivery vehicles [59]. Also, Secretary Schlesinger has publicly referred to the 'serious possibilities of replacing the existing (tactical nuclear) stockpile with nuclear weapons and systems more appropriate' to a variety of aims, including those of 'denying the enemy his military objectives without excessive collateral damage' and 'providing for selective, carefully-controlled employment options' [60]. Army Chief of Staff C. W. Abrams, appearing before the House Armed Services Committee in 1974, testified: 'We need, as a matter of urgent priority, an improved family of tactical nuclear artillery weapons to provide the field commander the precise ability to inflict damage on his enemy with greatly decreased collateral damage to non-military targets' [61]. Similarly, General A. J. Goodpaster has said that 'new weapons of lower yields and greater accuracy could increase military effectiveness while reducing possible collateral damage, thereby increasing their utility as well as the acceptability in NATO planning for employment in the NATO countries and the adjacent areas in which they would most likely be used' [62].

Miniaturization of TNWs can be achieved in various ways [63, 64, 65].

The assertion that mini-nukes, because of their limited destructiveness could be employed with little risk of escalation is not very convincing. Aside from the fact that 'the use of non-atomic explosive' is 'the best protection against escalation', Soviet military doctrine leaves little room for doubt that even the smallest TNWs 'would be regarded . . . by the Soviet forces' as 'not different in kind' from other nuclear weapons [66]. In other words,

> The use of 'small and clean' nuclear warheads by NATO is likely to be countered by the Warsaw Pact deploying its nuclear weapons, which, being less technologically advanced and less accurate will cause a much higher degree of destruction and radioactive fall-out. The hope that the Soviet side will not classify the 'mini-nukes' as nuclear weapons . . . seems unfounded; the change over from conventional into nuclear warfare has been regarded for too long as a major qualitative break ('fire-break') to be dismissed as a mere technicality [67].

Another argument against the deployment of mini-nukes is that they would probably be unacceptable to many West European governments, particularly FR Germany. Western European opinion has, as already mentioned, always opposed the US deployment of TNWs that raised the prospect of fighting a nuclear war on their territory and this argument will be even more valid after the signing of the SALT agreements giving parity between the strategic forces of the great powers.

Finally there is the question of the financial implications of development and deployment of new generations of miniaturized TNWs. According to some reports the USA has already spent about $10–20 million a year for research and development, while the cost of developing and processing a family of new, miniaturized TNWs is estimated to be $400 million annually over a period of several years [68], although it is expected to be even more.

It can be concluded with Enthoven's opinion that:

> One might say that the proposed mini-nukes combine the disadvantages of nuclear weapons with the shortcomings of conventional weapons. Their use would entail a high danger of escalation and mass destruction. Their costs are high in dollars and in operational flexibility. But like the conventional weapons their designers seek to emulate, their destructive power is comparatively low [69].

In addition to the proposals and arguments for miniaturization of nuclear weapons, there have been other proposals for changing the role of existing TNWs in Europe.

One of the most radical proposals is that calling for a US renunciation of the possibility of a tactical nuclear defence of Europe and the removal of all deployed US TNWs from Europe. The reasoning behind this proposal is that the contribution made by deployed TNWs to overall deterrence of aggression is at best marginal, and in any event is clearly outweighed by the escalation to strategic war that would, it is argued, inevitably follow the actual use of TNWs. As a consequence of this proposal, the defence of Europe must be entrusted exclusively to conventional forces. This view was quite prevalent within the Kennedy administration, but it gains weight again with the introduction of new conventional weapon systems such as zero-error bombs (smart bombs), missiles and anti-tank weapons (particularly those delivered from helicopters) which advances in laser, electro-optical, and infrared technology have made possible; greater mobility for infantry; better communication and surveillance (including the use of remotely piloted vehicles); greater use of modern sensors and electronic countermeasures etc. One model for this type of defence is presented in Chapter 9.

Bearing in mind political resistance to the idea of cutting substantially existing number of TNWs in Europe and pointing out the reasons for having a substantial force of TNWs in Europe, Enthoven has suggested that 'all the useful purposes can be more than adequately served by 1000 tactical nuclear weapons, especially when backed up as they are by Poseidon missiles with MIRV warheads (divided between surface-to-surface missiles, like Lance and Pershing, and artillery shells) [70], while former Assistant Secretary of Defense Paul C. Warnke has testified that the 'deterrent purpose of tactical (nuclear) weapons could abundantly be served by the maintenance of a few hundreds at most' [71].

One of the most elaborate and strongest arguments for an entirely new approach to tactical nuclear warfare based on the use of these weapons, particularly in Europe, was made in a paper by several members of the staff of the Los Alamos Scientific Laboratory [72]. After criticizing present TNW posture they proposed a strategy which is a radical departure from present plans to try in the first instance to defend Europe by conventional means, since it calls for the immediate use against the attackers of low-yield nuclear weapons for all but the most trivial incidents. A range of nuclear weapons, they say, should be designed for specific use in a possible European conflict situation. So far as the destructiveness of these weapons is concerned, 'NATO's political leadership wants no weapon explosion to expose people or their cultural heritage to indiscriminate destruction. This collateral damage can be largely controlled by the choice of target area and target defeat criteria'. But the military commander should have weapons capable of effective use for attaining his military objectives. 'Delivery accuracy, range and responsiveness of the delivery system, weapon yield and fusing, target acquisition, damage assessment and battlefield survivability of the system all play an obvious role in assuring effectiveness' [73]. This concept of European defence is described in Chapter 10.

In his study, Record elaborates four alternatives to the present US tactical nuclear posture. The first envisages merely a reduction in the number of US TNWs and is designed to simplify command and control and to decrease the instability inherent in what many observers believe is an excessive deployment. Alternative two contemplates the elimination of particularly vulnerable TNWs as well as a measure to reduce the vulnerability of remaining systems. Alternative three calls for the replacement of most currently deployed TNWs by new, miniaturized nuclear weapons, whose very limited destructive effects proponents claim would minimize collateral damage and preclude escalation to a wider nuclear war. Alternative four envisages a US renunciation of a tactical nuclear defence of Western Europe, followed by the removal of all US TNWs from the continent. Alternatives one and two appear both desirable and politically feasible, the following specific measures seem appropriate:

(a) Reduction of deployed warheads to 2000; (b) Limitation of warhead yields to not more than 10 kt nor less than 0.5 kt; (c) Termination of Quick Reaction Alert; (d) Elimination of artillery-delivered nuclear warheads and (e) Reliance on battlefield and long-range theatre missiles as the principal means of delivering tactical nuclear warheads [74].

VI. Conclusions

Recent debates and statements on strategy in the United States have once more confirmed the view that the developments in military technology and pressure from military circles are the main incentives for creating and adopting a new policy and vice versa. This assertion can be further testified by the steps undertaken during past few years in the USA such as: further development and upgrading of the accuracy of ICBMs and SLBMs; improvement of command, control and communications systems of existing nuclear missiles; multiple warheads on missiles; introduction of long-range cruise missiles; accurately-delivered low-yield nuclear warheads; introduction of miniaturized TNWs, particularly in Europe, which would be delivered by precision-guided short-range missiles etc. All these developments, backed up by counterforce strategy and the first-use policy, are reducing the population's fear of the consequences of a full-scale nuclear war and represent a shift to doctrines designed to fight limited nuclear war.

Although there are fewer data available on recent Soviet changes in policy and developments of new weapons, it is believed that the Soviet Union has also made remarkable improvements in ballistic missiles and in the development of accurate nuclear weapons.

As far as TNWs in Europe are concerned, it should be stressed that what has been true in the 1950s and 1960s may be even more true today. To sum up, TNWs cannot be substituted for conventional forces; TNWs cannot defend Europe, they can only destroy it, as has been demonstrated by limited 'use' of TNWs in war games; there is no such thing as limited nuclear war; once the threshold is crossed, the danger of all-out nuclear war is immense; limitation of collateral damage and casualties in the case of use of existing TNWs are proven impossible, particularly in Europe.

As far as the question of introduction of low-yield, precision-guided miniaturized tactical nuclear weapons is concerned, although they may seem attractive, particularly from the point of view of military tactics, any policy depending upon the use of TNWs is unacceptable because their use will blur the distinction between conventional and nuclear weapons and because they lower the nuclear threshold; the development of such a new generation of TNWs would be the most extreme financial burden because of the billions of dollars of expense that would be necessary to develop the new generation of small nuclear weapons; the development and replacement of existing TNWs with weapons that might have lower yield and greater accuracy and presumably fewer collateral consequences would erode rather than strengthen deterrence; in the event of actual conflict the availability of these weapons might lead to their premature and unnecessary use; they increase the prospects of eventual all-out nuclear war; they make nuclear war more thinkable and more acceptable; but once the threshold is crossed the process of escalation could become irreversible; also there is no evidence that the other side will use similar low-yield weapons, and it is even more likely that it will have to answer with 'dirtier' and less sophisticated TNWs.

As the parity between the USA and the USSR and the SALT agreements make the possibilities of fighting a tactical nuclear war confined only to Europe more likely and reduce the credibility of the US strategic deterrence, it would be in the best interest of European countries that the US withdraw all TNWs from Europe and that the defence of Europe should be confined exclusively to conventional weapons. Bearing all this in mind, it is absolutely indispensable that the 'firebreak' between conventional and nuclear war should be maintained.

References to Chapter 2

1. Pasti, N.., Opinions on NATO nuclear strategy, in *NATO's Fifteen Nations*, Vol. 13, No. 1, (February–March 1968), p. 2.
2. Halpern, M. H., *Defense Strategies for the Soviets* (Little and Brown, 1971), p. 5.
3. Newhouse, J. *et al.*, *US Troops in Europe: Issues, Costs and Choices*, (Brookings Institution, 1971), p. 45.
4. Lawrence, R. M., On Tactical Nuclear War: Part I, in *Revue Militaire Générale*, (January 1971), p. 46.
5. Miettinen, J. K., European security balanced by tactical nuclear weapons? Paper in *The 22nd Pugwash Conference on Science and World Affairs*, Oxford, England, (7–12 September 1972), pp. 3–4.
6. Hunt, K., *The Alliance and Europe: Part II: Defence with Fewer Men*, in Adelphi Papers, No. 98, (1973), p. 2.
7. Heisenberg, W., *The Alliance and Europe: Part I: Crisis stability in Europe and theatre nuclear weapons*, Adelphi Papers, No. 96, (1973), p. 34.
8. Department of the Army, Field Manual 100-30 (Test), *Tactical Nuclear Operations*, (Headquarters, Department of the Army, August 1971), pp. 11–2.
9. Cliffe, T., *Military Technology and the European Balance*, Adelphi Papers, No. 89, (1972), p. 1.
10. Record, J. and Anderson, T. I., *US Nuclear Weapons in Europe. Issues and Alternatives*, (The Brookings Institution, Washington D.C., 1974), pp. 3–7.
11. *World Armaments and Disarmament*, SIPRI Yearbook 1976, (Almqvist & Wiksell, Stockholm, 1976), pp. 24–25.
12. Enthoven, A. C. and Smith, W. K., *How Much Is Enough? Shaping the Defense Program* (Harper & Row, New YorS, 1971), pp. 118–9.
13. Akerman, N., *On the Doctrine of Limited War* (Lund, Berlingska Boktryckeriet, 1972), p. 73.
14. *Documents on American Foreign Relations, 1955* (Council on Foreign Relations, New York, 1954), p. 18.
15. York, H., *Race to Oblivion: A Participant's View of the Arms Race* (Simon & Schuster, New York, 1970); Bottome, E. M., *The Missile Gap: A Study of the Formulation of Military and Political Policy* (Rutherford, Fairleigh Dickinson University Press).
16. Kissinger, H., *The Troubled Partnership* (McGraw-Hill, New York, 1965), pp. 97–98.
17. Brodie, B., War in the nuclear age, in Brodie, B. (ed.), *The Absolute Weapons* (Harcourt Brace, New York, 1946).
18. Kaufmann, W. W., The requirements of deterrence, in Kaufmann, W. W. (ed.), *Military Policy and National Security* (Princeton University Press, 1956); Kissinger, H. A., *Nuclear Weapons and Foreign Policy* (Random House, New York, 1959); Osgood, R. E., *Limited War: The Challenge to American Society* (University of Chicago Press, Chicago, 1959); and Wohlstetter, A., The delicate balance of terror, *Foreign Affairs*, Vol. 37, (January 1957).
19. Nixon, R. M., A report to the Congress, US foreign policy for the 1970s, *A New Strategy for Peace*, (18 December 1970), p. 122.
20. Nixon, R. M., A report to the Congress, US foreign policy for the 1970s, *A New Strategy for Peace*, (25 February 1971), pp. 170–71.
21. Iklé, F. C., Can nuclear deterrence last out the century?, *Foreign Affairs*, Vol. 51, No. 2, (January 1973); Panofsky, W. K. H., The mutual hostage relationship between America and Russia, *Foreign Affairs*, Vol. 52, No. 1, (October 1973); Greenwood, T.

and Nacht, M. L., The nuclear debate: sense or nonsense?, *Foreign Affairs*, Vol. 52, No. 4, (July 1974); Davis, L. E., *Limited Nuclear Options. Deterrence and the New American Doctrine*, Adelphi Papers, No. 121, (1975), and *World Armaments and Disarmament*, SIPRI Yearbook 1974 (Almqvist & Wiksell, Stockholm, 1974), pp. 55–66.

22. Brennan, D., Strategic forum: the SALT agreements, *Survival*, (September–October 1972), pp. 216–7.

23. Iklé, F. C., Can nuclear deterrence last out the century?, *Foreign Affairs*, Vol. 51, No. 2, (January 1973), pp. 13–17.

24. Russett, B., A countercombatant deterrent? Feasibility, morality and arms control, in S. C. Sarkesian, *The Military–Industrial Complex* (Beverly Hills: Sage Publications, 1972), pp. 201–42.

25. May, M., Some advantages of a counterforce deterrence, *Orbis*, Vol. XIV, (Summer 1970), pp. 271–83.

26. Panofsky, W. K. H., The mutual hostage relationship between America and Russia, *Foreign Affairs*, Vol. 52, No. 1, (October 1973).

27. Schlesinger, J. R., Secretary of Defense, Remarks to the Overseas Writers Association, (10 January 1974); Press Conference, (24 January 1974); and *Annual Defense Department Report, FY 1975*, (4 March 1974), pp. 3–5, 27–45.

28. Panofsky, W. K. H., The mutual hostage relationship between America and Russia, *Foreign Affairs*, Vol. 52, No. 1, (October 1973).

29. Malinovski, (Marshall), XXIInd Party Congress, *Pravda*, (25 October 1961).

30. Sokolovski, V. D. (ed.), *Soviet Military Strategy* (The Rand Corporation, Santa Monica, 1963), p. 367.

31. Sokolovski, V. D. (ed.), *Soviet Military Strategy* (The Rand Corporation, Santa Monica, 1963), p. 299.

32. Sokolovski, V. D. (ed.), *Soviet Military Strategy* (The Rand Corporation, Santa Monica, 1963), p. 364.

33. Sokolovski, V. D. (ed.), *Soviet Military Strategy* (The Rand Corporation, Santa Monica, 1963), pp. 374–5.

34. Shtemenko, S. M., *Nedelia*, No. 6, (31 January–6 February 1965).

35. Lomov, N. A., The influence of Soviet military doctrine on the development of the military art, *Kommunist Vooruzhennykh Sil*, No. 21, (November 1965), pp. 16–22.

36. McNamara, R. S., Secretary of Defense, Joint Session of Senate Armed Services Committee and Senate Subcommittee on Department of Defense Appropriations on the Fiscal Year 1968–72 Defense Program and 1968 Defense Budget – January 23, (1967), p. 68.

37. Clifford, C. M., Secretary of Defense, Press Conference, US Military Headquarters Berlin, Germany, (12 October 1968).

38. Schlesinger, J. R., *The Theater Nuclear Force Posture in Europe* (see footnote, p. 149), p. 6.

39. Karkber, Ph. A., Nuclear weapons and 'flexible response', *Orbis*, Vol. XIV, (Summer 1970), p. 288.

40. Possony, S. I., NATO's defense posture, *Ordnance,* Vol. 14, (July–August 1969), pp. 41–45.

41. Enthoven, A. C. and Smith, W. K., *op. cit.*, Ch. 4.

42. *Ibid.*, p. 148.

43. Enthoven, A. C., US forces in Europe: how many? doing what?, *Foreign Affairs*, Vol. 53, No. 3, (April 1975), p. 526, and Schlesinger, J. R., Secretary of Defense, *Annual Defense Department Report, FY 1975* (1974), pp. 87–88.

44. Lawrence, R. D. and Record, J., *US Force Structure in NATO: An Alternative*, (The Brookings Institution, Washington D.C., 1974), p. 46.

45. Record, J. and Anderson, T. I., *op. cit.*, pp. 16–17.

46. Osgood, R. E., *NATO: The Entangling Alliance* (University of Chicago Press, 1962), p. 109.

47. Ridgway, M. B., *Soldier: The Memoirs of Matthew B. Ridgway* (Harper, 1956), pp. 296–297.

48. Blackett, P. M. S., *Studies of War* (Oliver & Boyd, London, 1962), p. 63.

49. Record, J. and Anderson, T. I., *op. cit.*, pp. 10–11.

50. Enthoven, A. C. and Smith, W. K., *op. cit.*, p. 128.

51. Brodie, B., *Strategy in the Missile Age* (Princeton University Press, 1959), p. 323.

52. Blackett, P. M. S., *op. cit.*, pp. 66–67.

53. McNamara, R., Hearings, US House of Representatives Committee on Appropriations Subcommittee, Department of Defense Appropriations for 1964, *88th Congress, First Section, Part 1*, (Superintendent of Document US Government Printing Office, Washington, DC, 1963), p. 102.

54. Record, J., To nuke or not to nuke: a critique of rationales for a tactical nuclear defense of Europe, *Military Review*, Vol. LIV, No. 10, (October 1974), p. 10.

55. Nixon, R. M., US Foreign Policy for the 1970s: A Report to the Congress, (3 May 1973), pp. 183–84.

56. Lawrence, R. M., *op. cit.* and Beecher, W., 'Clean' tactical weapons for Europe over the threshold, *Army*, (17 July 1972), p. 18.

57. Shreffler, R. G. and Bennett, W. C., *Tactical Nuclear Warfare* (Los Alamos Scientific Laboratory of the University of California, 1970, processed), p. 5.

58. *IHT*, 17 April 1972.

59. Martin, J., *Washington Post*, (24 May 1974).

60. Schlesinger, J. R., Secretary of Defense, Subcommittee in US Security Agreements and Commitments Abroad and Arms Control, International Law and Organization of the Senate Foreign Relations Committee, (4 April 1974, processed), pp. 3, 4.

61. General C. W. Abrams, Statement before the House Committee on Armed Services, (14 February 1974, processed), p. 22.

62. General A. J. Goodpaster, *New York Times*, (27 January 1974).

63. Beecher, W., *op. cit.*, p. 18.

64. Enthoven, A. C., *op. cit.*, p. 529.

65. Aron, R., Peace and war: *A Theory of International Relations*, (Doubleday, 1966), p. 495.

66. Hunt, K., *op. cit.*, p. 18.

67. Heisenberg, W., *op. cit.*, p. 11.

68. Beecher, W., *op. cit.*, p. 20.

69. Enthoven, A. C., *op. cit.*, p. 530.

70. Enthoven, A. C., Subcommittee on US Security Agreements and Commitments Abroad of the Committee on Foreign Relations, (14 March 1974, processed), p. 86.

71. Warnke, P. C., Subcommittee on US Security Agreements and Commitments Abroad of the Committee on Foreign Relations, (14 March 1974, processed), p. 63.

72. *World Armaments and Disarmament*, SIPRI Yearbook 1974, (Almqvist & Wiksell, Stockholm, 1974), pp. 66–69.

73. Bennett, W. S., Sandoval, R. R. and Shreffler, R. G., A credible nuclear emphasis defense for NATO, *Orbis*, Vol. XVII, No. 2, (Summer 1973), pp. 465–5.

74. Record, J. and Anderson, T. I., *op. cit.*, pp. 56–70.

Part Two. The issues

3. Tactical nuclear weapons: problems of definition and application

M. Milshtein

What is a 'tactical nuclear weapon' (TNW)? A single definition does not exist in countries representing the confronting alliances, nor in the countries allied with NATO. In what way are they different from strategic nuclear weapons? Why are they called tactical nuclear weapons? What is their main application? And finally, is it correct to use the term 'tactical' in respect to any nuclear weapons at all?

These questions are not idle ones, because without a correct answer to them, it is impossible to evaluate adequately the entire package of problems, and with regard to the TNWs, above all, the problem of their limitation and reduction. The elaboration of a single criterion for classification of TNWs should have been made a long time ago, for several reasons, including the stimulation of progress at talks such as the negotiations on the reduction of armed forces and armaments in Central Europe, where the TNW reduction problem is one of the questions under discussion.

It is easy to see why the study of the Brookings Institution on US nuclear weapons in Europe declares that 'a major problem in any discussion of TNW is the lack of a precise definition of the term itself' [1].

But it would be incorrect to say that western, including US, literature totally lacks a definition of this term. On the contrary, there is a host of definitions, but in most cases they are very uncertain and contradictory. The Brookings Institution study, mentioned above, gives several of these definitions.

One US military analyst defines TNWs as 'Nuclear weapons designed to support land forces'; another believes them to be weapons 'designed to influence the land or air defence battle *directly*, rather than indirectly (for example, by interdicting an enemy's lines of communications or by carrying a threat to his homeland' [1]. TNWs have also been defined as 'nuclear weapons now for limited tactical military purposes' [2].

The Brookings Institution study defines TNWs as 'all nuclear weapon systems other than strategic bombers (and the bombs they carry), nuclear cruise missiles and land and sea-based intercontinental ballistic missiles' [3]. Similar definitions could be given, but none of them provide a clear and all-embracing answer to the question 'what are tactical nuclear weapons?' A definition of what is the difference between tactical and strategic nuclear weapons is equally unclear.

It is said in one case that 'the main difference between strategic and tactical tactical nuclear weapons is the difference in range. Tactical nuclear weapons have a shorter range but are sometimes more powerful than strategic weapons' [4]

Overall, most western analysts take one of two approaches when defining the difference between tactical and strategic nuclear weapons – yield and range of delivery systems. But as the Brookings Institution study points out, neither of these two approaches was completely acceptable. This division reflects to a certain extent the general division in the USA of the art of warfare into strategy and tactics and, in keeping with this, the division of all weapons into strategic and tactical weapons. For example, the USA speaks of a 'strategic air force' and a 'tactical air force' and, correspondingly, of a 'strategic air command' and a 'tactical air command'. This implies that the weapons used in the hostilities zone are tactical and the ones employed on an intercontinental scale are strategic.

The term 'tactical nuclear weapons' was introduced into military vocabulary by the USA more than 20 years ago. At that time nuclear weapons were deployed by US armed forces outside the American continent, mainly in Europe, and the possibility arose of using nuclear weapons not only on a global scale but on the theatre of hostilities (TH) as well. This leads to the conclusion that the USA still includes among TNWs the nuclear weapons and means of their delivery that are deployed on the TH and are intended to be used in battle there.

The main criterion in the US definition is not the yield of the nuclear device (i.e. its TNT equivalent) and not the destruction capacity, but primarily the range and operational purpose (in this case its use in the TH). Thus the USA will usually define TNWs as all nuclear weapons whose range is not sufficiently for them to reach the USA proper for a strike on the targets in this territory.

According to official data, the tactical nuclear weapons, or nuclear weapons in a theatre of operations include [5]:

1. Land-based tactical nuclear weapons:
    atomic demolition munitions;
    nuclear artillery;
    surface-to-surface missiles (for example Pershing, Sergeant, Honest John and Lance, which will replace Sergeant and Honest John);
    surface-to-air missile.
2. Nuclear-capable tactical aircraft:
    nuclear-capable planes carrying nuclear bombs;
    air-to-air missiles;
    surface-to-air missiles.
3. Sea-based tactical nuclear weapons:
    nuclear-capable aircraft-carrier planes carrying nuclear bombs;
    air-to-air missiles;
    ship-to-air missiles.

So the so-called TNW arsenal is quite varied and embraces different means, with ranges from several kilometres (nuclear artillery) to several thousand kilometres (the F-4 or F-111 planes) and yields from several kilotons to more than one megaton [6].

The main difference between tactical and strategic nuclear weapons is that the TNWs do not have the means for intercontinental delivery, in particular

ICBMs, or strategic air force planes, nor do the TNWs usually have a multi-megaton yield. On the other hand TNWs include weapons that are purely defensive, like atomic demolition munitions and surface-to-air missiles.

The USA also makes a difference between strategic and tactical nuclear weapons according to their use. It is assumed that strategic nuclear weapons are intended for deterring (or waging) an all-out nuclear war, and that tactical nuclear weapons are intended for deterring (or waging) a nuclear war in the theatre of hostilities. In practical use, strategic and tactical nuclear weapons are not so sharply divided that they represent two independent and different weapons types. If nuclear weapons are evaluated according to their yield from the point of view, for instance, of the capability to strike at targets in the USA proper, then tactical nuclear weapons do differ greatly from strategic nuclear weapons. The reason is that TNWs are deployed on theatres far-removed from the continental USA and, therefore, cannot reach its territory, thus posing no threat or danger to the USA. Therefore at first glance one can make the following absurd assumption: if the USA felt the TNW could do the same functions for it as the strategic nuclear weapons, then the USA would probably ardently advocate a complete ban on strategic nuclear powers' arsenals.

But in view of the destruction and annihilation for the European states if TNWs are used in Europe, there is no great difference as far as these states are concerned between tactical and nuclear weapons. 'No one has yet been able to devise any reasonable set of scenarios for our European-based tactical nuclear weapons,' says a US magazine and declares that 'there is no rational doctrine for the use of the US land-based tactical nuclear weapons in Europe' [7].

Alain C. Enthoven, former US Secretary of Defense aide, who co-authored the book *How Much Is Enough?* writes that the various war games and studies in the 1960s showed that 'even under the most favourable assumptions, it appeared that between 2 and 20 million Europeans would be killed, with widespread damage to the economy of the affected area' [8] even if the strikes were made at a restricted number of targets with low-yield nuclear weapons. Such studies were done both in the USA and by NATO.

In operation Sage Brush, held in Louisiana, 275 TNWs of from 2 to 40 kt were 'detonated' in a limited military operation. The assessors 'reported that the destruction was so great that no such thing as a limited purely tactical nuclear war was possible in such an area' [9].

In 1955 NATO staged the Carte Blanche exercise in Western Europe. A total of 335 devices were 'exploded' in 38 hours, 268 of them on the territory of the FRG and the GDR. German casualties, not including those attributable to residual radiation, were estimated at between 1.5 and 1.7 million dead and 3·5 million wounded. These figures are some five times the total German civilian casualties caused by RAF and USAF bombing during World War II – 305 000 killed and 780 000 wounded. As the result many people decided that the use of TNW 'will not defend Europe, but destroy it' [10].

In the 1960s the US Defense Department studied the possibilities of using TNWs in Europe and concluded that even given the most favourable assumptions, colossal damage could be inflicted on the population and the economy even with limited use of TNWs. There was also 'a high risk of 100 million dead if the war escalated to attacks on cities' [11]. How can one speak about 'tactical' nuclear weapons here? Western analysts accept that in the Soviet Union the so-called TNW deployed in Europe also presents a serious threat. In the book *The Superpowers in a Multinuclear World* it is stated that

> The United States has approximately 420 F-4 Phantom II and 70 F-111 fighter-bombers stationed in Europe which are 'dual-capable' – aircraft built to carry both nuclear and conventional loads – planes, about another 75 A-6 and A-7 strike planes on two aircraft carriers with the Sixth Fleet in the Mediterranean ... This group of American aircraft [is] based so as to be capable of reaching the Soviet Union with nuclear weapons ... Moreover, there are Pershing missiles in West Germany with a 450-mile range which could strike into the Ukraine ... It has been estimated that up to 20 per cent of the industrial capacity and population of the Soviet Union could be destroyed by forward-based systems alone [12].

The forward-based systems, with the so-called TNWs as their core, also present a strategic threat to the Soviet Union. So theatre nuclear weapons with ranges that do not allow them to reach the US territory are weapons of a tactical nature for the USA alone.

The USA is now rapidly developing TNWs, especially the weapons used by land forces, by making them more accurate and more compact. The old weapons are allegedly being replaced by new models. According to US analysts this development makes the use of the TNWs in the theatre less dangerous, because strikes could be made on small targets with high precision by low-yield devices.

In this way, nuclear weapons have become more 'acceptable' and the probability of possible use rises. This, supposedly, makes them more important as deterrents. But in practice the more improved TNWs do not lessen the danger of nuclear war; they drastically increase the possibility.

The difference between conventional and nuclear weapons is being obliterated and some military commanders might find the possibility of using highly accurate nuclear weapons more attractive than using conventional armaments. In this context the serious danger that may arise with the development of new types of very accurate and low altitude operational–tactical cruise missiles cannot be forgotten. These missiles are dual-capable, i.e. they can deliver both conventional and nuclear ordnance. Experts say the cruise missiles are relatively cheap to produce, simple to operate and are 'attractive' for possible use.

This development of nuclear weapons and their delivery means may lead to proliferation of nuclear weapons and to an irreversible escalation from a 'tactical conflict' to all-out nuclear war even if they are used on a limited scale. Therefore, TNWs must not be improved, but be reduced and completely removed from all probable theatres.

The talks on reduction of armed forces and armaments in Central Europe are the only place today where practical questions on TNW reduction are being considered. The USSR insists, as it has from the very beginning, that both the TNWs and the air force – one of the main means for TNW delivery – be included among the items to be reduced. The USSR is consistently working for the reduction of nuclear weapons in Europe, which is in complete accord with its fundamental position on the limitation and reduction of nuclear weapons and on achieving nuclear disarmament.

Therefore it may be concluded that:

(1) The term 'tactical nuclear weapons' is a very conditional and vague term; it does not really reflect either the potential of these weapons or their specifications. Moreover it has only a one-sided meaning – they are tactical only with respect to the USA. As for the countries on whose territory these weapons may be used, they are essentially strategic weapons because their use would lead to disastrous consequences.

(2) A more correct term for such weapons (as applicable to the Central European theatre) would be 'theatre nuclear weapons). This term covers the nuclear weapons that are deployed, could be deployed, or are intended for use in the theatre of hostilities: It does not qualify the weapons' nature, yield range or destruction potential. The term covers all nuclear weapons in all armed forces in the theatre. But the definition also leaves certain 'grey zones', such as the US missile-carrying nuclear-powered submarines, provided for NATO by the US Command, or the British bomber force intended for use in the theatre. But these 'grey zones' do not alter the fact that the definition of TNWs as 'theatre nuclear weapons' seems the most suitable in relation to the Central European theatre.

(3) Modern tactical nuclear weapons, if used on any scale, however limited (territory, depth, the number and nature of targets and yields), can in the European theatre lead not only to colossal civilian casualties and loss of material resources, but also can be an ominous step towards the global nuclear disaster.

(4) The further development of TNWs through their miniaturization and increased accuracy is extremely dangerous and increases the probability of their use, i.e. such development may drastically lower the nuclear threshold. The 'usability' of nuclear weapons, their 'safety', 'advantages' and the possibility for 'temporary and limited' employment is increased, and the difference between the conventional and the nuclear weapons is obliterated. Therefore, this process should be halted before it is too late.

(5) The complete rejection of the so-called TNWs and their total removal from areas of possible confrontation is the only correct solution from the point of view of the European peoples' security. If it is not possible to achieve such a solution immediately, it must be approached stage by stage, firstly – to divide TNW reduction into several steps and, in the long run, to achieve their total abolition. Measures may be taken either within the system of reducing armed forces and armaments in Central Europe, or independently, i.e. within a framework of TNW reduction alone.

Other approaches are possible. The one important point is that all proposals must lead to a reduction in so-called 'tactical' nuclear weapons and reduce the possibility of a nuclear war breaking out.

References

1. Record, J., *US Nuclear Weapons in Europe: Issues and Alternatives*, (Brookings Institution, Washington, 1974), p. 3.
2. Record, J., *US Nuclear Weapons in Europe: Issues and Alternatives*, (Brookings Institution, Washington, 1974), p. 2.
3. Record, J., *US Nuclear Weapons in Europe: Issues and Alternatives*, (Brookings Institution, Washington, 1974), p. 6.
4. *The Defense Monitor*, Vol. 4, No. 2, p. 1.
5. Secretary of Defense James Schlesinger, *The Theater Nuclear Force Posture in Europe.* A Report to the US Congress in compliance with public law 93-365, (Washington, 1975), pp. 15–18.
6. Record, J., *US Nuclear Weapons in Europe: Issues and Alternatives*, (Brookings Institution, Washington, 1974), pp. 20–23.
7. *The Defense Monitor*, Vol. 4, No. 2.
8. Enthoven, A. C., 'US Forces in Europe: How Many? Doing What?' *Foreign Affairs*, Vol. 53, No. 3, April 1975, p. 514.
9. Blackett, P. M. S., *Studies of War*, (Oliver & Boyd, London, 1962), p. 63.
10. Schmidt, H., *Defence or Retaliation*, (Praeger, 1962), p. 101.
11. Enthoven, A. C. and Smith, W. K., *How Much is Enough? Shaping the Defense Program, 1961-1969*, (Harper & Row, New York, 1971), p. 128.
12. Kemp, J., *The Superpowers in a Multinuclear World*, (Lexington, 1974), p. 49.

4. Arms control and tactical nuclear forces and European security†

J. Coffey

In recent years there has been renewed interest in tactical nuclear forces‡ as evidenced by a spate of studies on force postures and capabilities, on concepts for the employment of tactical nuclear forces, and on the roles these can play in deterrence and defence, most notably in Europe [1]. To some extent this spurt of articles and papers represents the rising to the surface of an underground stream long submerged by NATO official doctrine, which has emphasized the conventional component of those forces needed for 'flexible response', and to some extent reflects new factors which could significantly affect European security. Among these is the recognition and acceptance of strategic parity between the Soviet Union and the United States, which has enhanced earlier doubts about the credibility of the US strategic deterrent; continuing improvements in WTO forces which seem to threaten the conventional military balance in Europe; technological innovations, which promise smaller, cleaner, more flexible and more accurate tactical nuclear weapons; and the (somewhat belated) discovery that NATO tactical nuclear forces are by no means suited to the role they are supposed to play. NATO existing delivery vehicles are in some instances too inaccurate, warheads too large and too 'dirty', and the

---

† This chapter is based in part on one chapter of *Arms Control and European Security: A Guideline to East–West Negotiations*, by J. I. Coffey, published by Chatto & Windus for the International Institute for Strategic Studies.

‡ The phrase 'tactical nuclesr forces' (TNF) refers to those weapons systems deployed in, or which can readily be deployed in, a theatre of operations and which are designed and intended primarily for use against military targets such as troop concentrations, artillery positions, attacking aircraft, unit headquarters, supply points, air bases and so on. Tactical nuclear forces can be further subdivided into: (*a*) tactical nuclear delivery vehicles (TNDVs) such as artillery, rockets, strike aircraft, and short-range missiles, including both land-based cruise missiles and those mounted on surface vessels or submarines; (*b*) defensive weapons suitable for use in tactical operations, such as antisubmarine rockets (ASROC) and surface-to-air missiles (SAMs); (*c*) the nuclear warheads used by these various weapons systems. (The TNDV and the warhead, with associated surveillance, guidance and communications equipment, constitute a weapons system.) This definition clearly excludes strategic nuclear forces (intercontinental ballistic missiles, submarine-launched ballistic missiles and heavy bombers) and is intended to exclude regional nuclear forces, such as medium and intermediate range ballistic missiles, medium bombers etc. Admittedly, there are 'grey areas' in that some of the weapons excluded can be (and may well be) used against military targets in a theatre of operations and could even be deployed there; however, this may not be their primary mission. See also Chapter 3, by M. Milshtein, of this book.

For a further discussion of definitions, see Jeffrey Record, with the assistance of Thomas I. Anderson, *US Nuclear Weapons in Europe, Issues and Alternatives* (Washington: The Brookings Institution, 1974), pp. 4–7. This monograph will hereafter be cited as *US Nuclear Weapons in Europe*.

forces as a whole vulnerable to sabotage, a pre-emptive strike, and/or seizure by attacking troops [2].

The proposals for taking account of these new factors range from those calling for creating a viable war-fighting NATO nuclear force, based on combining advanced nuclear technology with precision-guidance for warheads, through those suggesting a restructuring of the current forces into a smaller, less vulnerable, and more easily controlled deterrent, to those calling for virtually complete reliance on conventional war-fighting capabilities [3]. Despite the range of the reassessment, little attention has been paid to the impact of arms control† on European security in general and on tactical nuclear forces in particular. In fact, of all the articles cited, only that by Miettinen specifically discusses measures for limiting tactical nuclear weapons. Yet any agreement on mutual force reductions, even if it takes the form proposed by NATO, is bound to affect tactical nuclear capabilities [4]. Moreover, the USSR and the East Europeans are pressing hard at the Vienna Conference on Mutual Reduction of Forces and Armaments and Associated measures in Central Europe for cuts in nuclear-capable aircraft, rockets and missiles [5] which, if accepted, could lead to reductions in US and Soviet TNDVs and/or nuclear warheads, and could constrain both present and future European nuclear delivery capabilities. Thus, it is somewhat surprising that so little has been said about the effects of, and the desirability of, measures that would impose limitations on tactical nuclear forces.

There are a number of reasons for this comparative silence, one being that these forces would play a significant role in the event of a nuclear war. The Soviet Union (and its East European allies) have not only integrated tactical nuclear delivery vehicles into units at all levels from the division upwards but have evolved a doctrine calling for their large-scale use to shatter enemy defences and to prepare the way for exploiting operations by mobile ground and airborne forces [6]. A second is that these forces are regarded by NATO as hedges against a WTO conventional attack which threatens major loss of NATO territory or forces [7]. A third is that they serve both as a deterrent to any conventional attack and as an inhibitor to the use of nuclear weapons by an adversary. A fourth reason is that they do, in the eyes of some West Europeans, constitute a link with US strategic nuclear forces, thus assuring that the awesome prospect of a strategic nuclear exchange will enhance deterrence.

There are also a number of reasons why measures for the limitation of tactical nuclear forces should be examined. One of these stems directly from the fact, already noted, that arms control measures which could affect these forces are being discussed by NATO and WTO in Vienna. A second is that any widespread use of tactical nuclear weapons in Europe would, because of the high average yield of the warheads and the comparative inaccuracy of many launch

---

† As used here, the phrase 'arms control' refers to any measure limiting or reducing forces, regulating armaments, or restricting the deployment of troops and/or weapons which is taken pursuant to an understanding with another state or states, is intended to induce responsive behaviour, or is aimed at diminishing the likelihood of war on ameliorating its consequences.

vehicles, result in enormous losses in lives and property. While these side effects could be mitigated by improvements in weaponry, they could also be reduced by cutbacks in numbers of tactical nuclear weapons systems, constraints on their employment, and other measures for the limitation of armaments. A third is that the vulnerability of some delivery vehicles puts a premium on their early use and may, in time of crisis, lead to a premature decision to employ nuclear weapons or even to inadvertent nuclear war; here, too, arms control can help by leading to reductions in, or the redeployment of, vulnerable weapons by both sides. A fourth reason is that, given the nature and scope of other East–West interactions, such as those on SALT, on mutual force reductions, and on European security at large, it seems not unreasonable to engage in cross-alliance discussions of measures to restructure tactical nuclear forces in Europe, rather than to adjust force postures and weaponry in accordance with national or alliance concepts of security.

It should be noted that measures for the control of tactical nuclear forces will not *automatically* limit damage, minimize the danger of nuclear war, enhance European security, or improve relations between East and West. For one thing, the number and the destructiveness of tactical nuclear weapons systems in Europe is such that even massive reductions would still leave each side formidable capacities to inflict damage [8]. For another, unless controls over nuclear delivery vehicles are world wide, both NATO and WTO will retain the ability to reintroduce TNDVs into Europe, as well as to employ regional nuclear forces such as Soviet MR/IRBMs and medium bombers and British and French SLBMs, IRBMs, and strategic bombers.† The force postures of the two alliances are so different and their doctrines for the employment of tactical nuclear weapons so divergent that any single arms control measure is almost bound to have asymmetrical effects on their capabilities – and hence on their assessments of the military balance in Europe. Finally, some kinds of controls, such as those on nuclear-capable aircraft, could affect differently perceptions of security, in the case of East Europeans by relieving some of their anxieties about the likelihood and consequences of tactical nuclear war in Europe, and in the case of West Europeans by raising doubts as to whether deterrence was effective across a spectrum of possible conflicts.

This is not to argue against examining – or adopting – measures for the control of tactical nuclear forces, but only to point out that such an examination must be far-ranging and that the measures recommended must be assessed in terms of their political consequences as well as their military ones. This examination will start with an analysis of measures such as limitations on numbers of TNDVs and warheads, restrictions on their deployment, controls over new tactical nuclear weapons systems, and constraints on the uses of weapons in time of war. Next there will be an evaluation of these measures, with a view to recommending some which seem both feasible and potentially acceptable.

---

† In addition, the USSR could direct ship-borne or submarine-launched cruise missiles against targets on land, while the United States could utilize against similar targets the missile submarines in European waters which are allocated to NATO.

Table 4.1. NATO tactical nuclear delivery vehicles in Europe, mid-1976[a]

	M-109 155 mm howitzer	M-110/M-115 203 mm howitzer	Honest John[b]	(P) Pluton, (L) Lance or (S) Sergeant[b]	Pershing[b]	Strike aircraft[b]
Belgium	41	24	8	—	—	36 F-104G[c]
Canada	18	—	—	—	—	—
Denmark	72	12	—	—	—	25 F-104G 15 CF-104G
France	—	—	—	24(P)	—	75 Jaguar 120 Mirage IIIE
Federal Republic of Germany	600[d]?	80	70	20(S)	72	229 F-104G (including 85 from Naval Air Force) 60 F-4
Greece	18	12	8	—	—	38 F-4 15 F-104G
Italy	126	12	8	6	—	18 F-104G 54 F-104S
Netherlands	36	12	8	—	—	36 F-104G
Norway	18	—	—	—	—	22 F-104G
Turkey	72	96	18	—	—	18 F-104S 36 F-104G 40 F-4

United Kingdom	40	12 (*CE*)	8 (*CE*)	—	—	28 Buccaneer (*CE*) 42 Buccaneer (*WE* including 14 carrier based) 24 Jaguar (*CE*) 16 Jaguar (*WE*) 50 Vulcan (*WE*)
United States	306 (*CE*)	192 (*CE*)	—	36 (L) (*CE*)	108ᵉ (*CE*)	72 F-111E (*WE*) 158 F-4 II (*WE*)ᶠ 232 F-4 II (*CE*)ᶠ 60 A6/A7 (*SE*)ᵍ
Total NATO, including	**1 347**	**452**	**128**	**62**	**180**	**1 324ᵇ**
France	**(1 347)**	**(452)**	**(128)**	**(86)**	**(180)**	**(1 519)**

(*CE*) = Central Europe; (*SE*) = Southern Europe; (*WE*) = Western Europe (United Kingdom, France, Spain).

ᵃ Only those weapons-systems listed as nuclear-capable in *The Military Balance, 1976–1977* (IISS, London) and/or Cliffe, T., *Military technology and the European balance*, Adelphi Paper No. 89 (London: IISS, August 1972), pp. 41–42, are counted here, even though this excludes weapons such as the 175 mm gun and *Mystère* IVA which other analysts consider nuclear-capable.

ᵇ All figures for aircraft count only those operational in organized units: excluded are squadron reserves (3–7 aircraft per squadron) and war reserves. All figures for missiles count only launchers, and exclude those extra missiles carried in battery trains or maintained in war reserves.

ᶜ Only those aircraft in bomber or fighter-bomber squadrons are counted, here and elsewhere; thus nuclear-capable aircraft in fighter-interceptor or reconnaissance squadrons are omitted.

ᵈ This figure includes all the self-propelled 155 mm howitzers of the Bundeswehr, but none of its towed 155 mm howitzers.

ᵉ *The Military Balance, 1976–1977*, p. 73 lists only 36; since, however, the United States deployed 108 Pershings last year, and has offered at the Vienna Conference to remove 36, under certain conditions, the old figure seems more reasonable than the new one.

ᶠ These figures are based on equipment levels of 24–5 aircraft per squadron, rather than the 18 given for fighter-bombers by *The Military Balance, 1976–1977*, p. viii; however, the latter levels do not jibe with the total number of aircraft given in *The Military Balance*, p. 7, which has been deemed overriding.

ᵍ Carrier-based.

ʰ Not all of these are deployed in a nuclear role; hence, this figure may overstate actual NATO capabilities as distinct from potential ones.

Sources:
1. *The Military Balance, 1976–1977*, pp. 5–7 and 18–26.
2. Exchanges with the International Institute for Strategic Studies.

Table 4.2. WTO tactical nuclear delivery vehicles in Europe[a]

Country	203 mm Howitzer	FROG[b]	Scud[b]	Scaleboard[b]	Shaddock[b]	Strike aircraft[b]	
Bulgaria	—	24[c]	8[d]	—	—	—[e]	
Czechoslovakia	—	30	16	—	—	48	SU-7[f]
GDR	—	18	8	—	—	—	
Hungary	—	18	—	—	—	—	
Poland	—	45	24?	—	—	48	IL-28
						72	SU-7
Romania	—	27	—	—	—	—	
USSR	16?	93	92[d]	16?	?	90?	IL-28 / Yak-28 (CE)[g]
						30	IL-28 / Yak-28 (SE)
						165?	SU-7 (CE)[g]
						17?	SU-7 (SE)
Total	**16**	**255**	**152**	**16**	**?**	**470**	

(CE) = Central Europe; (SE) = Southern Europe.

[a] Only those weapons systems listed as nuclear-capable in *The Military Balance, 1973–1974* (IISS, London) and/or by T. Cliffe, *Military Technology and the European Balance*, Adelphi Paper No. 89 (IISS, London, August 1972), Appendix C, Table G, pp. 46–47, are counted here, even though this excludes weapons which other analysts consider nuclear-capable.

[b] All figures for aircraft count only those operational in organized units; excluded are squadron reserves (of 3–7 aircraft per squadron) and war reserves. All figures for missiles count only launchers, and exclude those extra missiles carried in battery trains or maintained in war reserves.

[c] Based on one FROG battalion of three launchers per division, from Erikson, *Soviet Military Power*, p. 75. Brigadier Kenneth Hunt, *The Alliance and Europe: Part II: Defence with Fewer Men*, Adelphi Paper No. 98 (London: The International Institute for Strategic Studies, Summer 1973), p. 24, lists four SSMs per Soviet (and presumably WTO) division. Dupuy, *op. cit.*, also gives slightly higher figures.

[d] Based on one Rocket Brigade at full strength (16 launchers) for each of the four Soviet Armies in the GDR, and one brigade at half strength for the three Soviet Armies in Poland, Czechoslovakia and Hungary, and for similar formations in the countries of Eastern Europe. This may be high, as *The Military Balance, 1973–1974*, p. 69, lists only 300 Scuds and Scaleboards for the entire Soviet Army, about one-fifth of which is deployed in Eastern Europe.

[e] Only those aircraft in bomber or fighter-bomber squadrons are counted, here and elsewhere; thus nuclear-capable aircraft in fighter-interceptor or reconnaissance squadrons are omitted.

[f] These figures were derived by assuming that one-third of the Czechoslovak, Polish and Soviet fighter-bombers were SU-7s, and the remainder were MiG-17s and MiG-21s. According to the *Army, Air Force, and Naval Air Statistical Record*, the three 'Northern Tier' countries alone have 360 SU-7s, and Hungary another 50; however, these figures seem high when matched against the breakout by types of squadrons in *The Military Balance 1973–1974*.

[g] Obtained by applying the percentages of light bombers from p. 91 of *The Military Balance, 1973–1974* against the 1250 aircraft deployed in Central Europe (*Ibid.*, p. 95).

Sources:

1. *The Military Balance, 1976–1977*, pp. 8–10 and 12–14.
2. Aviation Studies International, *Army, Airforce, and Naval Air Statistical Record* (London: January 1973) (for numbers of ground-based weapons systems in East European armies).
3. T. N. Dupuy and Wendell Blanchard, *The Almanac of World Military Power* (2nd ed., R. R. Bowker Company, 1972).
4. Friedrich Wiener, *Die Armeen der Warschauer-Pakt-Staaten* (Munich: J. F. Lehmans Verlag, 1971), esp. pp. 35, 112 and 115, for weapons systems in the Soviet Army.

Finally, a more detailed assessment will be made of the military and political implications of those measures recommended. This will be done in order to provide a basis for judging their impact on European security, on the ability of the various states to safeguard their vital interests against the use or the threat of force and on their confidence in that ability. If both the examination and the assessment focus largely on NATO and WTO, this is not to say that only members of these alliances are concerned about European security, but rather that they are the proper subjects for initial discussion since they alone, either directly or through their great power allies, have access to tactical nuclear weapons. (For instance, the French have an independent tactical nuclear capability based on the Mirage IIIE and other nuclear-capable aircraft and the Pluton short-range surface-to-surface missile (SRBM). The bulk of the tactical nuclear delivery vehicles in the hands of United Kingdom armed forces are of US design, and their warheads are consequently under US control, but the United Kingdom has produced weapons for their Buccaneer and Vulcan V-2 bombers, which have tactical missions. Other members of NATO can obtain nuclear weapons from the United States; in fact nine of these have TNDVs and at least five allow the storage of nuclear warheads on their territory (Table 4.1). All six of the East European members of the WTO have nuclear-capable delivery vehicles (Table 4.2), but so far as is known no nuclear warheads are stored on their soil).

I. Limitations on tactical nuclear weapons systems

As suggested earlier, cut-backs in tactical nuclear weapons systems could, if carried far enough, reduce the damage each side would suffer in the event of war. Long before that point was reached, these cut-backs could make disarming strikes less feasible (since the number of targets would not decrease as rapidly as the number of aircraft and other launch vehicles). Mass attacks designed to blast holes in enemy defences would become more difficult. Even ceilings on increases in nuclear forces, or freezes in present force levels, could affect to some degree the ability to wage nuclear war in Europe – and hence perceptions of the intent to do so.

Unfortunately, various measures for the limitation of tactical nuclear weapons systems may affect differently artillery, missiles, aircraft and other types of delivery vehicles, as well as the warheads which they carry. Each will impinge in varying degree on the uses to which tactical nuclear weapons can be put in the event of war, such as blocking the advance of enemy troops, interdicting enemy reinforcement and supplies, and suppressing that enemy's own air and missile forces. Moreover, the effects of particular limitations will depend not only on their nature but also on the geographic areas to which they are

Table 4.3. Deployment of tactical nuclear delivery vehicles by region, mid-1976

Region	Artillery	Rockets	VSRBM	SRBM	Strike aircraft	Total
Northern						
NATO	102				62	164
WTO (including forces inside the USSR close to Norway)	—	(8[a])	(8[b])	(12[b])	(25[c])	(53)
Central						
NATO	1 361	94	56	180	560	2 210
WTO	234	217	167	16?	771	1 405
Southern						
NATO	394	38	6	—	213 land-based 60 carrier-based	705
WTO	36	106	49	?	150	341
Western						
NATO (including France)	—	—	(24)	—	324 (519)	324 (543)
Total						
NATO	1 824	134	62	180	1 324	3 493
(including France)	(1 824)	(134)	(86)	(180)	(1 519)	(3 712)
WTO (including forces inside the USSR close to Norway)	270 (270)	323 (331)	216 (224)	16 (28)	921 (946)	1 746 (1 799)

[a] See note e to Table 4.2.
[b] See note f to Table 4.2 for Scud. It is assumed that 12 Scaleboard launchers are also deployed on the Kola Peninsula.
[c] Assumes that 270 of the 300-odd Soviet aircraft deployed on the Kola Peninsula belong to the Naval Air Force and the remaining 65 are distributed by type as are the total planes available to the Tactical Air Force.

applied. Given the asymmetrical deployment of the nuclear forces of NATO and WTO, what may be feasible or desirable in Central Europe is not necessarily so on either of the flanks. Before discussing particular consequences, it may be desirable to examine the broad implications of various types of limitations.

A freeze on numbers of tactical delivery vehicles might be useful in ensuring that new nuclear-capable airplanes are used to replace old ones, rather than to increase the number of strike aircraft, or that improved missiles cannot be allocated to a country not now possessing them without withdrawing equal numbers of the older weapons from another country. As suggested by these illustrations, however, a numerical freeze would not preclude qualitative upgrading, nor would it be very meaningful, partly because the number of TNDVs is so large, and partly because it is possible to transfer these from one part of Europe to another. While a freeze on the numbers in each region would preclude this, it would also preclude any changes in present deployment patterns, leaving Norway, for example, unable to counter a new threat to its security by either the permanent or the temporary introduction of tactical nuclear weapons. Despite this, some kind of freeze might help NATO by inhibiting a further build-up of WTO nuclear forces. It would be less advantageous to the WTO, which presumably would want to offset NATO's qualitative and quantitative edge either by improving its own forces or by reducing those of the Western Alliance.

An alternative would be to impose ceilings on tactical nuclear delivery vehicles, either over-all or by regions. If these ceilings were applied to total numbers of TNDVs, they would permit a build-up by WTO, since NATO has many more TNDVs than do the countries of the WTO (Table 4.1). If, conversely, ceilings were imposed on each type of weapon, at the lowest level now extant, this would require NATO to scrap most of its medium artillery, and the WTO to cut back by about one-third on the number of rockets in its divisions. If the ceilings were set at the higher of the two levels, this would allow – and, given the tendency to seek 'balance', almost require – increases in some categories of weapons which could be both expensive and disturbing; for example, the USSR might, in an effort to match Western capabilities, develop and distribute nuclear shells for their 152 mm gun/howitzers as it is reportedly now doing.

It is, of course, possible that the two sides could agree on asymmetrical ceilings by types, thus reflecting trade-offs among the various kinds of delivery vehicles. Conceivably, these ceilings could be set low enough to cause reductions in some categories of delivery vehicles and to enable less extensive build-ups in others, by the side at a numerical disadvantage. Certainly they could, and probably should, be different for the Northern, Central and Southern regions as well as for types of weapons, thereby giving limited flexibility in redeployment. Since the asymmetries in both types of weapons and in deployments are so great, and the purposes for which these weapons are intended are so different, it might be extraordinarily difficult to reach agreement on ceilings, of whatever level.

An alternative would be to accept the present levels as given and to make reductions from these, either over all or by type, and either symmetrically or asymmetrically. One immediate problem which this poses is that the two sides would have to agree on each other's current levels, and on means to verify that these are in fact correct; if this were not done, either side could claim force levels lower than actually existed, make smaller numerical reductions, and thereby gain an 'advantage'. Fortunately, tactical nuclear delivery vehicles in Europe are (with some minor exceptions) rather difficult to hide, the numbers deployed seem fairly well known, and the intelligence nets of both sides are probably sufficiently good to ensure against gross violations of agreements. Thus, it may be possible to set aside the problem of verification and to turn to other problems connected with reductions in tactical nuclear weapons.

The first of these is that symmetrical reductions will affect unequally the very asymmetrical forces of the two alliances. For example, equal numerical reductions on an over-all basis could further enhance the numerical advantage possessed by NATO; indeed, if carried to its ultimate end, such a measure could leave NATO with hundreds of TNDVs and the WTO with none – an outcome scarcely likely to enhance WTO security. Conversely, percentage reductions on an over-all basis would require NATO to give up at least two delivery vehicles for every one relinquished by WTO,† a ratio which could serve to tilt the present military balance. Moreover, under either approach each side could choose to keep its larger and longer-range delivery vehicles, thereby maintaining its ability to inflict damage – and practically guaranteeing that it would do so in the event of war.

This grim finding suggests either asymmetrical reductions in total weapons, with internal ceilings on some types of TNDVs, or reductions by type. Here too symmetrical reductions could have asymmetrical effects. By way of illustration, equal numerical reductions in nuclear-capable artillery could denude the Soviet forces in Europe without having any perceptible effect on NATO, while equal percentage cuts might mean that NATO would have to give up as many as 6 guns for every piece relinquished by the USSR. While the differences in most categories of weapons are not so great, either equal numerical or equal percentage cuts could significantly offset the present NATO edge of approximately three to two in strike aircraft. Since the USSR places less reliance on aircraft than on missiles for delivery of tactical nuclear warheads, it might benefit from either arrangement but it is questionable whether NATO would accept the first alternative (which would mean virtually stripping Central Europe of dual-capable war planes) and it is certain that it would not accept the second, which would affect adversely its capabilities for both conventional and nuclear wars.

This would also suggest asymmetrical reductions by type, were it not for three things. The first is the difficulty of evaluating the relative military utility (and political importance) of various types of weapons, and hence of determining trade-offs which are in accord with the principle of 'equal security' on which NATO and WTO have agreed. The second is that the very asymmetries mentioned earlier may mean that equal percentage reductions (and consequently unequal numerical cuts in different types of weapons) will largely offset each other, thereby creating automatic trade-offs. The third is that critical inequalities can be mitigated by selecting the areas within which reductions are to take place – a subject to which we shall next turn. ‡

---

† Even if one uses Record's (incomplete) figures for nuclear-capable artillery (in *US Nuclear Weapons in Europe*, pp. 22 and 39), NATO has roughly 2200 TNDVs to 1000 for the WTO.

‡ A fourth reason, which could become relevant if mutual reductions in conventional forces preceded or accompanied reductions in TNDVs, is that reductions in conventional forces may either require reductions in TNDVs (as in the case of dual-capable artillery) or may partially offset their consequences, as cut-backs in fighter-interceptors would in the case of strike aircraft. For further details see Coffey, J. I., Arms control and the military balance in Europe, *Orbis*, Vol. XVII, No. 1 (Spring 1973), pp. 150–151.

The major argument for looking primarily at Central Europe† is that this is the area within which the greatest number of weapons are deployed. It is the area about which the greatest concern is expressed, both in terms of the potential damage resulting from a tactical nuclear war and in terms of the likelihood that some incident, such as, for instance, the seizure of nuclear weapons in time of crisis by splinter groups within the FRG, may increase the likelihood of that war. It is also the area currently receiving the greatest degree of attention, both with respect to proposals for restructuring TNF and in the negotiations on mutual force reductions. Moreover, it is, as suggested previously, the area in which NATO's and the WTO's tactical nuclear forces are (save for artillery) most closely balanced (Table 4.3).

Unfortunately, measures focused on Central Europe would not alleviate concerns among the flank countries, some of whom seemingly feel equally threatened by a concentration of nuclear delivery vehicles in, say south-eastern Europe. While these concerns would logically argue for extending arms control measures to the flanks, both the nuclear balances there and other relevant factors are so different that different measures would certainly be required. Moreover, even an extension of force reductions to the flanks would leave uncovered parts of western Europe and all of the USSR, from which strikes could be launched against countries such as Norway and Turkey.

If one attempts to include these areas, other problems arise. In the case of western Europe, the most significant is probably French opposition to any arms control measures which could impinge on French forces, but even an extension to the United Kingdom might create difficulties, first by involving the United States and United Kingdom aircraft now stationed there and secondly by inhibiting the redeployment to that country of any planes withdrawn from other areas of Europe. As for the USSR, that state has been notably sensitive to measures involving forces within its borders which, moreover, include such diverse TNDV as rockets organic to tank divisions, missiles which are incorporated into army and front artillery, light bombers of the Tactical Air Force, surface vessels and submarines armed with cruise missiles. While the failure to include the USSR would give that country advantages in terms both of ready forces and of reinforcements, it is questionable whether the USSR would in any case reduce by agreement elements inside its frontiers, and certain that it would not do so unless these applied in some way to the United States.‡

So far we have been talking about limitations on nuclear delivery vehicles, and not on the warheads which these vehicles could deliver. Theoretically, cuts in warheads themselves could reduce fears of a nuclear attack and the de-

---

† In this context, 'Central Europe' equates to the 'NATO Guidelines Area (NGA)' for mutual force reductions, i.e. the Benelux countries and the FRG for the West and the GDR, Czechoslovakia and Poland for the East.

‡ This issue illuminates another, which is that reductions in tactical nuclear forces would not, even if they extended to the Soviet Union, get at the Soviet regional nuclear forces which are of such concern to the West, or at the British and French strategic nuclear forces which are seen by some in the East as their counterpart. However, any attempt to include these in an agreement on the reduction of nuclear delivery vehicles would raise a whole new set of problems – as, regrettably, have efforts to deal with them in SALT.

struction from such an attack. Whether they would in fact do this depends on the magnitude of the cuts, on what is done with these weapons, and on what happens to weapons stored outside Europe.

At the moment, the USA reportedly has about 7000 warheads for tactical nuclear delivery vehicles deployed in western Europe, while the USSR may have from none to 4000 stored in eastern Europe.† Obviously, reductions of 10 or 20 per cent in these stockpiles would make little difference; only as cuts reached 40 or 50 per cent would they begin to affect the ability of either side to maintain various types of warheads, to allocate these to different purposes, and to insure that they were readily available to units which might need them in an emergency. Even then, reductions of stockpiles in Europe would not be that meaningful, as each great power could re-introduce warheads on short notice, either in preparation for an attack or in response to a crisis situation. Furthermore, while it would be comparatively easy to verify that weapons were being transferred out of the area, it would be more difficult to check on the stocks remaining (since this would require intrusive inspection) and virtually impossible to preclude weapons from being moved back in again – at least across western frontiers [9].

If, therefore, reductions in nuclear warheads are to be meaningful, these must provide for the destruction of any weapons taken from the stockpiles in Europe. Even this would not markedly affect the military capabilities of the forces in Europe unless it were part of an over-all programme for cut-backs in nuclear warheads. While the partial denuclearization of Europe could be a contribution to that process, it may not be the most feasible or the most desirable point at which to start it.

On balance, therefore, it would seem best to abjure mutual reductions in tactical nuclear warheads – especially since these may turn out not to be mutual. It would also seem best to avoid attempts to control TNDVs through the imposition of successively lower ceilings, either on an over-all basis or by type of weapon, as distinct from freezing forces at existing levels. Even then, such a freeze could not extend to the Soviet Union, and should not apply to western Europe, since any attempt to do so might create more problems than could be solved. Any freeze should probably be applied to northern, central and southern Europe on an over-all basis, rather than by region or by type of weapon, thereby maintaining some degree of flexibility while precluding major force build-ups.

This leaves untouched the issue of possible reductions in tactical nuclear delivery vehicles. As suggested previously, any reductions should be limited

† The commonly accepted figure of 3500 nuclear warheads includes those for Soviet MR/IRBMs, medium bombers and naval aircraft (*Report of the United States Delegation to the Fourteenth Meeting of Members of Parliament from the North Atlantic Assembly Countries, held in Brussels, 11 November through 15 November, 1968*, (28 March 1969), p. 29, cited in Cliffe, *Military Technology and the European Balance*, p. 4), which would reduce the base figure to approximately 2000. However, the information is dated, and the USSR must certainly have added by now to its stockpiles of tactical nuclear warheads. There are, however no indications that it has stored any of these in eastern Europe and some hints that it has not. (See Record and Anderson, *US Nuclear Weapons in Europe*, pp. 37 and 38.)

initially to central Europe, since the tactical nuclear forces deployed there are more or less balanced, and since those left untouched (as would be the TNDVs inside the USSR and in western Europe) are somewhat distant from the zone of operations. Reductions should probably be made by type, in order to minimize the likelihood that each side would scrap its older and shorter-range weapons and keep its newer, longer-range and more lethal ones. And they should probably be made on an equal percentage basis, since this would be comparatively easy to negotiate, fair to both sides, and likely to result in relatively balanced reductions in military capabilities.

It should, however, be noted that the effect on capabilities would be comparatively small even if the percentage reductions were large. This is partly because even a relatively few nuclear weapons could have devastating effects on military operations, and more than a 'few' would remain. Partly it is also because tactical nuclear forces from outside the area could be brought to bear, either directly, by launches from ships and submarines in adjacent waters and from air bases and missile sites in neighbouring countries, or indirectly, by redeploying aircraft and missiles to stations from which these could strike at targets in central Europe. Partly it is because neither side depends wholly on tactical nuclear forces but can bring into play regional and intercontinental strategic nuclear forces, some of which are already targeted on central Europe.

II. Deployment restrictions

Restrictions on the deployment of tactical nuclear forces have purposes more or less similar to those that might inspire limitations on weapons. One of these is to reduce the threat which these might present by withdrawing them to rear areas or by precluding particular countries from installing them on their territories. Another is to make them less usable for some purposes, by increasing the distance between them and potential targets. A third is to make their presence or their use less dangerous, by establishing controls which would preclude inadvertent escalation. And a fourth is to do all this in ways which will not, hopefully, be less reassuring than at present – although admittedly assurances arising out of the automaticity of escalation may decrease, even as those related to beliefs in the unlikelihood of war may increase.

As in the case of limitations on tactical nuclear weapons systems, there is a range of deployment restrictions which could be applied. One could, for example, exclude tactical nuclear forces from whole countries (such as Turkey) or from large areas (such as the Baltic Sea), or one could exclude them only from parts of countries, such as a strip along both sides of the West German–Czechoslovak frontier. One could restrict the deployment of all nuclear-capable

† Record and Anderson estimate that the 684 US strike aircraft in Europe in 1973 – of which about 250 were deployed in central Europe – could be increased within 30 days to 1854 planes. (*US Nuclear Weapons in Europe*, Table 4–2, p. 24.) Similarly, the USSR could more than double the 600-odd nuclear-capable strike aircraft now deployed in central Europe without denuding their armies in the Far East of air support.

weapons, or only of some. One could restrict the deployment of warheads along with delivery vehicles or in lieu of restricting delivery vehicles; that is troops of the GDR could keep their SCUDs and forces of the FRG their Lances – but not the nuclear warheads for these missiles. Restrictions on the deployment of weapons could be accompanied with organizational changes, such as the removal of TNDVs from smaller units and their allocation to a separate support command, which would make less likely their involvement in, or employment during, the first stages of any conflict. Alternatively, or in addition, one could establish control procedures such as electronic locks on warheads, or separate communications links from higher headquarters to those guarding nuclear warheads, which would inhibit their unrestricted use by military commanders.

Deployment restrictions applicable to whole countries or to other large areas could, it is argued, ameliorate threats, relieve tensions, and perhaps create 'zones of peace'. They might also decrease the possibility that nuclear weapons would be directed against neighbouring countries, either in counterforce attacks or in retaliation for such attacks. And they could, as suggested earlier, reduce the dangers and the risks of escalation arising out of either low level decisions or of misperceptions and misunderstandings by political leaders.

Unfortunately, deployment restrictions must be related to the ranges of the weapons which can be brought to bear; the withdrawal from the Kola Peninsula of short-range FROG rockets will scarcely be meaningful to Norway if Scaleboard SRBMs remain behind. The establishment of restricted areas does not rule out their use against the countries of that area, since longer-range weapons can be employed to shatter defences, to block the dispatch of reinforcements, and to persuade a country that its only choice is to surrender. Furthermore, restrictions of this kind may ameliorate threats to one side but will not relieve tensions unless they apply to both. (For example, turning the Baltic into a 'sea of peace' by precluding the transit of nuclear-armed warships may appeal to the members of the WTO, but it will scarcely appeal to Denmark unless restrictions are placed on Soviet destroyers armed with cruise missiles as well as on NATO submarines carrying atomic mines.) Finally, restrictions on the deployment of tactical nuclear weapons to particular countries may reduce the ability of an alliance to respond to threats to the security of one of its members, or to buttress that member against pressures backed by threats of force, and hence may diminish deterrence.

Restrictions applicable to smaller geographical areas or to relatively narrow zones both promise less and risk less. They are likely to affect most significantly very short-range TNDVs, particularly suited for battlefield usage, rather than longer-range ones. Their preclusion can be psychologically reassuring, both because it reduces particular military capabilities and because it lessens the possibility of inadvertent war; at the same time it may leave untouched the longer-range forces of both sides, which some see as linked more directly with deterrence. Conversely, narrow denuclearized zones may affect the ability to conduct a forward defence, because the shorter-range weapons systems most suited for that defence will not be available. An attacker could not only use longer-range systems but could, since he can choose the time and place of attack, quickly reintroduce battlefield delivery vehicles, at selected points. The absence of these on the other side would reduce not only defensive capabilities but the automaticity of escalation, which some see as an important contribution to deterrence.

Such advantages as may accrue to the establishment of narrow denuclearized zones could be enhanced by organizational changes in the opposing forces, such as the establishment of a separate Nuclear Support Command, and the reservation to higher levels of decisions concerning the employment of this Command. One advantage of this is that it would facilitate adjustments in military doctrine to the existence of denuclearized zones, thereby minimizing problems with lower-level commanders. Another is that it would presumably facilitate control both over the authorization to use TNDVs and over the targets against which they might be used, and a third is that it might keep short-range nuclear vehicles out of areas where they would be subject to early capture or destruction, and hence inclined to fire in self-defence or as a way of justifying their existence.

Re-structuring forces in this way, whether or not it is accompanied by the establishment of nuclear-free zones, would mean some loss of military readiness because of the non-availability of TNDVs and the time required to obtain authorization for their use. This may lead the armed forces to increase their requirements for longer-range weapons, thereby re-establishing the threat which reorganization was supposed to remove. And in some cases, as with 155 mm artillery, it would be virtually impossible to remove nuclear-capable elements without drastically reducing the conventional firepower of the remaining forces.

An alternative procedure, which would minimize some of these disadvantages, would be to keep the weapons but not the warheads, that is to set up nuclear-free zones. Such an approach would preserve the conventional capability, where this was important, but would eliminate the possibility of inadvertent escalation through decisions by lower unit commanders. Since nuclear warheads could be reintroduced rapidly (by helicopter if necessary) their peacetime absence should not degrade drastically the deterrent effect of TNDVs nor, unless an attack came without warning, their military capabilities. This is, however, a double-edged advantage, in that the quick re-introduction of nuclear warheads makes their absence less reassuring than would otherwise be the case. Moreover, there are problems in determining that they are not in fact present in any nuclear-free zone. This would require some technical knowledge of the procedures for supplying nuclear warheads, and some intrusive inspection, without advance notice, approval for which might not always be forthcoming. On balance, therefore, the marginal utility of nuclear-free zones may be offset by the difficulties of verifying that they are in fact nuclear-free.

All this suggests that restrictions on the deployment of tactical nuclear weapons will be difficult to establish and in some instances even more difficult to verify. Among the measures which could reduce the risk of inadvertent war and/or diminish perceptions of threat there are several which seem potentially feasible:

1. The delineation of narrow zones along frontiers from which certain types of delivery vehicles, such as rockets or nuclear-capable artillery, would be barred. This measure would delay, if not reduce, the ability to conduct battlefield operations; however, this may be more of a handicap to the attacker than to the defender, who is not likely to be willing (or able) to resort immediately to the use of nuclear weapons†, and to the extent that it did disproportionately diminish conventional capabilities, the balance could be restored by introducing models of weapons which were not nuclear-capable – or by agreeing on reductions in conventional forces which would offset the loss in firepower;

2. The stationing in rear areas of other delivery vehicles, such as Pershing and Scaleboard missiles, so that these could not, without movement forward, strike at targets deep in the territory of the other side. And while such movement would not be precluded, notice could be required – with failure to give such notice significant in itself;

3. The redeployment to rear bases of nuclear-capable aircraft, which might increase response time but would also give additional protection against pre-emptive strikes. Since NATO nuclear-armed aircraft would not be employed initially against conventional assault, this should not adversely affect Western capabilities, and since WTO aircraft play only a limited nuclear role, this should not significantly alter the WTO's defensive posture.

Although it is conceivable that restrictions could be placed on other types of weapons, these are more likely to take the form of trade-offs than of symmetrical arrangements, or to result from negotiations on other issues; thus, restrictions on the deployment of nuclear-armed warships might come as part of any limits placed on naval forces in the Mediterranean. Wide zones from which TNDVs and/or nuclear warheads would be barred would seem so detrimental to the West as to be ruled out – though this judgement might change if the USSR were willing to include stretches of its territory such as the Kola Peninsula, which abuts on northern Norway.

III. Controls on new weapons

The introduction of new nuclear delivery vehicles, of improved warheads, and of up-graded theatre nuclear forces can have several consequences. For one, these can change the military capabilities of one side or the other, as would the introduction by WTO of quicker-reacting SRBMs and longer-range strike aircraft. They can affect the survivability of these forces in ways both favourable and unfavourable to the maintenance of a stable military environment; for instance, the quicker-firing missiles mentioned above would help WTO to

---

† In the case of NATO, warheads are mated with short-range delivery vehicles only for training purposes or in time of crisis. In the case of the WTO, warheads would presumably have to be brought in from the USSR.

weather a NATO first-strike, the longer-range aircraft would help it to launch one. In both these ways, the introduction of new weapons can affect not only the war-fighting capabilities of the two sides but their perceptions of threat and their sense of reassurance. They can induce interactions in research and development and in procurement, as in the case of surface-to-air missiles and ground-hugging strike aircraft. And they can cause each side to be worried about the modernization of the other's forces: NATO lest the nuclear-capable MiG-23 replace the Polish MiG-17s (which cannot carry atomic weapons), and WTO lest the F-4 IIs scheduled to replace the F-104 Gs in the Air Force of the FRG add markedly to that country's ability to wage tactical nuclear war.

This does not mean that all technological innovations are bad. In one sense, low-yield warheads, more accurate delivery systems, and better means of command and control can diminish collateral damage and can reduce the danger that an adversary will escalate because the wrong target was attacked with the wrong size nuclear warhead. It is, however, hard to get one side to accept improvements in another's arsenal, such as smart bombs or mini-nukes, and it may be even harder to persuade an adversary that he should introduce innovations which would similarly enable him to fight a more controlled nuclear war, much less that he should limit his innovations to those which will produce this result. (Moreover, such innovations may not only be costly but detrimental militarily; for example, 'clean' warheads may hinder the establishment of radioactive barrier zones, which could be effective obstacles to attacking units.)

Controls over the introduction of new weapons are hard to establish. Given the tendency to make continuous improvements on newer models of the same weapon, or to upgrade those that are returned to depots for re-furbishing, it is virtually impossible to preclude gradual alterations in weapons systems. It is possible to inhibit the development of new weapons when these reach the test stage, either by placing restrictions on their testing or by limiting the numbers which can be deployed. But it is hard to cover all the kinds of systems, from destroyers to air-to-surface missiles, which play a nuclear role. Once weapons have been introduced into inventory, it is possible to identify new types quite readily, but the multiple roles of some delivery vehicles, such as the Buccaneer and the Yak-28 medium bomber, means that one cannot restrict their deployment or their numbers without effects going far beyond the nuclear capabilities of the two alliances.

It would be hard to obtain an agreement not to install these new types in Europe, which is the main area of confrontation between West and East, and therefore, the theatre in which each tends to deploy its newest and best weapons. Additionally, each side may see deficiencies in its own posture and imbalances between the postures of the two sides which new weapons would do something to rectify. And each may see modernization as an answer to the problems posed by ever-increasing costs; for example the Lance SRBM, which is scheduled to replace the Honest John rockets and Sergeant missiles in the hands of US forces, not only requires smaller crews but is more effective, so that replacements need not be made on a one-for-one basis [10].

One could, of course, slow down the introduction of new weapons by prescribing a time span between systems; e.g., new SRBMs could be introduced only every x years. Alternatively, one could specify a rate of deployment into the European theatre such that it would take, say six years, from the introduction of the first new weapon to the replacement of all the old ones. One could also conceivably agree to introduce new weapons first in rear areas, as by deploying modern fighter-bombers in the United Kingdom rather than in the FRG, thereby reducing the stimulus to counter-action occasioned by its deployment. Such measures could be not only valid in and of themselves but also meaningful accompaniments to reductions in tactical nuclear delivery vehicles, whose effects might otherwise be offset – or more than offset – by qualitative improvements in weaponry.

Given, however, the difficulties of verifying such arrangements, they must of necessity be limited to important, relatively scarce, and highly visible weapons; checks on the yield or degree of radioactivity of nuclear warheads would be virtually impossible. Given, moreover, the incentives to maintain flexibility and the complexities of working out controls on new weapons, it is probably desirable to eschew formal arrangements to this end; instead, NATO and WTO might exchange views about the kinds of improvements in weaponry which each side finds disconcerting, and try to stimulate an awareness of the potential consequences of these shifts, so as to induce unilateral – but reciprocated – restraints on the introduction of new weapons.

Obviously, such measures would not affect significantly the military balance in Europe – which is perhaps one of their virtues. They could, though, affect perceptions of intent and thus enhance the sense of security of all Europeans. Equally importantly, they could contribute to that larger dialogue between East and West which is essential if peace is to be maintained, or even if the uses of tactical nuclear weapons are to be limited.

IV. Constraints on the use of nuclear weapons

The purpose of constraints on the use of nuclear weapons would be to make the best of a situation in which continued reliance on these weapons for deterrence and defence precludes their removal from the theatre or drastic reductions in their numbers, and where their utility in crisis bargaining makes countries unwilling to accept deployment restrictions which would inhibit rapid movement from one area to another. The idea would be to see whether other constraints can serve the general purpose of maintaining stability through deterrence, and of upholding interests in a crisis (or in a conflict) with a lesser degree of risk than may currently be the case.

The kinds of constraints on the use of tactical nuclear weapons (if any) that may be acceptable, depend in part on estimates of the likely threat. If the threat is one of the conventional assault, a country may hope to deter such actions without escalation by using nuclear weapons only on its own territory, while if the threat is that of political pressures against an isolated area such as West Berlin, that same country may not wish to forgo the possibility of employing nuclear weapons outside its frontiers. The acceptability of constraints will depend also on a country's interest in precluding escalation, as against threatening it; in the former instance, it might be willing to accept restrictions on even the battlefield use of nuclear weapons whereas in the latter instance, as already noted, it would want to be able to attack targets deep in rear areas, with large weapons if necessary. These variables make it extremely difficult to analyse the utility of constraints on the use of tactical nuclear weapons, much less to select those which countries might adopt, but the first is an essential prelude to the second.

One such constraint could be an agreement not to be the first to use nuclear weapons. It can be argued that a policy of 'no first use' does not weaken the deterrent to nuclear aggression, since if nuclear weapons are once employed, the policy no longer holds. It can also be argued that a policy of 'no first use' would not necessarily erode deterrence of conventional aggression, partly because some forms of conventional attacks will necessarily be met with conventional responses and partly because a declaratory statement about no first use can never be wholly credible. If all the major nuclear powers should adopt such a policy – as the Chinese have already done and as the USSR has from time to time in the past suggested – nuclear blackmail would be impossible. Moreover, if a conventional attack should take place, a policy of no first use would give the attacker no incentive to pre-empt and would minimize the danger of inadvertent escalation on the part of the defender [11].

There are, however, equally powerful arguments against this policy, especially in a European context. One is that a declaratory ban would not insure against the use of nuclear weapons, since WTO forces are geared to fight a nuclear war and under some circumstance might very well do so [12]. Even if this were not the case, a country facing defeat might use nuclear weapons, fearing that the consequences of defeat in a conventional war would be even more disastrous than nuclear war itself. Furthermore, a policy of no first use would undoubtedly mean some diminution in the deterrent value of nuclear weapons, at least in the eyes of the NATO allies, especially since any increase in conventional forces can hardly be expected. More important, a 'no first use' policy might not actually lead to any reductions in weaponry, since each side would still deem it necessary to maintain a deterrent to the employment of nuclear weapons by the other. For this reason, the policy might not reduce Soviet fears, which centre around the unauthorized use of nuclear weapons and around incidents leading to a situation in which a policy of 'no first use' might be reversed. For all these reasons, it might be preferable to look to policies which would reduce the likelihood of conflict rather than those which would govern whether it were nuclear or conventional, and to examine other and more feasible constraints on tactical nuclear weapons.

One such constraint would be to use nuclear weapons only at high levels of conflict, thereby avoiding inadvertent escalation, making sure of enemy intentions and perhaps avoiding the need to use these weapons at all. This policy, which calls for the establishment of a virtually impassable 'firebreak' between conventional and nuclear war, is designed to minimize the dangers of escalation but it may also reduce the utility of escalation in conveying to an adversary the intent of the defender to resist and his willingness to employ all measures essential to that resistance. Moreover, the concept of the firebreak is based on the assumption that parties who escalate through several levels of conventional conflict will stop short of the use of nuclear weapons, whatever the situation at the time when this seems to be the sole feasible option remaining. It is, additionally, potentially unstable in the sense that if the 'firebreak' is set too high this may open the way to all kinds of military adventures, whereas if it is set too low, it may be subjected to unsupportable pressures, which could result in its being crossed at just the wrong time. The question is not whether one should be prudent in deciding whether to use nuclear weapons and restrained in their employment: that goes without saying. Rather, it is whether it is desirable to say in advance that there are circumstances under which one would not employ those weapons, thereby sacrificing the benefits of uncertainty for the uncertain benefits of making a clear distinction between conventional and nuclear war.

A variant of the firebreak is the 'pause', in which the nuclear response to conventional aggression would be delayed not until the conflict had reached a particular level of intensity, but rather for a given period of time. This time would be used first of all to enable an aggressor to halt inadvertent or unauthorized incursions – if, indeed, assaults on a scale needed to penetrate the defences of either WTO or NATO could be so described. It would give the defender time to communicate and bargain with his adversary, a process which would certainly involve the potential use of nuclear weapons if all other measures failed. It would also allow for consultation, and hopefully for consensus, among those countries involved in any decision concerning the use of nuclear weapons. In short, this concept would call for exploring every possible avenue of dampening a conflict before authorizing the employment of weapons which could have potentially destructive consequences for both sides in the conflict.

All these are advantages which reasonable men may be loath to forgo, which is why the 'pause' is part of NATO doctrine for the defence of Western Europe. There are, unfortunately, also some disadvantages to the concept. One is that the 'pause' may give an adversary time to complete a *coup de main* and to confront the defender with the options of acquiescing, of attempting to eject aggressor forces by conventional means, or of employing tactical nuclear weapons after all. Another is that the 'pause' may be so prolonged – especially if the NATO allies attempt to reach agreement on the number, the timing, and the targeting of any nuclear strikes – that the front may break. Even if this does not happen, the aggressor may advance to positions wherein he is less vulnerable to the battlefield employment of nuclear weapons or which would be useful 'bargaining chips' should he decide to seek a negotiated solution. Another disadvantage is that the 'pause' leaves to the aggressor the option of launching a nuclear first strike which, even against alert forces, may be meaningful. And a final disadvantage is that even a 'pause' may not enable the defender to keep control of the situation, since this depends on who uses what weapons against which targets.

Obviously, the tighter the degree of control over warheads, targets, and procedures for releasing nuclear weapons, the less the likelihood of *some* form of unauthorized use, such as employment in self-defence by military commanders. If these controls extend also to the armed forces of allies who do not themselves possess nuclear weapons, they can also preclude unauthorized use by nationals of those countries, even if *their* political authorities give consent. However, few control systems are tight enough to rule out all possibilities that junior officers may decide on their own to employ nuclear weapons, as would appear from the discussion of the control system employed in the French *force de frappe* [13]. Nor will they preclude misunderstandings and miscalculations by political leaders, whoever they – or he – may be. Thus, controls over the employment of tactical nuclear weapons safeguard against some causes of inadvertent nuclear war – but not against all.

Moreover, at least three reservations must be made about the effectiveness of such controls on the maintenance of a stable military environment in Europe. One is that controls which are too tight and direction which is too centralized may somewhat degrade the deterrent effect of nuclear weapons, in that the uncertainties of an adversary about the decisions of several independent political authorities is always going to be greater than uncertainties about the decision of one. Another is that centralized direction and control may erode to some extent the sense of security of allies not possessing nuclear weapons, who may be unwilling to entrust their national security in time of crisis to a foreigner, however sympathetic he may be and however involved his country may be. A third reservation is that the usefulness of controls in dampening or de-escalating conflicts depends not only on the effectiveness of the controls but on what uses are made of nuclear weapons once the decision to use them has been made.

One approach would be to enhance the ability to fight a nuclear war at a relatively low level and within relatively restricted areas, thus climbing the 'escalation ladder' one or two rungs at a time. This approach would emphasize short-range weapons such as artillery and rockets rather than longer-range ones such as missiles and strike aircraft. It would call for the employment of delivery vehicles based within the area rather than those deployed outside it, so that an adversary would not be forced to extend the conflict if he tried to neutralize these weapons; in this context, 155 mm howitzers are to be preferred to Pershing SRBMs, and Pershings to F-111E fighter-bombers based in the United Kingdom. Small-yield warheads delivered with great accuracy would be preferred to larger-yield ones, even though the latter might be more effective militarily. The aim in all cases would be to show equal determination to check any aggression and to avoid intensifying or extending the conflict unless this were essential to secure the first objective.

This approach also suggests restraint with respect to the selection of targets. Attacks on military targets rather than civilian ones will help to indicate that the nuclear response is aimed at defeating an adversary, not at punishing him, and strikes against advancing units rather than against supporting weapons would make clear a desire not to extent the conflict. So would the use of 'iron bombs' against air bases, missile sites, and other places where enemy TNDVs are deployed, since the use of nuclear weapons against these targets might lead the enemy to believe that a disarming strike was under way and to launch his own missiles. While there are some costs attached to forgoing the use of nuclear weapons in counter-force strikes and in interdictory operations, a large-scale tactical nuclear war in Europe could be so disastrous to both sides that it may be preferable to attempt to settle it by bargaining, rather than by a fight to the finish.

Part of this bargaining process would undoubtedly involve the areas within which nuclear warheads are delivered. As indicated earlier, attacks deep in an enemy's rear are more likely to escalate the conflict than are shallow ones, both because of the kinds of targets likely to be struck and because the purposes of these strikes may be misinterpreted. Possibly the best way of convincing an opponent that one would prefer not to escalate the conflict would be to employ nuclear weapons only on one's own territory. This policy could combine a meaningful war-fighting capability with a comparatively low risk of escalation, especially if reliance were placed on purely defensive weapons, such as atomic demolitions. It could give enemy soldiers an incentive to 'lean forward in their foxholes' rather than to attack, since they would be safe from nuclear strikes until they had crossed the border. It would avoid punishing innocent people, such as the civilians in enemy territory, and it could minimize collateral damage, both by providing an incentive to exercise care in targeting and, hopefully, by enabling the removal of one's own people before nuclear weapons were employed.

Unfortunately, restricting the use of nuclear weapons to one's own territory could create problems. One of these, which is also true of other restrictive measures, is that it would give the opponent the option of launching a first strike counter-force attack – albeit against alert and dispersed tactical nuclear forces. Another is that it would sacrifice the advantages which could accrue from strikes against supporting weapons, such as artillery and missiles, and that it would reduce the effectiveness of interdictory and suppressive attacks, despite the accuracy attainable by utilizing laser guidance and homing devices on bombs and ASMs. A third drawback is the psychological difficulty of detonating nuclear weapons in one's own country (and in all likelihood, against some of one's own people) without striking at targets in enemy territory. The final one is that such a policy may increase fears that nuclear war in Europe could become acceptable, at least to the greater powers, with the result that the strategic deterrent would lose much of its effectiveness in protecting the interests and the territories of Europeans.

One could, of course, also limit the sizes of warheads which would be employed against particular targets; in fact, this would be an essential complement to a policy of using nuclear weapons on one's own territory and a highly desirable one to any emphasis on battlefield operations. One difficulty with such a measure is that it could be very costly, running perhaps to thousands of millions of dollars [14]. Another is that it would, if it were to be effective, require accompanying improvements in launch vehicles and in guidance systems, which would be even more costly, and which could upgrade war-fighting capabilities at all levels – a move which would scarcely fit the common concept of arms control. It would also be almost impossible to verify, even with intrusive inspection, since some nuclear bombs have adjustable yields, and would, like some other measures proposed, require radical readjustments in the operational doctrine and force postures of the members of WTO, who may not be prepared to make such adjustments.

This suggests that some measures to inhibit escalation and to limit damage from a nuclear exchange may have offsetting disadvantages, as would seemingly be the case if nuclear weapons were employed solely on one's own territory. Others, like a policy of 'no first use', may weaken the effectiveness of the nuclear deterrent and still others, like the maintenance of a 'firebreak' or the imposition of constraints on warheads, not only may be difficult to implement but also may run counter to current military doctrine, and hence be unacceptable. This leaves open only comparatively limited and short-term measures, such as delays in initial use and limited employment on the battlefield, as contrasted with extensive attacks against targets in rear areas.

The first appears both feasible and desirable, though it might require tight controls over release procedures and perhaps the establishment of a separate tactical nuclear command. The second also seems feasible, although costly, since it presumably would require some adjustments in both NATO and WTO nuclear postures. Whether it would be acceptable to WTO, whose (Soviet-inspired) doctrine calls for strikes throughout the zone of operations, or to NATO, whose European members see attacks in enemy rear areas as threat-ening strategic nuclear war, and thereby enhancing deterrence, is perhaps questionable. Whether either side would, in time of war, refrain for long from hitting at the adversary's TNDVs, wherever deployed, is even more doubtful.

These rather sombre judgements argue against formal constraints on the uses of tactical nuclear weapons. They do not, however, rule out unilateral efforts to minimize the possibility of inadvertent nuclear war (as does the Soviet practice of keeping warheads inside the USSR and the NATO concept of the 'pause') or to inhibit escalation, as could restricting initial uses of nuclear weapons to battlefield targets – preferably on one's own soil. They positively encourage discussions across alliances of the pros and cons of various measures, with a view to reshaping the doctrines and the procedures of both sides in ways which would enhance the stability of the military environment in Europe.

V. Arms control and tactical nuclear weapons: the first stage

The preceding discussion suggests that the asymmetries in force structures, in military doctrines, and in geographic positions of NATO and WTO will make it hard for them to reach agreement on measures for the control of tactical nuclear weapons. Moreover, these weapons are of such importance to both alliances that neither will readily accept curbs on its nuclear arsenal; in fact, the Western allies have, as previously noted, sought to exclude from consideration at Vienna both atomic warheads and tactical launch vehicles in Europe. This argues for confining any initial measures to those that do not greatly alter the military balance and which do not require extensive revisions in military doctrine. It also argues for postponing those kinds of measures which, whatever their utility in promoting crisis stability or in avoiding inadvertent war, may arouse fears concerning the intentions of other countries, whether allies or adversaries. On the contrary, the aim should be to alleviate fears, whether these derive from estimates of military capabilities or from forecasts of future behaviour. With this in mind, let us see whether we can steer between the Scylla of doing too little and the Charybdis of undertaking too much.

If these criteria just listed are to govern, there are a number of measures for the control of tactical nuclear weapons which, whatever their theoretical utility, should not be considered at this time. One would be the extension of any constraints on tactical nuclear delivery vehicles to the countries at the western tip of Europe or to the USSR, since this would enormously complicate the problem of defining those constraints, would require extensive measures of verification, and would probably be unacceptable. Another would be that of establishing controls over the number of nuclear warheads in Europe, as distinct from employing some of these for bargaining purposes. While controls are theoretically possible, their effects would be markedly asymmetrical, verification would be extraordinarily difficult, and they might be militarily meaningless (or even destabilizing) unless they extended to the homelands of the great powers.

A third measure that might be deferred would be that of establishing large new nuclear-free zones or setting up wide areas from which all tactical nuclear delivery vehicles would be barred†. Such a measure would have far-reaching consequences, and, while it might be applauded by the members of the WTO, would be unsettling to those countries in NATO which have previously rejected this approach. For quite different reasons, which relate primarily to their technical difficulty, controls over the introduction of new weapons should also be set aside. Some restrictions on the use of nuclear weapons, such as a policy of 'no first use', employment only on one's own soil, or restraint until conventional operations have reached high levels, are all so prejudicial to deterrence and so difficult to implement in time of war as to suggest caution in their application.

---

† Some areas in Europe are already nuclear-free, either because the countries concerned have decided not to produce nuclear weapons (as has Sweden) or because their status militates against the acquisition of nuclear weapons, as does that of Austria.

There are, however, certain things which could well be undertaken now. One would be to initiate East–West discussions on the utility of 'pausing' before nuclear weapons are employed and on ways of limiting any initial use to targets on or near the battlefield. (If nothing else, these discussions might test the validity of the NATO concept of 'flexible response', a concept which not only is alien to Soviet doctrine but also is seemingly seen as infeasible [15].) Similar discussions could also be held concerning the effects of introducing new weapons on perceptions of threat and on estimates of intent, on the kinds of reassurances that might ameliorate the consequences of deploying these weapons, and on unilateral practices that might bolster these reassurances, such as phased introduction, beginning in rear areas.

One could also attempt to impose restrictions on the deployment of some types of TNDVs:

1. By delineating narrow zones on the territories of states which belong to the two alliances from which artillery and rockets would be barred, if not in central Europe then in areas such as the Norwegian Finnmark or Thrace, where this could easily be done, where the political consequences would be important and where the military impact would be minimal;

2. By moving to rear bases nuclear-capable aircraft and SRBMs such as Pershing and Scaleboard.

Finally, one might attempt also to negotiate reductions of, say 20 per cent by types, in TNDVs currently deployed in central Europe.

Taken together, these measures could accomplish a number of desirable objectives. For one, they could, by removing from frontier zones artillery and rockets and by strengthening the concept of the 'pause', reduce the danger of inadvertent nuclear war. For another they could, by limiting initial strikes to the battle area, diminish collateral damage and minimize the risk of undesired escalation. They could also, by reducing some TNDVs and redeploying others, indicate peaceful intent, alleviate fears and, hopefully, enhance perceptions of security both East and West.

Since security does not rest solely on assurances of peaceful intent, it might be well to add that these measures would have little effect on the defensive capabilities of either NATO or WTO. True, the removal of dual-capable artillery and rockets from border zones could affect marginally the capacity for immediate strikes against battlefield targets, but neither side is normally prepared to launch such strikes instantaneously, and NATO's policy is not to do so. Moreover, it does not necessarily follow that strikes against battlefield targets have to be made by TNDVs in the battle area; with the development of precision-guided munitions (or 'smart bombs') either aircraft or SRBMs can carry out this mission [16]. If such launch vehicles are otherwise employed or are held in reserve, rockets and artillery pieces could be moved back within range very quickly.

Initial concentration on battlefield targets should not handicap the defence; in fact, some Western analysts have pointed out that such a concentration may be essential if advancing WTO forces are to be halted before they reach the Rhine [17]. It would, of course, leave untouched the bulk of the adversary's

TNDVs, thus forfeiting the advantages supposedly attaching to a pre-emptive counterforce strike; however, given the size, diversity, and degree of dispersion of both WTO and NATO tactical nuclear forces, these advantages may be more apparent than real. In any case, such an option could always be exercised (as could that of interdictory attacks on bridges, marshalling yards and other choke points) should strikes against battlefield targets fail to halt an enemy.

The reason for this is very simple: tactical nuclear capabilities on both sides would remain high, even after a 20 per cent cut in TNDVs in central Europe. As far as battlefield nuclear weapons are concerned, NATO would lose about 20 rockets, about the same number of missiles and from 200 to 300 nuclear-capable artillery pieces (depending on whose estimates one takes), while WTO would give up about 40 rockets, approximately 30 missiles, and perhaps 50 203 mm howitzers, which may or may not be nuclear-capable (Table 4.3). Cuts of this magnitude might require NATO to thin out its nuclear support across the Central Front and force the WTO to be more selective in the employment of battlefield nuclear weapons, but they would not significantly alter the present situation. Each side would have more than enough weapons to wage tactical nuclear war, should it choose to do so, and although the cut-backs in dual-capable artillery might affect NATO's conventional capabilities, the M-109 self-propelled howitzer could be replaced by other models or could be offset by similar cut-backs in WTO medium artillery under mutual force reductions.

As far as weapons suited for interdictory and suppressive strikes are concerned, NATO would have to give up about 25 Pershings and 110 strike aircraft, and the Warsaw Treaty Organization about 80–90 strike aircraft and possibly a few Scaleboard SRBMs. Again, cuts of this magnitude would not affect the nuclear capability of either side, since each would still have hundreds of weapons remaining, each could bring in reinforcements prior to, or during the first stages of, an attack, and each has other nuclear delivery systems available, such as the UK Vulcan bombers and the USSR TU-16s.

Admittedly, cuts in strike aircraft could, to some degree, affect the ability to conduct conventional operations. The effects on NATO could, however, be minimized if US strike aircraft were redeployed to the United Kingdom or to Spain rather than to the United States, a distance considerably greater than that over which Soviet planes would have to withdraw. They could also be offset for both sides (unless agreements to reduce conventional forces precluded this) by replacing indigenous nuclear-capable aircraft with those not able to deliver atomic bombs. They could be eased by phasing reductions over a three-year period, during which time attrition and/or the replacement of older aircraft by (fewer numbers of) more effective newer ones should result in lower inventories.

To sum up, both alliances could significantly reduce their tactical nuclear delivery vehicles in Europe (perhaps by greater percentages than those analysed) without affecting the military balance and without destabilizing consequences; in that sense, the common goal of equal security at lower cost may be attainable. Reductions such as those discussed should also lessen perceptions of threat: in the case of WTO because they would begin the phase-down of weapons systems with which some members are concerned, in the case of NATO partly for this reason, and partly because *any* cuts may help to change estimates of Soviet intentions. Finally, perceptions of security should be maintained, in the case of the Warsaw Treaty Organization because its overall military power would not be significantly affected and in the case of NATO because the link to the strategic deterrent (through nuclear-capable delivery systems) would still remain.

The other measures proposed could also contribute to these outcomes, especially if agreement were reached on the establishment of narrow weapons-free zones and on changes in doctrine which emphasized initial strikes against battlefield targets; indeed, even unilateral measures to this effect could be useful in alleviating perceptions of threat. Moreover, the initiation of talks on these subjects (and on self-restraint in the introduction and deployment of new weapons) could further alter perceptions, and perhaps help improve East–West relations.

This does not mean that all problems will be solved, and all Europeans satisfied. The measures proposed are but partial and incomplete, in that the potential for waging tactical nuclear war in (and over) central Europe remains largely unchanged. Moreover, none of the measures proposed will directly affect the countries of northern and southern Europe – save perhaps for the establishment in those regions of limited weapons-free zones. Nor will these measures offset those nuclear forces in western Europe and in the USSR (or the US) which pose considerably greater military threats than do TNDVs in central Europe. Unfortunately, it is not feasible, either in any foreseeable arms control negotiations or in this analysis, to deal with all possible problems.

Even more unfortunately, almost any measures (including the modest ones just described) may create problems rather than solve them. Some people in western Europe may look askance at *any* cuts in NATO nuclear forces, upon which they rely to offset WTO superiority in conventional troops. Others may see arms control agreements as giving the USSR a voice in NATO decisions, a voice which would be even stronger if these agreements provided for discussions of military doctrine and of concepts for the employment of nuclear weapons. Others may see these arrangements as foreclosing the option of creating a *European* tactical nuclear force – as, indeed, the Russians would like to do. Accordingly, many in the West may resist controls over tactical nuclear weapons, just as many in the East may resist adjustments in conventional forces, even while pressing for the dissolution of the NATO Nuclear Planning Group and for other constraints on NATO nuclear forces. These attitudes are understandable, in that they reflect the common belief that one's own strength enhances security and promotes stability while the power of an adversary is a threat to the peace. They are also detrimental, in that they tend to perpetuate that reliance on armaments which is a significant contributor to tensions between East and West and a major cause of uneasiness in Europe.

As suggested previously, one must begin modestly, with reductions which do not markedly alter military capabilities and with deployment restrictions which can convey reassurances disproportionate to their military implications. Only if such measures do in fact alleviate fears (and other circumstances are favourable) can one expect further progress towards the control of tactical nuclear forces. Given, however, the manifold concerns about those forces, and the potential consequences of their use (or misuse), every effort *should* be made to progress further, that is, to impose constraints on their employment, to remove some or all of them from particularly sensitive areas, to trade off those weapons systems which are perceived as threatening (or to substitute for them others which are less threatening), and to curb the forces and the warheads deployed outside Europe. In this context, one can truthfully say that the proposals advanced in this chapter should be regarded as a beginning, not as an end.

References and Notes to Chapter 4

1. The monograph by Record and Anderson is itself one indication of this, as is a briefer but equally comprehensive piece by Grey, C. S., Deterrence and defence in Europe: revising NATO's theatre nuclear posture, *Journal of the Royal United Services Institute for Defence Studies*, (December 1974), reprinted in *Strategic Review*, Vol. 3, No. 2 (Spring 1975), pp. 58–69. Both of these refer to other articles and monographs, notable among which are those by Lawrence, R. M., On tactical nuclear war, *Revue Militaire Générale*, (1971), Vol. 1, pp. 46–59 and (1971), Vol. 2, pp. 237–261; Cohen, S. J., Tactical nuclear weapons and US military strategy, *Orbis*, Vol. XVI, No. 1, (Spring 1971), pp. 178–193; Heisenberg, W., *The Alliance and Europe Part I: crisis stability in Europe and theatre nuclear weapons*, Adelphi Paper, No. 96 (The International Institute for Strategic Studies, London, (Summer 1973); David, R., Libre opinion: dissuasion et défense nucléaire, *Défense Nationale*, (March 1974), pp. 31–34; and Laurence Martin, Theatre nuclear weapons and Europe, *Survival*, Vol. XVI, No. 6, (November/December 1974), pp. 268–276.
2. These points are made both by Martin, Theatre nuclear weapons and Europe, *Survival*, Vol. XVI, No. 6, (November/December 1974), p. 269 and by Record and Anderson, *US Nuclear Weapons in Europe*, pp. 51–53. See also Lowenstein, J. S. and Moose, R. M., *US Security Issues in Europe: Burden Sharing and Offset, MBFR and Nuclear Weapons, September 1973*, A Staff Report prepared for the use of the Subcommittee on US Security Agreements and Commitments Abroad of the Committee on Foreign Relations, United States Senate (Washington: GPO, 1973). This study will hereafter be cited as *US Security Issues in Europe*.
3. These and other proposals are discussed both by Grey, Deterrence and defence in Europe (see Note 1) and by Record and Anderson, *US Nuclear Weapons in Europe*.
4. The NATO proposal calls for asymmetrical reductions in NATO and WTO ground combat forces, leading to eventual numerical parity in such forces in Central Europe. While it eschews any reductions in nuclear-capable aircraft, or in other tactical nuclear delivery vehicles as such, other than on a 'one-time' and limited basis it would, by removing organized units of division size, result in some cuts in TNDVs, particularly on the part of WTO. *The New York Times*, (16 September 1973), p. 4.

5. For the WTO proposal see Khlestov, O. (head of the USSR delegation in Vienna), Mutual force reductions in Europe, *World Economics and International Relations*, No. 6, 1974, reprinted in *Survival*, Vol. XVI, No. 6, (November/December 1974), especially pp. 294 and 296.

6. Sidorenko, A. A., *The Offensive (A Soviet View)*, translated and published under the auspices of the United States Air Force (Washington: GPO, 1973), especially pp. 40–41 and 58–64. A good summary of Soviet nuclear battlefield tactics will be found in Cliffe, T., *Military Technology and the European Balance*, Adelphi Paper No. 89, (The International Institute for Strategic Studies, London, August 1973), pp. 31–34.

7. Martin, L., Theatre nuclear weapons and Europe, *Survival*, Vol. XVI, No. 6, (November/December 1974), p. 270. This view is supported officially by the United States Secretary of Defense, James R. Schlesinger, in his *First Annual Report to the Congress on the United States Defense Posture*, (3 March 1974, mimeograph), pp. 37–38.

8. According to Record and Anderson, *US Nuclear Weapons in Europe*, warhead yields for US tactical nuclear weapons in Europe extend from approximately 0.1 kt to over 1 Mt, and for Soviet weapons from several kilotons to 500 kt, (pp. 20 and 37, respectively). Indeed, it is estimated that the megatonnage of US nuclear warheads stockpiled in Europe is equivalent to that of the US Minuteman ICBM force, i.e., over 200 Mt: Cohen, S. J., Tactical nuclear weapons and US strategy, *Orbis*, Vol. XVI, No. 6, (Spring 1971), p. 186.

9. Office of National Security Studies, Bendix Aerospace Systems Division, *Verification of a Nuclear Freeze in Europe* (Ann Arbor, Michigan, 1965).

10. *Statement of Secretary of Defense Melvin R. Laird before the House Armed Services Committee on the FY 1973 Defense Budget and the FY 1973–1977 Program*, (17 February 1972, mimeograph), p. 88.

11. For a further discussion of these points see Ullmann, R. H., No first use of nuclear weapons, *Foreign Affairs*, Vol. 50, No. 4, (July 1972), pp. 669–683.

12. Wolfe, T. W., *Soviet Power and Europe, 1945–1970*, (Johns Hopkins Press, Baltimore, 1970), pp. 453–457.

13. *The International Herald Tribune*, 18 June 1973, p. 1. See also *US Security Issues in Europe*, p. 18, which implies that European pilots of Quick Reaction Alert aircraft could, once take-off was authorized, drop nuclear bombs whether or not the US so wished.

14. Record and Anderson, *US Nuclear Weapons in Europe*, pp. 66–67.

15. Wolfe, T. W., *Soviet Power and Europe, 1945–1970*, (Johns Hopkins Press, Baltimore, 1970), pp. 456–458.

16. According to one report, the US Army is seeking funds for the development of a highly accurate Pershing II warhead with a yield of about one kiloton. *The Washington Post*, (24 January 1974), quoted in Record and Anderson, *US Nuclear Weapons in Europe*, p. 52.

17. See Grey, C. S., Deterrence and defense in Europe, (see Note 1), pp. 65–66, and Cohen, S. T., Tactical nuclear weapons and US military strategy, *Orbis*, Vol. XVI, No. 1, (Spring 1971), pp. 185–186.

5. Tactical nuclear weapons in Europe: implications for East–West relations

W. Multan

The question of tactical nuclear weapons in Europe has a variety of extremely important and interesting aspects and can be examined in diverse contexts. The present chapter will be confined to examining the presence of TNWs in Europe and the political implications of this fact for the further development of East–West relations, as well as the prospects for building durable security on this continent. Most people will agree that tactical nuclear weapons have for many years been an extremely significant element in East–West relations, since they influence the perception of the state of endangerment on both sides of the demarcating line between NATO and the Warsaw Treaty Organization.

The particular topicality of this question is underlined by the existence of parity in strategic nuclear forces between the Soviet Union and the United States as well as by the progress of political *détente* and the widespread expectation that *détente* will be extended to the military field.

The past few years have witnessed several interesting analyses of various components of the problem of tactical nuclear weapons stockpiled on both sides of the dividing line. Of great interest were the comparative studies of the NATO and the WTO TNWs, with special emphasis on quantitative and qualitative data, concepts of use, possible effects of use, etc. This chapter presents several conclusions ensuing from these analyses, and, on the basis of these offers some suggestions which might stimulate thinking towards the search for a further improvement in East–West relations, an increase in mutual trust and in the sense of security.

As early as 1949, following a concept of General Omar Bradley, Chairman of the Joint Chiefs of Staff, the Truman Administration started efforts aimed at launching the production of TNWs and bringing them to Europe. This took place in 1953, at that time in the form of aerial bombs and 280 mm artillery projectiles. It was argued that tactical nuclear forces were cheaper than conventional forces. At that time, there were no plans for using TNWs independently. In the overall doctrine of massive retaliation, the role of TNWs was to annihilate the remnants of enemy resistance centres, following a massive strike with nuclear forces. Both the massive retaliation doctrine and the related concept of employing TNWs were to enable NATO states to pursue a definite policy toward socialist countries, without risking realiation on the part of the latter. E. T. Cohen and W. C. Lyons write: 'However, it should be appreciated that these TNWs were developed in the late 1940s and early 1950s, when the USSR had developed little strategic nuclear capability, let alone any tactical nuclear forces. Under such conditions the West could be certain that mass destruction would be limited to the enemy and his territory' [1]. The WTO's response to this doctrine was the development of a defence concept in conditions of the enemy's use of nuclear weapons [2]. Doubtless, the fact that only one side possessed nuclear weapons must have influenced the lines of development of conventional forces by the other side.

Subsequent developments confirmed that the deployment of US tactical nuclear weapons in Europe did not contribute to the diminishing of tension in East–West relations. On the contrary, tension was intensified. Nor did the US move increase the sense of security on this continent. It did, however, initiate a qualitatively new stage in the development of the arms race on a global scale. In the late 1950s and early 1960s Poland submitted a number of proposals aimed at checking the nuclear arms race in Europe. Although these proposal were never to become the subject of negotiations, mainly because of the negative attitude of the FRG government and the US Administration, nevertheless the idea of nuclear-weapon free zones has found application in Antarctica and Latin America and met with interest in Africa, the Middle East and South-East Asia. Finally, the concept of nuclear-weapon free zones in general has been worked out by a team of government experts, under the auspices of the Geneva Disarmament Committee.

The problem of what should be regarded as tactical and what as strategic nuclear weapons [3] has been considered in Chapter 3 of this book. One of the ways of distinguishing is the destructive force of nuclear warheads. An overwhelming majority of experts hold that nuclear warheads should have an explosive force of less than 500 kt, 100 kt, or even less. However, some tactical forces in Western Europe are equipped with warheads of up to 400 kt, or far more than the yield of the warheads of some strategic missiles [4]. Moreover, the two bombs dropped in 1945 on Japan, with an explosive force of 14 and 21 kt, were of strategic importance for that country. At present, more than ten thousand nuclear warheads with an average destructive force of 40 kt each are aimed at Europe [5]. Simulation games held by NATO and the US Defense Department in the late 1950s and early 1960s showed that the use of nuclear weapons in Europe would entail casualties running into millions (see p. 173). York estimates that the explosion of a megaton bomb in Europe would cause the death of one million people [6]. Hence the continuing validity of the assertion that nuclear weapons would not be able to defend anyone in Europe, but would certainly annihilate everybody. Discussions on the use of nuclear weapons for tactical purposes frequently overlook the consequences of an extremely dangerous phenomenon – radioactive pollution. Its effects are of long duration and affect above all the civilian population [7].

At the root of the development of US TNWs in Europe was the necessity to counteract the alleged deficiencies in the numerical strength of forces. According to an overwhelming majority of experts, however, the existence of TNWs steeply increases the demand for reinforcements for combat units. According to US calculations, even in the first phase of a nuclear attack, the losses will run as high as 50 per cent of the initial strength [8]. Thus, the presence of TNWs in Europe presses for an increase in conventional forces rather than prevents it.

As far as the range of delivery vehicles is concerned, it is considered that tactical weapons include those incapable of reaching the territories of major nuclear powers. According to this classification, TNWs comprise missiles and aircraft deployed at a distance from the USSR territory which exceeds their range. This method is applicable also to the US territory. Practically, some types of missiles and aircraft (Pershing missiles, F-4 aircraft) deployed in West Europe and classified among tactical weapons have a sufficient range to reach targets in Soviet territory. Moreover, if the basic problem is considered of the existence of great asymmetries between the NATO and the WTO in such areas as the range, delivery capabilities and accuracy of various types of vehicles, and even differences between vehicles belonging to the same type, the whole question of comparability of tactical nuclear weapons in the possession of the two military alliances becomes exceedingly complex.

In this chapter, all nuclear warheads are treated jointly with delivery vehicles deployed in Europe, without distinction between tactical and strategic weapons.

As is admitted by many western authors, the NATO states did not have in the past, and do not have now, a clear-cut concept of TNW employment [9]. Under the present doctrine of flexible response TNWs are to play, theoretically, a double role. In case of a failure of conventional forces, TNW are to be resorted to. The concept also provides for a pre-emptive tactical nuclear attack 'when the consequences of conventional defeat would be . . . major loss of NATO territory or forces. . .' [10].

The second function of tactical nuclear weapons already assigned by the Kennedy Administration is to provide assurance for the USA's European allies that the United States is ready to use nuclear weapons in defence of the West. The western concept for the employment of TNWs was in the past, and is now, subject to critical analyses by western theoreticians. A considerable part of these analyses basically questions the practicability of tactical nuclear warfare. The very concept gives rise to essential doubts. This chapter will not consider differences in the philosophy of the use of TNW by the NATO and the WTO states.

Many authors argue that, given the nuclear-strategic parity between the Soviet Union and the United States, there is no reason for the existence of the NATO concept of an armed conflict in Europe, in which solely conventional and tactical nuclear forces would be employed. They reason that once an exchange of nuclear strikes has begun – irrespective of the yield of weapons – a global conflict with the employment of strategic weapons would be inevitable [11]. To the Soviet strategists, 'it is beyond doubt that in the present international situation and with the present development level of war technology, any armed conflict is bound to turn into a global nuclear war if *nuclear powers are involved in this conflict*' (Author's italics) [12]. In other words, Soviet strategists regard as unlikely an armed conflict with the participation of the United States and the USSR limited to the European continent, or to predetermined types of weapons. Further, a nuclear conflict in Europe might be the outcome of an earlier clash between the US and USSR in another region of the globe, e.g. in the Middle East, in the North Atlantic, etc. [13]. Even if it were possible, however, to limit a future nuclear conflict to the European continent and to tactical nuclear weapons, the losses would be so great that one could hardly speak of either side's victory.

To sum up this part, it should be stressed once again that the size of the tactical nuclear arsenal in Europe is so big, and the concepts of its use so vague, that any military conflict with the participation of states having access to this arsenal would entail losses incomparable with the assumed objectives of combat actions, and, consequently, such a conflict would be unequivocally unreasonable. On the other hand, miniaturization of TNWs does not offer a way out of this situation, since it poses the danger of an armed conflict's passing from the phase of employment of conventional weapons to the phase of nuclear weapons.

The terrible losses that would ensue from a nuclear conflict constitute the basic problem preoccupying the minds of not only politicians, but also strategists in Europe. This dilemma has also led for some time now more and more people to question the significance of presence of TNWs in Europe as an element of the West's sense of security, based on the military potential of western states. These views provide the basis for suggesting changes in the concept of the use of TNWs by the NATO states. Some of these authors explicitly, and others implicitly, indicate that practically all of these weapons are unnecessary, particularly in view of the parity in nuclear-strategic weapons between the world's two biggest nuclear powers, the United States and the Soviet Union. Tangible progress is urgently needed in imposing some controls on the development of the latter kind of weapon.

On the other hand, there is a constant search for technical and technological solutions as well as concepts for TNW use which would make it possible to attain the intended objectives without disastrous consequences for the whole of the continent. So far this search has not been rewarded with a major success. Moreover, looking back at the dialectic of armaments development following World War II, one can hardly expect a radical solution to this dilemma. For it seems that any technical, technological and conceptual innovations introduced by one side provoke a definite reaction of the other side, thus leading to a diminished sense of security on both sides. A solution should therefore be sought, not in improvements in tactical nuclear weapons and concepts of their use, but in measures gradually narrowing down states' rivalry in this field. For the very presence in Europe of a huge TNW arsenal is undeniably a source of continuing tension and lack of security, and makes improvements in East–West relations more difficult.

Having made a thorough comparative analysis of the quantity and quality of conventional armed forces and armaments in the possession of NATO and the WTO, many western sources admit that there is some kind of equilibrium because each side commands a sufficient conventional potential to prevent the other side from gaining a major advantage in a strike with forces of the same kind [14]. In other words, each side possesses a sufficient minimum of conventional forces to ensure its own security [15]. Following a comparison of the NATO and the Warsaw Treaty conventional forces in Europe, Record concluded: 'a successful conventional defence of Europe seems not only feasible but also well within the grasp of the forces now deployed on the continent' [16]. Under such circumstances, parallel and gradual reductions of the TNW arsenals by both sides would not impair the vital security interests of either side.

The existence of such an equilibrium ensuring a sense of security to both sides may be the starting point for considering, and later negotiating, measures aimed at gradual limitation of the possibilities of the use and possession of nuclear weapons in Europe [17].

Considering the above and taking into account the interest expressed since the mid-1950s by WTO member states in introducing restrictions on tactical nuclear weapons in Europe, and also considering the development of *détente* in East–West relations, there now seems to be a more favourable atmosphere than ever before for efforts aimed at: (*a*) diminishing the danger of the use of TNWs; (*b*) checking their further quantitative and qualitative development; and (*c*) gradual elimination of TNWs from Europe. This should be the order in which to examine and solve principal elements of the complex of measures ultimately aimed at diminishing the danger of a nuclear conflict. On this point there is agreement with L. T. Coldwell, who writes of the importance of negotiating a more stable situation as far as TNWs are concerned [18].

The present state of East–West relations appears to be favourable for starting serious talks to work out an agreement on preventing a nuclear war on the European continent. Such an agreement – the parties to which would be the four nuclear states in Europe – could be an adaptation of relevant provisions of the USSR–US agreement of 30 September 1971 to the demands of a European multilateral agreement. In view of France's standpoint, such an agreement could be negotiated and signed outside the framework of the Vienna talks on armed forces and armaments reductions in central Europe.

Another international act of decisive importance for further progress in checking the quantitative and qualitative development of TNWs and for an increased sense of mutual security would be the signature of a multilateral agreement prohibiting: (*a*) the use of nuclear weapons in Europe; or (*b*) first use of these weapons in Europe. An introductory stage of such a prohibition could be an agreement on non-use of nuclear weapons in Europe against states, in whose territory such weapons are not deployed. Non-nuclear states could undertake to prevent the use of nuclear weapons either from their territory, or by means of installations at their disposal. Moreover, the guarantees of nuclear powers contained in Security Council resolution no. 255 (concerning nuclear weapons) could be modified to include the explicit statement that nuclear states would not use nuclear weapons against states which do not possess such weapons, or from the territory where such weapons are not to be found. This group of agreements, as directly stemming from Principle Two of the decalogue of Principles contained in the Helsinki Final Act (refraining from the threat or use of force) could embrace the whole territory of Europe. It should provide a good basis for negotiating measures impeding the further development of TNWs and permitting gradual elimination of these weapons, initially from central Europe, and later from the whole of the continent.

Following the signature of these agreements, further agreement on freezing the number of nuclear warheads and delivery vehicles in central Europe should not prove impossible to negotiate, both in view of the readiness of the major parties to assume such an obligation and because of the existence of possibilities for verifying the implementation of such an agreement. As far as delivery vehicles are concerned, initially the freeze would embrace vehicles *exclusively* for delivering nuclear warheads. The second stage of the freeze would also encompass dual-capable vehicles. Here again, experience gained during the SALT negotiations might prove instrumental.

Major difficulties would be encountered in limiting the qualitative development of TNWs. In this case it would be necessary to prohibit exchange of both warheads and delivery vehicles for new and better ones. SALT experience could be rather less useful here. However, given the goodwill of the parties concerned, to find a solution it should be possible to this problem as well.

The stage which would follow the freeze would be the *gradual elimination* of tactical nuclear weapons from the European continent. This basic phase of nuclear disarmament could be achieved on the following principles:

1. The advisability should be considered of giving priority to gradual withdrawal of nuclear warheads from central Europe, leaving delivery vehicles at an undiminished (frozen) level. In case of danger, warheads could be brought back relatively easily and quickly. Their removal to more distant regions in the period of normalization would have a directly positive impact on military *détente* in Europe.

2. At a certain stage of talks on the reduction of nuclear warheads, there would begin a reduction in central Europe of vehicles serving *exclusively* to deliver such warheads. Later emphasis would be placed on reduction of dual-capable vehicles.

The territorial scope of agreements on a freeze and reduction of TNWs would initially coincide with the area defined in the Vienna negotiations. If successfully implemented in this area, the agreements could be extended to cover other regions.

The very process of freezing and reducing nuclear stockpiles in Europe and the security of states of that continent could be guaranteed by the nuclear-strategic umbrella of the USSR and the USA. This would reflect the military contribution of these two countries to European security.

As already mentioned, negotiations on these disarmament measures concerning nuclear weapons could be held in the framework of the Vienna talks, or outside it. However, it seems likely that *real chances for conducting such talks would be created only by positive results in the negotiations now in progress in Vienna.*

The basic question in planning and adopting any measures concerning a freeze and reduction in TNWs would be full respect for the principle of undiminished security of the two military-political groupings in Europe as well as individual states, whether they are members of either grouping, or remain outside it. This is obvious, since the principal objective of any freeze and for reduction measures is to ensure greater security to all European states.

Conclusion

Although tactical nuclear weapons were partially included in the Vienna talks on the reductions of armed forces and armaments and associated measures in Central Europe, nuclear weapons constitute such an important element in East–West relations and in all the problems of European security that they should be given special attention.

The picture presented here is far from being an complete one, for it involves a great many complex problems difficult to solve even with the goodwill of both sides. It omits a number of complex questions of a political, military and technical nature. It also overlooks the essential problem of controls. Despite its general character it is contriversial in some points. However, it was not intended to offer a perfect solution, but only to direct the discussion toward a problem which constitutes one of the most important components of East–West relations and of European security in general

This chapter concludes with a quotation from Leonid Brezhnev's speech made in mid-1976 in Berlin: 'Europe has entered an essentially new epoch, radically different from all that which was before. Lack of understanding of this fact would for Europeans be tantamount to meeting a disaster halfway'; and further on: 'It is no easy task to disarm arsenals and, strictly speaking, the nuclear arsenal which Europe is today. It is important, however, that this really be initiated' [19].

A long-range vision of the future of Europe is needed if a positive contribution is to be made to a more stable security on our continent. What is needed is perhaps a radical shift in perceptions and a new appreciation of contemporary realities, especially with regard to the whole existing system of security.

References and Notes to Chapter 5

1. Cohen, S. T. and Lyons, W. C., A comparison of US–allied and Soviet tactical nuclear force capabilities and policies, *Orbis*, Vol. XIX, No. 1, (Spring 1975), p. 72.
2. Miller, M. J., Jr, in *Ordnance Magazine*, (May–June, 1970).
3. Record, J., *US Nuclear Weapons in Europe. Issues and Alternatives*, (The Brookings Institution, Washington, DC, 1974), p. 1.
4. This is illustrated by the fact that Pershing missiles with which the Bundeswehr is equipped, can deliver from 60 to 400 kt warheads, while Minuteman III, Polaris A3 and Poseidon missiles, classified as strategic, deliver from 50 to 200 kt warheads.
5. In addition to the 7200 nuclear warheads stockpiled in Europe, the United States has another several thousand tactical nuclear warheads stored in the territories of non-European states (see chapter 1). These warheads, just like the reserves kept in the United States, can be brought to Europe within a very short period of time.
6. York, H. F., The nuclear 'balance of terror' in Europe, *The Bulletin of the Atomic Scientists*, (May 1976).

7. The fact has been emphasized by, among others, Gen. H. Trettner, who said that the effects of radioactive pollution are almost wholly overlooked by the organizers of NATO war games. See: Trettner, H., Atomare Gefehrsfeldwaffen für Mitteleuropa?, *Revue Militaire Générale*, (February 1971).

8. Record, J., US Tactical Nuclear Weapons in Europe: 7000 Warheads in Search of a Rationale, *Arms Control Today*, Vol. 4, No. 4, (April 1974).

9. One should agree with M. J. Miller, who says that: 'The Soviet Army and the United States Army are both prepared to fight a land nuclear war, but neither is quite sure how to conduct such a war', in *Ordnance Magazine*, (May–June, 1970). Also Martin, L., Theatre nuclear weapons and Europe, *Survival*, No. 6, (November–December 1974); Cohen, S. T. and Lyons, W. C., A comparison of US–allied and Soviet tactical nuclear force capabilities and politics, *Orbis*, Vol. XIX, No. 1, (Spring 1975), pp. 72 ff.; Collins, A. S., Tactical nuclear warfare and NATO: viable strategy or dead end? *NATO's Fifteen Nations*, No. 3, (June–July 1976).

10. The theatre nuclear force posture in Europe: Report to Congress by Secretary of Defense Schlesinger, *Survival*, No. 5, (September–October 1975).

11. The view that every use of tactical nuclear weapons would lead to a global conflict has been expressed by, among others, Polk, J. H., The realities of tactical nuclear warfare, *Orbis*, (Summer 1973).

12. Wojenna Strategia [War Strategy], Sokolowski, W. D. (ed.), (Wydawnictwo MON, Warsaw, 1964), p. 240; Kortunow, W., Istoriczeskaja obrieczennost politiki gonki wooruzenii. *Mieżdunarodnaja Żiźń*, No. 9, (1976), pp. 3–14.

13. Miettinen, J., Time for Europeans to debate the presence of tactical nukes, *The Bulletin of the Atomic Scientists*, No. 5, (May 1976).

14. In mid-1973 wide-scale comparative studies of the NATO and WTO military potentials were concluded. They were conducted by the US Defense Department in close collaboration with intelligence services. The *Washington Post* of 7 June 1973, wrote on this subject: 'The purpose of the study was to measure relative East–West fighting capability more accurately and realistically than had been done before'. Naturally, detailed results of the study have not been made public. The conclusions of the study published by the newspaper unequivocally undermine the traditional – and widely exploited by propaganda – thesis about the WTO's superiority over NATO conventional forces. The newspaper wrote about the study: 'It challenges not only many previous US military estimates but also the official line of NATO's European hierarchy. It suggests, in fact, that to defend against the most likely – though not the most severe – threat posed by the Warsaw Pact ground forces requires less than what we (NATO) have'. The study says that NATO ground forces would be capable of effectively withstanding, without reinforcements, a strike by WTO forces for at least 90 days, i.e. a period long enough to start negotiations and prevent a nuclear conflict. Also as far as air forces are concerned, the study has not found justification for the thesis of WTO's superiority; rather, a reverse thesis would be more likely. See also: Canby, S. L., Damping nuclear counterforce incentives: correcting NATO's inferiority in conventional military strength, *Orbis*, Vol. IX, No. 1, (Spring 1975); Enthoven, A. C. and Smith, W. K., *How Much Is Enough? Shaping the Defense Programme 1961–1969*, (Harper & Row, New York, Evaston and London, 1971); Coffey, J. I., Arms control and the military balance in Europe, *Orbis*, Vol. XVII, No. 1, (Spring 1973). A similar assessment of the balance of power in Central Europe can also be found in: Secretary of Defense James R. Schlesinger, Annual Defense Department Report FY 1975/1974/, pp. 87–9; Statement by US Defense Secretary D. Rumsfeld in *US News and World Report*, (16 March 1976).

15. NATO states and France have more troops in central Europe and spend more on military purposes than WTO states, *Strategic Survey*, (1975), p. 64.

16. Record, J., *US Nuclear Weapons in Europe: Issues and Alternatives*, (The Brookings Institution, Washington, DC, 1974), p. 1.

17. For some time now some western authors writing on various variants of the development of the military situation in Europe have been mentioning the variant of basing defence on conventional forces. See for example, Collins, A. S., Tactical nuclear warfare and NATO: viable strategy or dead end? *NATO's Fifteen Nations*, No. 3, (June–July 1976), pp. 72 ff.

18. This is indicated by, among other things, the ninth round of the Vienna negotiations. In this respect the Final Document issued by *The Conference of European Communists and Workers' Parties* (29–30 June 1976), Berlin, GDR, is worth notice.

19. *Trybuna Ludu*, (30 June 1976).

6. The irrationality of current nuclear doctrines

Frank Barnaby

I. Irrationality of use of tactical nuclear weapons in Europe

Soviet and American views differ on when and how tactical nuclear weapons would be used in a European war. NATO's plans for the defence of Europe, though, are based on an awesome bluff [1]. On the one hand, NATO says that a significant attack by the Warsaw Treaty Organization (WTO), even if non-nuclear, would be countered with nuclear weapons. But this would set in motion a series of events leading to the destruction of most European cities, and the death of most of its people [2].

Most European political leaders (and probably most of its military leaders, also) would not be willing deliberately to initiate such a chain of events. But the public admission of such unwillingness would, it is said, 'undermine the credibility of the deterrent'. The belief is that this would increase the probability of an attack.

Nevertheless, there doubtless are (or will be) some political and military leaders who actually *would* destroy Europe in order to save it. And if such people should have their fingers on the trigger at the crucial moment, they might bring about a nuclear holocaust. In short, NATO's bluff could be called and Europe could be utterly destroyed.

Whether or not a nuclear war will take place in Europe will be determined at least as much by NATO deployments and doctrines as by those of the WTO forces. However, what will happen to western Europe when a nuclear war takes place will be almost entirely determined by what the Soviets actually do with their nuclear weapons (and, of course, vice versa). The larger NATO weapons are mainly intended for use against targets in eastern Europe – and such NATO weapons as may be exploded in western Europe will, hopefully, be targeted so as to limit collateral damage as much as possible.

Take, for example, the Soviet medium-range ballistic missile equipped with a megaton warhead. Six hundred such missiles are deployed. These contain about 10 times as much explosive power as all other tactical types combined.

It is impossible to know just how the USSR plans to use these weapons, but it is relatively easy to say what kind of targets they would be effective against [2]. Because of their great power, their relatively poor accuracy (a few kilometres), their peculiar range (covering all Europe, and not much more, from European USSR) and their vulnerability (they are more vulnerable than more modern types to a pre-emptive attack), they would be most effective in early strikes

213

against governmental and major military command centres, major communications centres, large transportation centres like harbours and railyards, manufacturing facilities, and civil and military airports. These targets are mostly in or near large cities. In fact, a list of them would be, for all practical purposes, simply a list of the large cities of western Europe, and a lot of smaller ones besides.

We can, therefore, be reasonably sure that the majority of these 600 Soviet missiles are at this moment targeted against the centres and suburbs of the cities of western Europe. They are likely to be launched in the initial stage of a nuclear war, perhaps immediately after the 'bluff' is called.

The bombardment of western Europe by just these 600 ballistic missiles (or an important fraction of them) could easily kill virtually the entire urban population by *blast alone*. In addition, if a large portion of the warheads were exploded on or near the ground, then a major fraction of the rural population could also be killed by nuclear fallout – and so could a very large number of people outside Europe [2].

The use of even a fraction of NATO's nuclear weapons would, of course, obliterate eastern Europe. In short, if there were nuclear war in Europe, society as we know it would cease to exist.

Not only is the threat to use current types of tactical nuclear weapons in Europe not credible but neither is the doctrine of deterrence by massive assured destruction, on which NATO and WTO strategies are ultimately based. Assured destruction depends on the existence of strategic nuclear weapons. There is, therefore, a fundamental relationship between tactical nuclear warfare in Europe and the strategic nuclear arsenals of the USA and the USSR.

II. Irrationality of strategic nuclear doctrines

The policy of deterrence by assured destruction is obsolete, immoral and incredible. Some of the serious deficiencies in the policy have been described by Herbert York, who observes that while deterrence may be the best strategy now available to the USA 'it is a terrible strategy, and our highest-priority, long-run objective should be to get rid of it altogether' [3]. The main advantage of this policy is that, in contrast to a counterforce strategy, it requires fewer rather than more nuclear weapons. The stockpile now relied on to provide assured destruction is from 10 to 100 times as murderous and destructive as it need be to satisfy that purpose. 'Therefore', York says, 'our highest priority objective for the immediate future should be to reduce greatly the current level of "overkill" even while we still maintain the strategy of deterrence'. The best that is usually claimed for nuclear deterrence is that it 'works' and that it is 'stable'. The first of these claims is highly speculative and, in any event, unprovable. But York believes that the current nuclear balance is stable and has been for some time.

Moreover, he believes that the present balance is stable in two different ways. First, it possesses what is called 'crisis stability'; that is, in a military crisis, one side cannot add much to its chances of survival by striking first, and so there is no strong inducement to do so. The current nuclear balance is also reasonably stable in the 'arms race' sense; that is, there does not appear to be any way for one side to achieve an overwhelming advantage over the other side by quickly acquiring any feasible quantity of some weapon, and so again there exists no really strong inducement to do so.

In York's opinion, the problem with deterrence is that if for any political or psychological or technical reason deterrence should fail, the physical, biological and social consequences would be completely out of proportion to any reasonable view of the national objectives of the USA or the USSR.

> Some authorities have proposed that we confront these awful possibilities by undertaking huge, complex programs designed to cope directly with a massive nuclear attack. Such programs usually include the installation of a so-called thick system of antiballistic missiles combined with very extensive civil defense and post-attack recovery programs. In detailed examinations, however, the main elements of such proposals have always been judged to be either technically unsound, or economically unfeasible, or socially and politically unacceptable, and so no such programs are currently underway or even being seriously considered.
>
> In brief, for now and the foreseeable future, a nuclear exchange would result in the destruction of the two principles as nations regardless of who strikes first. This is what is usually meant by the phrase 'Mutual Assured Destruction'.
>
> It is most important in any discussion about international affairs or the current military balance to have clearly in mind what the current technical situation means: the survival of the combined populations of the superpowers depends on the good will and the good sense of the separate leaderships of the superpowers. If the Soviet leadership, for whatever reason, or as a result of whatever mistaken information, chose to destroy America as a nation, it is unquestionably capable of doing so in less than half an hour, and there is literally nothing we could now do to prevent it. The only thing we could do is to wreak on them an equally terrible revenge. And, of course, the situation is the same the other way around.
>
> No one can say when deterrence will break down, or even why it will. Indeed, if the leadership of all the nuclear powers always behave in a rational and humane way, it never will. But there are now five nuclear powers and there will be more someday, and if any of them ever makes a technical, political or military nuclear mistake for any reason, real or imagined, then there will be a substantial chance that the whole civilized world could go up in nuclear smoke. This is simply too frightful and too dangerous a way to live indefinitely; we *must* find some better form of international relationship than the current dependency on a strategy of mutual assured destruction [4].

So far as the size of the force currently devoted to mutual assured destruction (the matter of 'overkill') is concerned, York explains that informed opinions about how many weapons are really needed vary over an extremely wide range. For example, shortly after leaving the post of Special Assistant to the President for National Security Affairs, McGeorge Bundy wrote 'In the real word of real political leaders – whether here or in the Soviet Union – a decision that would bring even one hydrogen bomb on one city of one's own country would be recognized in advance as a catastrophic blunder; ten bombs on ten cities would be a disaster beyond history; and a hundred bombs on a hundred cities are unthinkable'. For a very much higher estimate, York quotes calculations made in the early 1960s.

> In order to quantify the question, it was assumed that 'assured destruction' meant guaranteeing the deaths of 25% of the population and the destruction of a majority of its industrial capacity. From that, it was calculated that as many as 400 bombs on target might be needed.
> As an intermediate estimate, we may turn to what the French and British have actually done to produce what they evidently think is a deterrent force. In each case the number of large bombs devoted to that purpose seems to be something less than one hundred [4].

In York's view, Bundy was right: that from one to ten are enough when the course of events is being rationally determined.

> In the case of irrational behavior, there is no way of calculating what it would take. The case of irrational behavior is, therefore, of little interest in connection with the question of how big the deterrent force should be; rather, the matter of irrational behavior only enters into questions about when and how deterrence will fail, and about whether a policy based on deterrence is of any political value at all [4].

Why is it, if one or ten, or maybe a few hundred bombs on target are all that are needed to deter, that the USA possesses more than 20 000? And why so much total explosive power? Similarly, why has the Soviet Union deployed a comparable nuclear force?

These numbers are *not* the result of a careful calculation of the need in some specific strategic or tactical situation. They are rationalizations after the fact.

> One method for doing so is called 'worst case analysis'. In such an analysis, the analyst starts with the assumption that his forces have just been subjected to a massive preemptive attack. He then makes a calculation in which he makes a series of very favorable assumptions about the attacker's equipment, knowledge and behavior, and a similar series of very unfavorable assumptions about his own forces. Such a calculation can result in an arithmetic justification for a very large force indeed, provided that we really believe there is a chance that all the many deviations from the most probable situation will go in one way for them and in the other way for us.
> An additional argument for possessing many more weapons than are needed for deterrence involves a notion called 'Damage Limi-

tation'. The idea is that a part of our force should be reserved for attacking and destroying those enemy weapons that for some reason were not used in his first, preemptive strike. Besides the obvious technical difficulties with such a scheme, it is counterproductive for political reasons. In today's world, the internal politics of each of the two superpowers requires them to maintain strategic forces that are roughly equal in size. That in turn means that if one side builds a large force for 'damage limiting' purposes, the other side will build a roughly equal force which will inevitably be 'damage producing'. Such a chain of events obviously leads from bad to worse. Furthermore, the kind of forces needed for this so-called 'damage limiting' role are technologically identical to those needed for a first strike, and so such a strategy is obviously dangerous for that reason also.

In brief then, even if we accept for the time being the need for a policy of deterrence through mutual assured destruction, the forces now in being are enormously greater than are needed for that purpose. And again, if we recognize that deterrence can fail, and if we admit to ourselves the consequences of such a failure, then we see that greatly reducing the current degree of overkill is both possible and essential [4].

This reduction in the numbers of nuclear weapons should be the first step in comprehensive programme of nuclear disarmament.

Even if the present strategic balance is stable now it probably will not remain so for long, mainly because of qualitative improvements in strategic weapons. Some of these developments may well enhance perception about 'winning' a nuclear war. And this is why they are so destabilizing.

The most dangerous current development in strategic nuclear weapons is the continuous improvement of the accuracy of warhead delivery. This accuracy is normally measured by the Circular Error Probability (CEP) – the radius of a circle, centred on the target, within which 50 per cent of the warheads aimed at the target will fall. The most modern US ICBM warhead, for example, probably has an accuracy of about 350 m at a range of 13 000 km. The Mark 12A warhead, developed for the Minuteman III ICBM, will have an accuracy of 200 m. This will be a first-strike weapon, able to destroy enemy ICBMs in their silos.

III. Inadequacy of any policy based on nuclear weapons

If the use of current types of nuclear weapons in Europe is not a credible strategy, what should be done? One solution which has been suggested [5] is to provide a range of nuclear weapons specifically designed for use in a possible European conflict situation. So far as the destructiveness of these weapons is concerned, 'NATO's political leadership wants no weapon explosion to expose people or their cultural heritage to indiscriminate destruction. This collateral damage can be largely controlled by the choice of target area and target defeat criteria'. But the military commander should have weapons capable of effective use for attaining his military objectives. 'Delivery accuracy range and responsiveness for the delivery system, weapon yield and fusing, target acquisition, damage assessment and battlefield survivability of the system all play an obvious role in assuring effectiveness'.

A future NATO military posture based on a number of considerations is visualized.

(1) The US SIOP [Single Integrated Operational Plan] has some indeterminate residual value in deterring irrational forms of Warsaw Pact aggression or escalation, but this role should not be articulated in declared policy.

(2) No US capability for other than defensive war should be retained on European soil.

(3) Any NATO forces deployed in Europe that are solely deterrent (i.e. punitive or retaliatory) should be provided and controlled by Europeans.

(4) The defensive capabilities of the NATO force should be tailored to the characteristics of each prospective battle area in which these capabilities may be exercised. The force should be constructed in each case around US-supplied and US-controlled nuclear weapons and should be designed to defend NATO at its borders.

(5) The design of the force should accommodate to economic pressure to reduce the cost of acquiring and maintaining forces. In particular, the size of the US manpower commitment in Europe should be reduced. Since the NATO force will depend largely on nuclear weapons to defeat massed attacks by Warsaw Pact armor, our goal should be eventually to restrict the US role to command and control of these weapons. The Europeans should be left to provide those elements of the force that would cope with other than massed armor attacks. This goal seems to be a clear application of the Nixon Doctrine to the defense of Europe. Also, emphasis should be placed on procuring weapons that minimize acquisition and maintenance costs.

(6) The force must be usable and effective without exceeding collateral-damage constraints. We offer two thoughts on the control of collateral damage, defined as unintended destruction which should be minimized or avoided if possible: (a) To make a nuclear-emphasis defense acceptable to our allies, the goals for minimizing expected collateral damage must be set at levels at least equal to and preferably even lower than would be associated with a conventional defense, (b) Concern about collateral damage stems largely from the presence of high-yield weapons (greater than a few kilotons) in our NATO stockpile and from vivid memories of wars as long engagements that ravaged the continent. We contend that a nuclear-defensive war by NATO, fought in proximity to the border with low-yield weapons and discriminating delivery systems, would result in a short conflict. Under such conditions, collateral damage is far less an issue, and the criteria set forth in (a) are reasonable.

(7) We are fated to pursue the brinkmanship game of strategic deterrence (assured destruction) in the NATO theater until NATO adopts this new defensive stance. The problem is far more one of policy and doctrine – and US leadership – than it is one of technology. There are advantageous changes possible in both weaponry and forces, but the first step is a recognition that a new approach is necessary. As

NATO comes to this realization, there is much to trade the Warsaw Pact in MBFR negotiations: our theater-range weapons for theirs, for example. We need not develop expensive new 'bargaining chips'; both sides have many of them already. Because Europe is the issue, this is a problem for the multilateral MBFR talks, not for bilateral SALT II negotiations. In any case, our negotiating position does not become worse by NATO's acquiring an effective defence [5].

At first sight, this solution may seem attractive, particularly from the point of view of military tactics. *But there are fundamental weaknesses in any policy depending upon the use of tactical nuclear weapons.* Few would be confident that this or any other feasible policy would, in practice, prevent armed conflict involving the major powers from escalating to an all-out nuclear war so long as any nuclear weapons, regardless of type, are used. And there is no guarantee, or even likelihood, that the opponent will adopt similar tactics. In particular, the introduction of very low-yield (less than a kiloton) nuclear weapons would blur the present distinction between conventional and nuclear weapons. It is of *paramount importance* that an absolute 'firebreak' should be maintained between nuclear and conventional war.

It is doubtful that nuclear weapons have had the political utility often claimed for them, since their existence has indeed greatly decreased world security and most particularly the security of those countries actually having nuclear weapons.

No nuclear policy based on deterrence is credible to rational persons. Such policies must be seen for what they are – mere rationalizations for the development and deployment of the weapons made available by military technology.

The commonly held view that the very destructiveness of nuclear weapons precludes the outbreak of nuclear war is incorrect. Even if 'rational behaviour' is assumed, nuclear war is unlikely to occur only if it is believed that neither side can win. If one power perceives a chance of winning, then there is a risk that it will decide to strike while it has the advantage. Moreover, in the event of a serious crisis, the side placed at a disadvantage may, if it believes a nuclear war inevitable, attack first in the hope of reducing the damage it thinks it is bound to suffer.

At present, and for the foreseeable future, a general nuclear war could not be 'won' by either side. As a result of timing and reliability considerations, small numbers of the most vulnerable components of the strategic forces – the bombers and the land-based ICBMs – would stand a chance of surviving an initial attack, even if attacking forces were much improved over existing ones. More important, despite the vast amount of money being spent on anti-submarine warfare research, nuclear submarines will almost certainly remain invulnerable to a mass attack. In addition to the strategic nuclear forces disposed by the two sides, there are of course the thousands of land- and sea-based medium- and short-range nuclear weapons.

Because of the massive deployments of nuclear weapons of all types by both sides, the likelihood of either side escaping unscathed following an initial attack is virtually nil. But no one knows whether the present technological situation will continue for decades. There are various developments, some of which are now being pursued by both the USA and the USSR, which tend toward the acquisition of a first-strike capability. These are likely to produce periods of extreme instability and heightened risk of nuclear war.

Some of the catastrophic consequences of such a war were described on 5 September 1974 by Dr Fred C. Iklé, Director of the US Arms Control and Disarmament Agency before the Council of Foreign Relations, Chicago. Iklé explained that new information has become available over the past years as a result of accidents and chance discoveries which cast new light on the effects of nuclear war. He lists six examples.

(1) In 1954, the United States exploded an 'experimental thermo-nuclear device' on a coral reef in the Marshall Islands. It was expected to have the power of about 8 million tons of TNT. But actually it exploded with about double the yield predicted – 15 million tons of TNT. And it produced much more fallout than expected. An area of more than 7000 square miles was seriously contaminated. Radioactive debris showered down on a Japanese fishing boat 40 miles from outside the pre-announced test area. About 100 miles downwind from the explosion, Rongelap atoll unexpectedly received serious fallout, so that inhabitants there had to be evacuated. One section of the atoll received about 6 times the lethal dose. And the US Government promptly issued a notice expanding the danger area to about 400 000 square miles or roughly eight times the area previously designated as the danger zone. This experience furnished a dramatic lesson in the difficulty of predicting fallout.

(2) The same thermonuclear test unexpectedly drove home to us some of the human meaning of fallout, largely an abstraction to most of the world at the time.

Soon after the explosion, a sandy ash showered down on crew members of the Japanese fishing boat I mentioned, settled in their hair, and on their skin. The crew, having no idea about the nature of this strange substance from the sky, kept working. But before long, the awful symptoms of radiation sickness began to be felt.

At Rongelap atoll it was two days before people on the island were evacuated. By that time they had received about one fourth the lethal dose of radiation. Fortunately, they had not been at the northern end of the island, where the fallout would have brought quick death. But children were later found to have serious permanent thyroid injury, which would retard their growth. Just recently, a young man who was exposed in that test while still in his mother's womb, underwent surgery at Cleveland Metropolitan General Hospital. Growths were removed from his thyroid gland.

This brought to 28 the number of residents of Rongelap who have had such surgery.

(3) The third unexpected discovery made us aware how nuclear explosions can bring about massive disruptions to worldwide communications. This type of disruption could have seriously impaired the ability of governments and military commanders to receive attack warning and maintain control. In 1958, the United States exploded two nuclear devices high above Johnson Island in the Pacific. High frequency radio communications which crossed the sky 600 miles from the detonation point were unexpectedly lost. Some interruptions

lasted minutes; others many hours. The disruption resulted from complex interactions among effects produced by the explosion: the shock wave's disruption of the ionosphere which normally reflects radio signals back to earth, radiations from debris, and ionization of the atmosphere. The reasons for the unexpected disruption were explained – but only well after the event.

(4) The fourth chance discovery made our experts focus on the distant damage to electronic equipment and computers that nuclear detonations can cause. Given that our engineers, happily, had never seen a nuclear war, they were used to worrying primarily about heat and blast damage, familiar to them from Hiroshima and Nagasaki and from subsequent weapons tests. But meanwhile, the British had discovered that the electromagnetic pulse produced by nuclear explosion could destroy critical command and control links and computer memories beyond the range of blast damage. The British, having a smaller test program than our own, assumed we must be aware of this vulnerability. We weren't. Only through coincidence was knowledge of this effect relayed to our own experts.

(5) The fifth discovery alters our assessment of the vulnerability of missile forces that are protected in underground silos such as our Minuteman. As you know, there is continuing concern that our Minuteman missile force might become vulnerable to a sudden attack, hence lose its deterrent value. For years, simplistic calculations have been used – the kind of calculations that a teacher can put on half a blackboard – to show that accurately aimed multiple warheads, so-called MIRVs, would inevitably increase this vulnerability. Then, the complexity of the real world was rediscovered. It was found that through a phenomenon dubbed 'fratricide' some of these warheads might destroy or divert each other before they could destroy the intended target. In this case, the discovery suggests something reassuring: our simple calculations may have exaggerated the vulnerability of our missiles.

(6) The sixth and last example concerns a new uncertainty about what nuclear war might do to people and to the very environment on which life depends – an uncertainty that has gone unnoticed for 25 years. This is the possibility that a large number of nuclear explosions might bring about the destruction, or partial destruction, of the ozone layer in the stratosphere that helps protect all living things from ultraviolet radiation.

We do know that nuclear explosions in the earth's atmosphere would generate vast quantities of nitrogen oxides and other pollutants which might deplete the ozone that surrounds the earth. But we do not know how much ozone depletion would occur from a large number of nuclear explosions – it might be imperceptible, but it also might be almost total. We do not know how long such depletion would last – less than one year, or over ten years. And above all, we do not know what this depletion would do to plants, animals, and people. Perhaps it would merely increase the hazard of sunburn. Or perhaps it would destroy critical links of the intricate food chain of plants and animals, and thus shatter the ecological structure that permits man to remain alive on this planet. All we know is that we do not know.

IV. Conclusions

The current debate on nuclear policies underlines the inherent dilemma produced by the existence of large nuclear arsenals (tactical or strategic). All nuclear doctrines must have severe shortcomings, mainly because they cannot reduce the probability of the use of the weapons to an acceptably low level.

It has now been officially stated that strategic nuclear deterrence has, in practice, been so crude that the only option has been a massive all-out strike – in the words of former US Secretary of Defense James Schlesinger 'massive pre-planned strikes in which one would be dumping literally thousands of weapons' on the enemy. (The highly sophisticated (but pseudo-scientific) deterrence theories, involving bargaining, limited retaliation and the like, worked out with enormous intellectual effort by strategic analysts have, therefore, had absolutely no effect on official nuclear policies). But bad though this situation is, the consequences of changing to a more flexible strategy, are even worse.

The only sensible policy for the nuclear-weapons powers is to reduce their nuclear arsenals as a step to the total abolition of nuclear weapons.

References to Chapter 6

1. *New Scientist*, Vol. 70, No. 1003, p. 516.
2. York, H., *Ambio*, Vol. 4, No. 5–7, (1975).
3. A discussion of nuclear policies is given in: *World Armaments and Disarmament, SIPRI Yearbook 1974* (Almqvist & Wiksell, Stockholm, 1974), Stockholm International Peace Research Institute) and *World Armaments and Disarmament, SIPRI Yearbook 1975* (Almqvist & Wiksell, Stockholm, 1975), Stockholm International Peace Research Institute).
4. York, H., *Deterrence by Means of Mass Destruction*, Pacem in Terris Conference III, (Washington, October 1973).
5. Bennett, W. S., Sandoval, R. R. and Shreffler, R. G., A credible nuclear emphasis defense for NATO, *Orbis*, Vol. XVII, No. 2, (Summer 1973).

7. 'Mini-nukes' and enhanced radiation weapons

J. Miettinen

I. Introduction

The first use of nuclear weapons in Japan in 1945 prompted Japan's capitulation because the weapon was new, its effects were awesome and poorly understood so that they caused a psychological shock and Japan had neither defence nor retaliatory capacity against it.

For a moment it seemed that the 'ultimate weapon' had been invented – a weapon which would give to its owner the world hegemony. However, this belief was soon shaken. In 1949 the Soviet Union exploded its first nuclear bomb and in 1953 its first hydrogen bomb, only nine months after the first similar test by the USA. At this time these two powers were already in top gear in the nuclear arms race – in the race which still continues today.

In the 1950s they both also brought the so-called tactical (or battlefield or theatre) nuclear weapons to Europe – the USA about 1953, the USSR about 1957. The numbers and types of these weapons steadily increased until the end of the 1960s when political *détente* began to emerge between these two powers and in Europe. This political process culminated in the Final Act of the Conference on European Security and Co-operation (CESC) signed in Helsinki on 1 August 1975 [1].

To be able to continue, political *détente* requires progress in military *détente*. Therefore, when agreement was reached on starting the CESC it was agreed that parallel negotiations would take place on mutual reduction of forces and armaments in Europe [2]. These negotiations began in Vienna on 31 October 1973, and are called the 'Vienna negotiations'.†

These negotiations have now proceeded for three years without visible results. There has been some progress, however. The negotiators now have better understanding of the other side's capabilities and goals. Furthermore, unity and organization within each alliance may have improved, and this also can have a certain stabilizing effect. During the process the negotiations have become less technical and more political, which also is good. Apparently, there now exists great technical readiness to achieve some results as soon as the political atmosphere is favourable.

---

† The official title is negotiations on 'Mutual Reductions on Forces and Armaments and Associated Measures in Central Europe' (MURFAAMCE).

Tactical nuclear weapons have played only a minor role in the Vienna negotiations, so far. The Warsaw Treaty Organization (WTO) has proposed from the beginning that similar reductions – totalling *c*. 16 per cent – would be applied to nuclear units and launchers as to other troops and armaments [3]. NATO first wanted to limit the discussions to ground forces, but on 16 December 1975 it proposed to withdraw in the first phase 29 000 US troops and 1000 nuclear warheads together with a number of launchers including 54 F-4 Phantom fighter bombers and 36 Pershing missiles against 68 000 Soviet troops and 1700 Soviet tanks [4]. In this way tactical nuclear weapons were first included in the negotiations. This package proposal was not accepted as such by the WTO, which made a counter proposal on 19 February 1976 [5]. In it the WTO proposed in the first phase 2–3 per cent reductions (counted as totals for each alliance) of US and Soviet troops, plus 200–300 tanks, 54 nuclear-capable interceptors and some numbers (e.g. 36) of tactical missiles, with the same numbers for the USA and the USSR. A number of nuclear air defence missiles could also be included. This was not approved by NATO [6]. So far the focus of these negotiations has been on general purpose ground forces and tanks rather than on nuclear forces. One of the reasons has been the continuous modernization of the US nuclear forces in Europe.

This chapter will describe briefly the present arsenals of tactical nuclear weapons in Europe, their command, planned use (doctrines) and modernization trends and evaluate their significance to the military balance in Europe.

II. Categories of nuclear weapons

The tactical or battlefield or theatre nuclear weapons

For these, the following numbers are usually given: for NATO 7000 and for the Warsaw Treaty Organization 3500 tactical nuclear weapons (TNWs). These numbers cannot be accurate because of continuous deployment, but they do symbolize the capabilities. The battlefield category mainly consists of:

(*a*) nuclear artillery,
(*b*) short-range battlefield missiles, and
(*c*) aerial bombs and air-to-ground missiles,

which all can be used in offensive and defensive roles. The terms 'mini-nukes' and 'enhanced radiation weapons' mentioned in the title refer to yields and ratios of weapons effects, not to the delivery system (see pp. 227–229). They may be developed for all these three categories.

In addition, there exist nuclear mines and nuclear air-defence missiles which are mostly intended for defensive roles. Reduction of all these weapons is being negotiated in Vienna.

The medium range weapons 'strategic' to Europe

These include on the western side the French and UK SLBMs and fighter bombers, the French MRBMs, the US Forward Based Systems (FBS) and the US SLBM submarines submitted to a NATO role; on the eastern side they include the Soviet I/MRBMs and light and medium bombers. The numbers of warheads are in thousands. These weapons are not treated in any arms control negotiations.

The long range nuclear weapons

These weapons are mainly intended for a *global strategic role*. They include the ICBMs, SLBMs and strategic bombers, some long range bombers, the strategic cruise missiles under development and possibly, a small number of naval cruise missiles which can be used against coastal targets in some conditions. Any of these weapons can in some circumstances be used for strategic or interdiction missions in the European theatre although they are primarily intended for maintaining the central balance. They are discussed in the SALT negotiations.

III. The US theatre nuclear weapons

Present arsenals

The USA modernizes its nuclear weapons on all levels. The goals were well described by the US Secretary of Defense James R. Schlesinger in his Defense Budgets FY 1975 and FY 1976 and 197T. They include on all levels: higher accuracy, smaller collateral damage and therefore smaller yields for point targets, improved security, better mobility for mobile launchers, decreased vulnerability, faster and safer command, control and communication, greater flexibility and better war-fighting capability. According to these trends the number of US tactical nuclear weapons decreased from over 10 000 (average yield 20 kt) in 1970 to 7000 (average yield 4 kt) in August 1974. Reduction of another one thousand warheads including 54 Phantom F4 fighter bombers and 36 Pershing missiles (corresponding to *c*. 270 warheads) was proposed as part of a package in Vienna on 16 December 1975, but this package was not accepted by the WTO. The USA is, however, likely to continue the reduction unilaterally (or at least adopt a policy of replacement by smaller yield varieties). Weapons likely to be removed include:

    1. the F4 Phantoms, which will be replaced during 1976 and 1977 by the F15s, which are all devoted to non-nuclear roles;

    2. some of the Pershings, as proposed in Vienna;

Table 7.1. US and NATO tactical nuclear weapons in Europe [7]

Type of tactical nuclear weapon	Number of delivery vehicles	Number of warheads or bombs	Maximum range
Atomic Demolition Munitions (ADM), atomic mines	not applicable	300	not applicable
Artillery, M-109 155 mm M-110 203 mm	1 010	3 030	10–18 miles
Surface-to-surface missiles Honest John and Sergeant (being replaced by Lance);	175	175	25–85 miles
Pershing	108	324[a]	variable, 60–450 miles
Tactical aircraft[b] US aircraft, F-4 and F-111	316	1 020	500–1 000 miles combat radius
Other NATO aircraft	612	1 224	240–600 miles combat radius
Surface-to-air missiles (SAM)	720	720	60 miles
Land-Based Antisubmarine Warfare (ASW) aircraft	380	380	1 800 miles combat radius
Totals	**3 621**	**7 173**	

[a] The US maintains approximately 216 of these in reserve.
[b] These do not include 98 nuclear-capable aircraft on the two US aircraft carriers regularly stationed in the Mediterranean Sea.

3. the nuclear mines or 'atomic demolition munitions' (ADMs) (about 300 in Europe), the prechambering of which has encountered political obstacles in the FRG; some of them will be replaced by soil-penetrating missile warheads (Pershing and airborne) under development;

4. some of the about 700 nuclear anti-aircraft missiles; with the present hit probabilities of air defence missiles, use of nuclear warheads for such a purpose would be militarily unnecessary and politically infeasible; and

5. most of the heaviest aerial bombs.

The USA is evidently aiming to have at the end of the 1970s a few thousand rather low yield tactical nuclear weapons most of which probably are planned to be used in defensive mode in the territory of friendly allies. The launchers used will be mainly aircraft and artillery with a small number of tactical missiles. For these weapons low yields of thermal energy, overpressure and delayed radiation (fallout) are imperative and it is here that the 'mini-nukes' and enhanced radiation weapons may play a very important role.

'Mini-nukes'

'Mini-nuke' is a catchword the use of which is presently banned by the US Administration because of the political repercussions it caused in western Europe in 1973–4. It has been used of very small fission weapons. In 1973 the NATO nuclear planning group defined mini-nukes as 'fission weapons having a yield equivalent of 50 tons of TNT and the accuracy (CEP = 50 per cent hit probability) of 1 m' [8]. In other sources yields of from 10 to 100 tons are mentioned. Since the smallest yields of the tactical nuclear weapons presently deployed are about 0.5 kt (except for the smallest ADMs) it is a matter of taste whether new varieties having yields between 0.05 and 0.5 kt are called mini-nukes or not. It has been proposed that mini-nukes could be used early in the battle without a risk of escalation.

It has been stressed that there is nothing new in the radiation effects of these weapons, the only difference being that a 100 t mini-nuke has a 10 times smaller yield than a 1 kt 'proper' tactical nuclear shell. The percentage distribution of energy is the same in both cases (shock 50, thermal 35, initial radiation

Table 7.2. Radius of non-radiation effects of small nuclear weapons (radius in metres[a])

1 kt = 155 mm howitzer shell		
0.1 kt = typical 'mini-nuke'		D

Airburst, optimal height

Shock	1 kt	0.1 kt
25 lb/in^2	250 m	110 m
5 lb/in^2	700 m	320 m
Thermal radiation		
2° burns	800 m	250 m
1° burns	1 130 m	360 m

[a] By linear approximation from Glasstone's data [9].

Table 7.3. Radius of radiation effects of small nuclear weapons (radius in metres[a])

Airburst, optimal height

Initial γ-radiation	1 kt	0.1 kt
1 000 rem	630	320
100 rem	1 050	630
10 rem	1 550	1 050
Neutrons (fast)		
1 000 rem	640	390
100 rem	930	640
10 rem	1 260	930

[a] By linear approximation from Glasstone's data [9].

5 and delayed radiation [part of it eventually forming fallout] 10 per cent). In principle this is true, in practice not quite. For a 1 kt weapon the radius of thermal effect (2° burns, 800 m) is wider than that of a lethal dose of neutrons (1000 rem, 640 m) while for an 0.1-kt mini-nuke the thermal effect radius (2° burns, 250 m) is *smaller* than that of neutrons (1000 rem, 390 m), respectively (Tables 7.2 and 7.3). Thus, the smaller the yield the higher the percentage of soldiers killed by initial radiation and also of *those not killed but submitted to a high dose of initial radiation.* The consequences of this will be discussed later.

The enhanced radiation weapons (ERW)

The biological effect of the enhanced radiation weapons is primarily due to fast (14 MeV) neutrons. Since no pure fusion weapon has been invented yet, there must be a fission initiator, too, in the enhanced radiation weapons, but evidently it can be relatively small. Precisely how small it can be has not been revealed so far. The smallest critical masses of ^{235}U and ^{239}Pu likely to be achievable today are of the order of a few kilograms or 1 kg, respectively. If a large amount of ^6LiD is added to the weapon to undergo fusion, it must be

Table 7.4. New US radiation casualty criteria recommended for use in the targeting process [10]

(1) **Immediate permanent incapacitation (IP)**
   A: 17 000–19 000, mean 18 000 rads. Personnel will become incapacitated within five minutes and will remain so until death which will occur within one day.
   B: 7 000–9 000, mean 8 000 rads. Personnel become incapacitated within five minutes and for physically demanding tasks remain so until death which occurs in one to two days.

(2) **Immediate transient incapacitation (IT)**
   2 500–3 500, mean 3 000 rads. Personnel will become incapacitated within five minutes of exposure and will remain so for from 30 to 45 minutes. They will then recover but will be functionally impaired until death, which will occur in four to six days.

(3) **Latent lethality (LL)**
   650-rad band (800–500 rads). Personnel will become functionally impaired within two hours of exposure. They may respond to medical treatment and survive this dose; however, the majority will remain functionally impaired until death in several weeks.

The corresponding radii of damage R_D of a 1 kt low airburst are:

Criterion	Dose, rads	R_D, metres†
(1) IP, A (physically undemanding)	18 000	400
IP, B (physically demanding)	8 000	500
(2) IT	3 000	640
(3) LL	650	760

† There is some discrepancy between these values and those in Table 7.3.

difficult to avoid most of the fissionable material undergoing fission in the high neutron density environment produced by the fusion. The fission yield, therefore, would be of the order of tens of kilotons. The total yield, for the weapon to be *relatively* very clean, must be quite high, therefore. However, it may be possible to have a low-yield enhanced radiation weapon, if one can separate the fission and fusion materials by some distance, arrange the geometry so that the supercriticality is achieved only for a short moment when a small burst of heat and neutrons (e.g. *c*. 0.2 kt) is released, and reflect these unidirectionally to the fusion mass simultaneously changing the supercriticality to subcriticality; then it may be possible to have a small enhanced radiation weapon – e.g. 0.2 kt fission and a few kilotons fusion. But is it necessary to have small enhanced radiation weapons? These weapons are intended to kill personnel within tanks attacking in a formation, and since the distances between individual tanks are likely to be of the order of from 100 to 200 m, it is probably advantageous to have a rather wide coverage, say 1–2 km diameter, for the total coverage of an assaulting armoured spearhead. A 1-kt fission weapon already has an effect radius of 500 m for immediate permanent incapacitation (IP), 8000 rads, for unprotected personnel [10] (Table 7.4). One kiloton fission weapon has approximately the same lethality against troops in the open as seven artillery battalions (five 155 mm and two 8 inch battalions) firing improved conventional munitions in a single volley. Against troops in tanks the radiation dose would be reduced by 80 per cent but yet it would have from 20 to 30 times wider lethal coverage than a single artillery volley of the size mentioned. This is why there is demand for enhanced radiation weapons, in which about 80 per cent of energy is released as fast neutrons (*c*. 14 MeV). Such neutrons penetrate steel easily and have a high biological effectiveness (about ten times higher than gamma-rays). Therefore, by using such weapons the radius of immediate incapacitation of personnel in tanks could be increased to some 800–1000 m, but relatively high fusion yields would be needed (this is about the limit at which the neutrons are absorbed in the atmosphere whatever the yield). Therefore, we can assume that in enhanced radiation weapons, the fission yield is probably in the low kilotons, the fusion yield in tens of kilotons.† Thus these are no mini-nukes.

Nuclear artillery

Nuclear artillery seems to play the central role in the US modernization of tactical nuclear weapons.

(1) *155-mm howitzers*. Presently NATO has 684 dual-capable 155-mm towed howitzers M-109 SP which have a range of 15 km, and for them a total of 2000 nuclear shells. Of these 360 are US tubes, 324 NATO tubes. The shells all have a yield of 2 kt [11]. These shells contain plutonium-239 as the fissile

---

† Articles published in the USA in summer 1977, during the neutron bomb press debate, mention a yield of 1 kt (e.g. *Newsweek* (4 July 1977), p. 29).

material. The CEP of the present guns is 40, 100 and 172 m at short, medium and long ranges respectively.

(2) *8-inch howitzers.* The US and NATO forces also have altogether 326 8 inch self-propelled howitzers M-110, which are corps level artillery. Their maximum range is about 17 km, CEP the same as above, and the yield presumably about 1 kt. There are 1000 nuclear shells available in Europe for these guns. They contain uranium-235 in highly enriched form (oralloy = 93.9 per cent ^{235}U) as the fissile material. The US Army also has self-propelled 175 mm guns (maximum range of 32 km) for which it has no nuclear shells and an unknown number of towed M-115 8 inch howitzers with dual capability [11].

(3) *Modernization.* There are now different programmes for modernization of this artillery. In his FY 1975 Defense Budget Secretary Schlesinger applied funds

> to replace with a new, improved tube all of the existing tubes of our self-propelled 8 in howitzers and 175 mm guns, both of which are mounted on the same chassis . . . The new 8 in tube when firing the new rocket-assisted projectile now completing development would have the same accuracy as the current 8 in howitzer has at 17 km but with considerably increased range. Moreover, the new tube will have a longer life and also will be able to fire the current 8 in conventional round as well as the 8 in nuclear round. With these improvements the US heavy artillery will be a better match for current Soviet heavy artillery, particularly in counter-battery operations. Furthermore, the replacement of the 175 mm gun tubes with the new 8 in howitzer tubes will greatly simplify the ammunition logistics problem [12].

Thus, even the current 175-mm guns will be able, after the tube change, to fire the present conventional and nuclear 8 in shells, as well as the new 8 in nuclear shells (W-75) being programmed for the years 1978–80. They will cost $400 000 apiece and the total number to be ordered will be above 1000 because the present 175 mm cannons will also be able to fire these shells after the tube change. The new 8 in nuclear shells will contain plutonium instead of oralloy. Since the oralloy is valued at $1000 million for 1000 shells, there will be a great net saving which itself will prompt the modernization.

All existing towed 155 mm howitzers will also be replaced by new towed extended range 155 mm (XM 198) howitzers that are just completing development. These will be able to fire not only the current 155 mm conventional and nuclear rounds but also the new 155 mm rocket-assisted projectiles with which they will have a maximum range of 18 miles (29 km) instead of the present 15 km. The new precision-guided, rocket-assisted shells for the 155 mm and 8 inch howitzers are the so-called Cannon-Launched Guided Projectiles (CLGP). These are intended for hard point targets, like tanks. 'The CLGP is a projectile equipped with a semi-active laser guidance system. It is fired from an artillery tube, using conventional fire direction techniques. As the projectile arrives over the target area, either a ground or an air observer equipped with a laser designator illuminates the target with a thin laser beam. The CLGP, with its semi-active laser guidance system, homes in on the laser energy reflected from the target, thus giving it a very high kill capability even against moving hard targets such as armoured vehicles'. For tanks, conventional explosives will be enough, of course. The Pentagon also ordered in 1972 2000 new nuclear shells (W-74) for these howitzers to replace the old ones mentioned above. It is uncertain when this deployment will take place. The new 155 mm shells will have two yields, one of which is 'substantially higher than the other and higher than the present yield'. Since the present yield is probably 2 kt, the new must be several kilotons.

The artillery shells are probably the prime candidate for the delivery system of mini-nukes and enhanced radiation weapons. Artillery has good target acquisition capability, rapid responsiveness, it is easy to control by front commanders, its range and accuracy are sufficient for the tasks envisaged for these new munitions and the guns are available in large numbers everywhere (and indistinguishable from non-nuclear ones). Because of these qualifications nuclear artillery is considered to be the most suitable delivery system for targets close to friendly troops, i.e. the ones for which the mini-nukes and enhanced radiation weapons have been developed.

Of other US plans for miniaturization of its tactical nuclear arsenal, the precision-guided missile warhead of 1 kt yield for the Pershing missile is perhaps the most important one.

IV. Soviet theatre nuclear weapons

An estimate of the Soviet arsenal is presented in Table 7.5 [14]. It can be seen that the Soviet Union has somewhat more nuclear surface-to-surface missiles

Table 7.5. Soviet tactical nuclear weapons in Europe and IRBM/MRBMs in Western Soviet Union [11]

System	No. deployed (July 1974)
Artillery rockets T5 (NATO code FROG 1-7)	(600)
Short range missiles T7 (NATO code SCUD A)	
(NATO code SCUD B)	(300)
(NATO code Scaleboard)	
Medium range cruise missile (NATO code Shaddock)	(100)
MRBM (US code SS-4 Sandal) 500†	(500)†
IRBM (US code SS-5 Skean) 100†	
Nuclear mines and aerial bombs‡	n.a.
Artillery M-55 203-mm gun/howitzer	n.a.

† Of the total 600 of M/IRBMs about 100 are believed to be in Eastern Soviet Union, 500 in the Western parts being targetable to Europe. These weapons are obsolete; a new 2400-mile mobile or silobased SSX-20 has been tested and evidently will replace them. It has three MIRV warheads. Total number to be deployed is said to be 1200 missiles, about half of them mobile [18].
‡ Number of nuclear-capable Soviet aircraft is estimated at 1000 [13].

(1000) and two times more nuclear capable aircraft (700) than the USA in Europe [15].

The dual-capable Soviet T5 (FROG) and T7 (SCUD) missiles have been constantly modernized during the last ten years the range and mobility of the newer varieties being better than those of the older ones. The reliability also is assumed to have improved, and possibly the accuracy. The number of T7s in front and army levels has gone up by 50 per cent in recent years [16].

The interesting question is whether the Soviet M-55 203 mm gun/howitzer, deployed in large numbers since 1950s, is dual-capable. It has a long range, over 30 km. The IISS has held this possible for the last few years [11]. NATO also seems to believe that the new Soviet self-propelled howitzer (152 mm, on an SA-4 transporter) designated by NATO 'M-1975', may be dual-capable.† Its estimated range is 17.5 km. If these artillery pieces do indeed have a nuclear role, then the number of Soviet tactical nuclear warheads in Europe may be much larger than the usually assumed figure of 3500. Many Western observers contend that the number of Soviet theatre nuclear warheads has been growing during the 1970s (e.g. John Ericson [16, 17]).

The major focus of modernization, however, seems to be on the Soviet IRBMs. According to information from a source close to the Pentagon [18] the replacement of the SS-4s and SS-5s in about 540 silos is just beginning in complexes aimed at Western Europe and China, the old varieties being replaced by *c*. 1200 SS-20s, some of which will be placed into silos, some kept mobile. SS-20 is a new MIRVed mobile IRBM using the first two booster stages of the SS-16 ICBM. Both are carried on the same transporter-launcher and have three MIRV warheads. The SS-20 has a maximum range of 2400–4000 km. Its accuracy is said to be much (even ten times) greater than that of SS-4 and SS-5. What is worrying the USA is that the SS-20 is believed to be capable of quick transformation to the SS-16 ICBM, which would increase the Soviet ICBM number by 1000 over that allowed by SALT.

The implications to Europe of this modernization were set out recently by Fred Iklé, Director of the US Arms Control and Disarmament Agency (ACDA) when he asked in a speech in Los Angeles why the USSR deems it necessary to add the SS-20 when it already possesses regional superiority in Europe [18]. Addition of 3000 accurate large yield warheads to the nuclear potential aimed at western Europe will certainly have implications in the Vienna negotiations although the weapons are located outside the negotiation area.

The Soviet Union is also increasing the flexibility with which it can use nuclear weapons. By replacing older tactical aircraft with modern dual-capable fighters and fighter-bombers such as the swing-wing varieties Su-17, Su-19 and MiG-23, code-named in the West Fitter C, Fencer A and Flogger, it has increased range and accuracy. The number of delivery systems has increased and theatre-wide command, control and communication system improved [19].

---

† According to a NATO source, both these artillery pieces are indeed dual-capable (*International Herald Tribune* (22–23 October 1977)).

V. Command, control and security

Although the United States and Soviet Union have given launchers of nuclear weapons to their allies, they keep the warheads under the custody of their own troops. This is the so-called 'double key' system. The USA applies the same system also for weapons in its own use (two persons are always needed in order to make the weapon usable). All US warheads are further provided with mechanical locks and an electronic switch called Permissive Action Link (PAL) [20, 21]. The PAL-system can be operated only by a code given to the commander by radio after the Presidential release of use. The PAL-system has been continuously improved since 1962 when it was first applied. Presently, PAL D and PAL F are in use. PAL F makes use of multiple codes which are practically unforgeable. The security of the weapons against unauthorized use has thereby greatly improved.

The request for use may come from a NATO Government or a NATO commander [12]. It is immediately forwarded to all NATO Governments and the NATO Defence Planning Committee (DPC), which is formed by representatives of thirteen NATO countries (all but France and Greece). The decisions of the Governments and DPC are forwarded to the US Government and its decision to NATO Council and other NATO Governments. That is a lengthy chain of political decision making which may take several days. There exists, however, a short-cut. The Supreme Allied Commander in Europe (SACEUR) is always a US general and in an urgent case he may obtain the release of the weapons directly from the US President, who only needs to negotiate with other NATO Governments if time allows. The allies do not have a 'key' (or veto right) to the use of these weapons. The Soviet weapons are released for use by the Politbureau, the French weapons by the French President.

It may well be that the peacetime risks of unauthorized use are low by the above arrangements. In alert situations, however, thousands of warheads are taken from their safe depot and distributed to troops. They are transported,

Figure 7.1. Release of nuclear weapons in NATO [12]

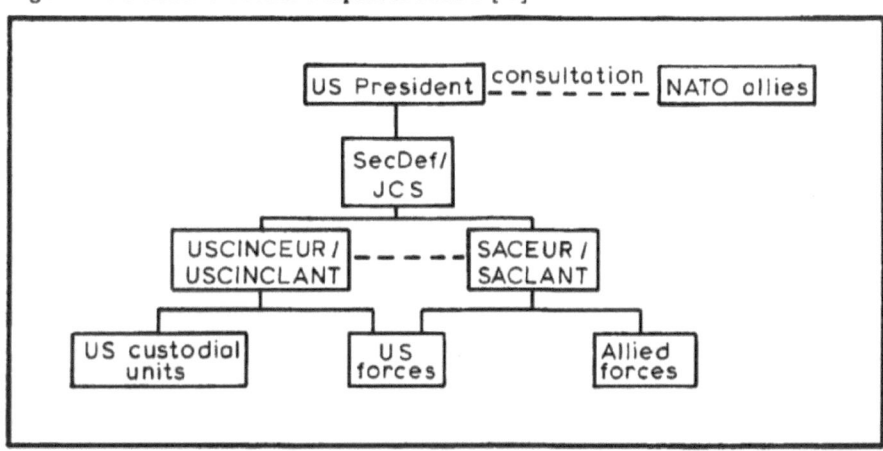

first to the units and then with the units when these are taking their wartime locations. All kinds of risks grow exponentially in such situations. Risks of accidents, capture by saboteurs or commandos or by allied troops in case of political coups do continue to exist.

In the battlefield the employment of theatre weapons is planned in 'packages' of specified number and yields for a specified area. For a given 'package' release is given within a limited time-frame, typically hours. The 'package' is employed within the time-frame in a short time span – typically tens of minutes – to achieve a shock effect and to synchronize the use with other military actions.

VI. The French tactical nuclear systems

France has two types of tactical nuclear weapons, aerial bombs and the tactical missile Pluton [11]. In 1972 France began to provide its Air Force with the tactical nuclear weapon AN-S-2. About 90 aircraft, 30 Mirage-III-E, 30 Jaguar, and 30 Navy bombers have been provided with this tactical weapon, the yield of which is announced to be 10–80 kt. Although these aircraft are designed for a tactical role, their ranges, Mirage 500–700 km and Jaguar 650–1650 km, are wide and make possible deep interdiction. In 1977 the French Navy aircraft, type 'super-Etendard', on two carriers 'Foch' and 'Clemenceau', will also have AN-S-2 nuclear bombs. These aircraft have a wide range, 800–2800 km depending on the load and the height of flight. Early in 1974, France began to provide a tactical missile SSM-Pluton to its Army. Two corps with headquarters in Metz and Baden-Baden will each get three 'regiments' of Pluton. Each of these regiments has three 'batteries' of two launcher-transporters. These two corps will have 36 missiles each, i.e. altogether 72 missiles. Finally, the French Army will have 120 missiles. The Pluton missile can have a yield between 10 and 25 kt. Its maximal range is 120 km and it detonates at a height of 300–400 m having a damage area of $c.$ 10 km^2. If, after about two thirds of its trajectory, the missile is likely to have an error greater than 1500 m, it destroys itself automatically.

In addition to these tactical weapons, France has 18 IRBMs and four ballistic missile submarines each having 16 SLBMs (in total six such submarines may be built) [11]. Although these weapons are considered 'strategic', as part of the 'Force de Frappe', they can be used for operational tasks, too.

The French tactical missile Pluton can only fire to the Federal Republic of Germany from their present location within French boundaries. Its role is described to be either purely escalatory, not as a defensive weapon but as part of the 'Force de Frappe' or, 'to launch a heavy, massive and decisive strike upon an enemy which threatens French boundaries and hereby to break its assault instantaneously'. (General Marty). General Marty also describes the use of Pluton as an intermediate step of escalation and as an offensive weapon 'to destroy the enemy by an offense in a chosen terrain'. All in all, the French tactical weapon is more part of the Force de Frappe than a real war-fighting tool because of the highly political nature of the French nuclear doctrine.

234

VII. The UK tactical nuclear weapons

The British Army on the Rhine (BAOR) has two artillery regiments provided with Lance missiles and dual-capable 8 in self-propelled howitzers, but these are NATO weapons under the two-key arrangement, the warheads being in the custody of the American custodial units [11].

In addition, the United Kingdom has six strike squadrons with 50 Vulcan B2 bombers and four strike squadrons with 56 Buccaneers which are nuclear-capable [11]. Particularly the Vulcan bombers were formerly considered strategic, but presently, being relatively obsolete and probably not able to penetrate the Soviet air defences, they might be best designated as long-range theatre weapons. In any case, their targeting has been integrated to NATO plans.

The United Kingdom also has four SLBM-submarines which must be considered to be strategic [11]. They are also under the same targeting programme as the US SLBMs, but in the Nassau agreement in 1962 when the USA gave the Polaris A-3 missiles for the above UK submarines, it was stated that 'except when Her Majesty's Government decides that supreme national interests are involved, these forces are used for the defence purposes of the Western alliance in all conditions' [translation].

Thus, in practice all British nuclear weapons are under NATO or US control rather than under national control, a sharp difference from the French nuclear forces.

VIII. US and Soviet doctrines and tactical nuclear weapons

In order to evaluate the trends in programmes for modernization of weapons in general and the role of the possible mini-nukes and enhanced radiation weapons in particular we must know their planned use, and how they fit in the doctrines. This is a difficult and contradictory field, since both sides have a declared defensive posture but in fact their armaments and purely military postures are clearly offensive.

The US doctrine

This doctrine was enunciated clearly and with emphasis by Secretary of Defense James R. Schlesinger in *The Theater Nuclear Force Posture in Europe* published in May 1975 [12]. The NATO military posture consists of a 'Triad': conventional forces, theatre nuclear forces and strategic forces. The present NATO strategy of 'flexible response' was approved in 1967 by NATO and is stated in its Military

Committee Document 14/3. The doctrine emphasizes a spectrum of military capabilities ranging from conventional warfare through use or threat of use of theatre nuclear weapons to those of strategic nuclear weapons. It is designed to cause to the enemy great uncertainty as to which response might be selected. Significant changes which have occurred since 1967 are, e.g.:

1. The achievement by the Soviet Union of parity of strategic forces which places greater emphasis on the deterrent role of conventional and theatre nuclear forces.

2. Continued improvement of the conventional forces of both sides.

3. New technology for improvement of both nuclear and conventional forces.

4. The arms control negotiations (SALT and MURFAAMCE) by which greater stability may be achieved.

'The role of theater nuclear forces is to deter and defend against theater nuclear attacks; help deter and, if necessary, defend against conventional attack; and help deter conflict escalation', Schlesinger writes. 'For deterrence the WTO must perceive that sufficient NATO theater nuclear forces can survive initial conventional and nuclear attacks and, in conjunction with surviving conventional forces, blunt WP armored attacks and attack remining WP theater forces'. Special stress is placed on the ability of theatre nuclear forces to prevent the enemy armoured forces from massing. If deterrence works, there will be no war. 'If deterrence fails, NATO forces must be able to achieve these objectives and reverse the tactical situation'. The threat of escalation to strategic level may also be used to attempt to terminate the conflict.

Schlesinger particularly stresses that NATO cannot rule out the first use of theatre nuclear weapons† because 'it is possible to envision significantly worse circumstances than those (normally used) planning assumptions in which NATO conventional forces are unable to hold under conventional attack'. The first use should be on the one hand 'clearly limited and defensive in nature' . . . but on the other hand 'delivered with sufficient shock and decisiveness to forcibly change the perceptions of WP leaders and create a situation conducive to negotiations' This is achieved through a 'package' of maybe 20–40 nuclear warheads typically released within a short time-span, a few tens of minutes, as described earlier (p. 234). How the above contradictory demands ('limited . . . but shock') can be fulfilled remains unclear. Schlesinger remarks that 'conventional forces cannot substitute for an adequate theatre nuclear force' but stresses the need to reduce collateral damage 'since the tactical use of nuclear weapons may involve detonation on NATO territory'. If tactical weapons are 'delivered with sufficient shock and decisiveness' to stop the enemy as Secretary Schlesinger describes, it may well be that the Warsaw Treaty countries see this use as more 'political' than 'tactical' even though it would be intended to be 'tactical' and 'defensive' and only low yield weapons would be used in fact. Even though the NATO first use would cause a shock to the

---

† A proposal to foreclose the first use of nuclear weapons in Europe was made by the WTO in December 1976. NATO responded at a short notice by rejecting the proposal.

opponent, it may be that it has anticipated it in planning its operation. The NATO first use must now seem rather credible to the Warsaw Treaty countries in connection with any military conflict in Europe. Rather than deterring the enemy, the NATO first use threat may convert the enemy's conventional attack to a nuclear attack to gain the advantage of surprise.

Thus, the NATO plan is to use theatre nuclear weapons with restraint against frontline targets only and applying many restrictions, particularly regarding yield, height of detonation, and risk of collateral damage. Mini-nukes and enhanced radiation weapons serve well these demands, particularly if detonated sufficiently high and in small numbers only, but they are not sufficiently different from other TNWs to be free of their other limitations, like their political and escalatory nature, and they are worse than the normal fission weapons in increasing the genetic risks and other delayed effects (see p. 243).

The Soviet doctrine

The Soviet military posture in Europe also contains a 'triad' of forces: conventional, theatre nuclear and strategic nuclear, but the operational Soviet military doctrine apparently does not subscribe to a strategy of 'graduated response'. Their doctrine, training and forces all indicate readiness for conducting a fast, short war in Europe either with a conventional shock army or with the use of theatre-wide, large scale nuclear strikes. Primarily, the nuclear weapons are maintained for deterrence, but if it fails, if nuclear weapons have to be used, they must be used only in connection with a large offensive operation carried out with speed, in concert with all available forces. The use is to be massive, directed against the enemy's whole territory – its political, administrative, industrial and military centres – simultaneously and in depth. The aim is to cause complete disorder in the entire territory of the enemy. Primarily, the Warsaw Treaty countries are to be defended by counter-attacks. The armoured forces of the Warsaw Treaty Organization are postured to exploit their massive nuclear attacks with rapid massive penetrations. The doctrine stresses great speed, surprise, maintenance of continuous pressure, holding of initiative. These can be best guaranteed if the enemy can be put into complete disorder and this is to be done by theatre-wide nuclear strikes if it cannot be achieved by conventional means.

This Soviet view was originally presented in the 1962 classic book by Sokolowskiy on Military Strategy and in a third edition published in 1968 [22] which now is somewhat outdated, but similar views are presented in works by Sidorenko [23], Savkin [24], Novikov and Sverdlov [25] and Shovkolovitch et al. [26]. Some opinions about a more flexible use, however, have been presented in a few recent articles. According to the Soviet doctrine nuclear weapons make simultaneous surprise strikes possible in great depth. An advantage is obtained only if the effect of these strikes is immediately exploited by deeply penetrating airborne and fast mobile operations. In nuclear war, it is necessary to maintain the initiative to gain advantage. Otherwise, there will develop only a tremendously destructive war of attrition. This does not mean, however, that the Soviet Union would particularly cherish tactical nuclear weapons†. The brunt of modernization has been on conventional forces in recent years (artillery, tanks, air force, personnel carriers etc.) although one of the main reasons for the great deployment of armoured personnel carriers probably was the anticipation of nuclear environment. The Soviet doctrine and forces are best characterized today by 'dual capability'.

Comparison

Thus, we can see that the US and Soviet doctrines are very different and in many respects almost opposite. For the USA the tactical weapons are primarily weapons of war-fighting, defence in a friendly country and means to delay escalation to the strategic level and to reinforce the direct defence under the strategy of 'denial'. The NATO allies like to think of them primarily as a 'link' to the strategic deterrent. To the Soviet Union they are primarily a deterrent, but if war has to be fought, they are the 'plough' to clear an avenue for a massive offensive. This very asymmetry makes the onset of their use extremely dangerous and decreases the credibility of NATO plans to control the escalation unilaterally.

IX. On introduction of mini-nukes

As described on p. 227 mini-nuke is a catchword used for fission weapons of from c. 0.05 to 0.5 kt yields.

Purely militarily mini-nukes have the advantage that they can be used in close proximity to friendly troops; they do not produce fallout and thus do not contaminate the terrain. Nevertheless, great doubts remain about the advisability of their deployment and about the use of TNWs in general.

First, even the smallest mini-nukes would change the nature of war. They are unequivocally distinguishable from non-nuclear weapons and even though the yield would be 'very small', it would be difficult to estimate accurately in the confused, excited world of the battle, and the enemy might perceive it larger than it really was. All nuclear weapons have a thoroughly political nature which has two kinds of repercussions: their political nature puts formidable inhibitions against their use. Simply, it may be that the political leadership will not allow their early use, i.e. their use at the time when it would be most advantageous. For instance, the Government of the Federal Republic of Germany has not allowed

---

† The proposal of the WTO for an agreement on non-first use of nuclear weapons mentioned above corroborates this view.

prechambering of the Atomic Demolition Munitions to forestall any too early escalation of a crisis to the nuclear level. It is difficult to believe the nuclear war would be initiated because of a small, local threat which could be fought by a small scale use of tactical nuclear weapons.

On the other hand, even a small scale use of these weapons by the opponent, particularly if it has a demoralizing effect, may lead to great pressure for retaliation in order to neutralize the tremendous advantage the other side would otherwise get. Both sides really seem to consider that initiation leads to retaliation.

Mini-nukes sometimes are justified on the very dubious basis of their assumed 'linkage' effect to the strategic level.

Although perhaps political in the eyes of the opponent, mini-nukes would nevertheless be to the allies symbols of nuclear warfighting rather than 'links' and thus undermine the psychological 'linkage' to the strategic level.

Mini-nukes would also severely undermine the Non-Proliferation Treaty (NPT). As pointed out previously [27], this treaty obliges the non-nuclear signatories which might have contemplated acquiring small tactical nuclear weapons not to acquire them but, through its article VI it also obliges all signatories 'to pursue negotiations . . . for . . . nuclear disarmament. . .'. Development of a new generation of nuclear weapons (mini-nukes or enhanced radiation weapons) would clearly violate the spirit if not the letter of the above article. The worst thing is that mini-nukes would be more likely to be used against the non-nuclear states than the bigger varieties. Thus, the non-nuclear signatories of the NPT could justly consider themselves to have been deceived.

Introduction of mini-nukes might well compel even such countries as Yugoslavia, Switzerland and Sweden to rethink their NPT policy (see, e.g. [28]) not to speak of the non-signatories. In any case the introduction of these new varieties of bomb with their new effects would force all countries to start a lot of new research on nuclear weapon structure and nuclear weapon effects [8], which in itself is not desirable.

X. Some reasons demanding non-use of nuclear weapons in Europe

It is possible to present a number of persuasive sounding reasons to maintain tactical nuclear weapons in Europe and to miniaturize and modernize them for better war-fighting capability – and no one has presented such reasons better than James Schlesinger in his Theatre Nuclear Posture thesis. However, there also exist a multitude of good reasons why they must be eliminated.

Biological reasons

Nuclear weapons are classified as dubious weapons of war [29]. Planning of even limited use of these weapons (war limited to Europe) puts tens or even hundreds of millions civilians under the threat of nuclear radiation, the consequences of which to future generations are incalculable. It is particularly misleading to plan to introduce the miniweapons or enhanced radiation weapons under the guise of 'small' and 'clean' weapons because these weapons would maximize the genetic effects. The European politicians who reluctantly and intrigued by uncertainty may approve their introduction hardly understand their true nature.

Mini-nukes and enhanced radiation weapons kill by initial radiation, mainly fast neutrons. Huge doses. 8000–18 000 rems, are 'recommended' for immediate permanent incapacitation of the enemy (Table 7.4, page 228). To achieve such doses to personnel within tanks about two times higher doses would have to be used outside the tanks.

In radiation breeding, plants and microbes are usually submitted to semi-lethal dose to maximize the genetic effect. Soldiers would be exposed exactly similarly by the miniweapons and enhanced radiation weapons.† For each million soldiers killed, two million would get a genetic dose increasing the number of deleterious genes in their progeny during many generations many-fold, about 30 rems being the doubling dose. These weapons therefore could inflict a grave deleterious effect on the future generations of Europeans. Today's highly mobile war fought by smaller or larger tactical nuclear weapons would probably roll back and forth over the two Germanies which would become most inflicted. All tactical nuclear weapons are dubious not only because of their mass-destructive nature but because of their inherent ability to produce late effects extending to future generations, but the mini-nukes and enhanced radiation weapons are even more dubious than the normal fission weapons because their lethal radius of initial radiation greatly extends their lethal radius of thermal radiation and over-pressure, while the situation is the reverse with yields above 1 kt of the standard fission weapon.

The whole idea of 'proper selection of radiation casualty criteria' to 'optimize' the results on the battlefield can only induce feelings of revulsion, exceeded only by those produced by chemical weapons, which probably constitute the ultimate misuse of science. No 'deterrent' which has incalculable consequences to future generations should be introduced to the battlefield under the guise of a 'clean' and 'small' weapon. Such a 'deterrent' is more incredible than credible.

---

† The radius of the circle around the air-detonation of an 0.1 kt miniweapon, within which unprotected soldiers get a lethal dose, say, greater than 1000 rem from fast neutrons, is about 390 m and the area 478 000 m² (see Table 7.3). The radius of the circle within which unprotected soldiers get a dose greater than 100 rem is 640 m and its area 1 287 000 m². Between these two circles remains 809 000 m² where the soldiers would get doses between 100 and 1000 rem. With proper medical care and possibly with a prophylactive dose of a radiation protective medicine like cysteamine many of these soldiers would have a fair chance to stay alive. Thus, roughly twice as many soldiers would get a non-lethal but optimal 'radiation breeding dose' as would receive a lethal dose.

It is the duty of the radiobiologists and other scientists who can understand the true nature of these new types of nuclear weapons to point it out to the military and to the politicians who plan to introduce them to Europe.

Military reasons

The main military justification for low-yield tactical nuclear weapons is that they are needed to blunt a massive armoured penetration. In reality, the enemy probably would not mass before it had destroyed the major part of its opponent's nuclear weapons which could threaten his troops. First, there would develop an initial nuclear dual of theatre weapons, then a highly mobile, chaotic nuclear battle in which both sides would suffer losses of up to 50 per cent in the forward divisions. This battle could roll back and forth in central Europe for a few days or a week, then a brief pause might accrue because of the hopeless chaos, turmoil, and frustration. Thereupon the one of the combatants who would still have reserves and who would first regain the manoeuverability necessary to exploit the confusion of his opponent might have some tactical advantage from the apocalyptic first two phases. Skill in forming and using reserves would then become decisive. Thus the use of tactical nuclear weapons greatly increases the need of manpower instead of reducing it.

It even seems unlikely that there can be any such thing at all as an organized nuclear battle. The military order and organization would collapse because of an unbearable psychological burden on the soldiers. The nuclear 'leg' is therefore an extremely uncertain 'leg' of a military posture. It should not be relied upon at all, because in reality it perhaps could not be used.

Destructiveness

The best known fact regarding nuclear weapons is their extreme destructiveness. The purpose of miniaturization is to reduce collateral damage. The USA is evidently planning to have an arsenal of a few – possibly from four to five – thousand mostly low-yield nuclear weapons in Europe. It may well be that no decision about introduction of mini-nukes or enhanced radiation weapons has yet been made. If they are introduced, it is probable that only a low number of them will be deployed. There are many targets in a theatre wide nuclear battle which cannot be destroyed by a small yield. The Soviet Union probably does not yet have any low-yield nuclear weapons at all. Even its nuclear artillery shells (if it has such) are likely to be in the kiloton range. The reason is simple. Its present offensive-bound doctrine does not require any low-yield weapons. If nuclear weapons are detonated according to a preset plan and high airbursts are used, medium yield weapons can be used for deep interdiction without impeding one's own offensive. Central Europe is extremely urbanized. Its extensive cities sprawl out and merge together so that in wide areas there is hardly any opening to be found for an offensive through the 'urban wall'. The Soviet doctrine says about employing tactical nuclear weapons in an urban area [26]:

> Prior to reaching the decision to employ nuclear weapons (in that case when they have already been employed during the course of action) it is necessary to take into consideration the importance of the city as a military, administrative, political and economic center and the attitude of the populace of the city toward our army.

And regarding the importance of urban areas for flank protection:

> Nuclear strikes can best be delivered against the edge of a city or within it where the enemy will hold his reserves or forces prepared to deliver a blow at the flank of our attacking troops.

In the light of the huge asymmetry between the US and Soviet doctrines and tactical nuclear arsenals the wish to minimize collateral damage to civilian structures and population may remain wishful thinking even though NATO would restrict its initial use of tactical nuclear weapons to the lowest possible yields. The opponent cannot do the same.

In the long run even the USSR would probably develop these new weapons if the USA deployed them and if they do offer marked advantages in combat. However, there is no guarantee that the Soviet Union would change its doctrine.

Political reasons

The European Security Conference was successfully concluded in 1975. It removed most direct causes of war in Europe and as a whole the *détente* symbolized by it certainly has increased the security in Europe although the progress has not been in all aspects as fast as many would like it to have been. The military situation in central Europe has been painstakingly analysed in Vienna. There seems to exist a rough balance of conventional forces in central Europe, as far as it is measurable because of the asymmetries (see p. 244). There cannot be continuous improvement and normalization of the political relations if the nuclear postures are continuously developed. Therefore, the theatre nuclear weapons must also be included in the potential agreement. A change is also required in the existing 'deterrence' doctrines. Maintenance of deterrence with a first use option poisons the mutual relations and makes normal trust impossible. NATO justifies its first-use option by saying that it is purely defensive and only will be used if the other side attacks. But it is not at all sure that the war begins with a regular attack. Most of the last hundred wars did not! The war may well begin, for instance, with a clash in the North Atlantic which first leads to a military build-up which then compels one side to pre-empt. Simply, a nuclear posture with a first-use option is both offending and dangerous. All plans for use of nuclear weapons in Europe must be foreclosed, otherwise normal relations cannot develop between nations in Europe.

Psychological reasons

That tactical nuclear weapons still exist in Europe today is probably mainly due to the fact that extremely few people know much about them and understand their inherently dangerous nature. In a crisis, the demand for knowledge about these weapons will grow exponentially amongst the general population which will suddenly realize that their whole existence depends upon the use (or rather the non-use) of these dubious weapons. Such a realization may change the credibility of the use of these weapons and since their deterrence effect is based upon the credibility of their use, the deterrence may fail at a critical moment. Therefore, a stable defence cannot be based upon such dubious weapons.

Ethical reasons

Mankind has some experience of the use of an inherently illegal weapon in war: the chemical weapon in World War I. This sad experience led to the Geneva Protocol. World opinion, particularly among veterans, was convinced that any form of chemical warfare is evil. It is vital that the general population in Europe realizes that this is true also of nuclear weapons before it is too late. Japan is the only country which has had to experience the use of nuclear weapons. The Japanese people are deeply convinced that these weapons must never again be used. In other countries the general population has been lulled to indifference regarding nuclear threats by 30 years of non-use of these weapons and, particularly, by the last almost ten years of *détente* and SALT negotiations. Without the pressure of public opinion the political leaders of the two great powers have negotiated agreements about increase of strategic weapons instead of agreements on their decrease.

The nuclear weapon is, because of its inherent destructiveness, its indiscriminate nature and its genetic effects extending to future generations, a 'forbidden' weapon which must never be used.

Therefore, those who try to make it usable by miniaturization and modernization are on a wrong path. Politicians and laymen should not be deceived by the 'cleanliness' of the enhanced radiation weapons and mini-nukes, particularly.

XI. *Measures to eliminate the risks of nuclear war in Europe*

Strengthening of the Vienna negotiations

These negotiations ought to be brought to conclusion. The figures for ground and air forces have been revealed. There exists a rough conventional parity†. It would be possible to have a compromise on a troops reduction of about 15 per cent, at least one only including conventional armaments. If an agreement on inclusion of nuclear weapons cannot be reached in the first reduction they should not be allowed to prevent the conclusion of the first round in Vienna. The proposed reductions of a few per cent of the existing nuclear arsenals – 54 fighter bombers and 36 missiles with corresponding nuclear warheads – are so little that they would not have any great military significance anyhow. It may be that Vienna, with its present framework, is not a proper forum for negotiating significant reductions of theatre nuclear weapons because they are a part of the total nuclear balance in Europe which also includes the medium range nuclear systems aimed into the reduction area and these systems are not the object of negotiations. But yet, they are very important to the military balance in Europe. Thus, a new forum probably would be needed for negotiating the elimination of all nuclear weapons and plans of their use in Europe. All European states perhaps ought to participate in it. In addition to those states participating in the Vienna negotiations, France would be needed because of its nuclear weapons and doctrines. Neutral and non-aligned states would be needed because the risks of nuclear war threaten them equally with the allied states – and they are not even protected by a nuclear 'umbrella'. Furthermore, they may be needed to achieve the goal. For instance, the European Security Conference probably would never have reached a successful conclusion without the inexorable diligence of the neutrals in seeking for just compromises. Members of military alliances are gravely biased in matters of nuclear weapons. Neutrals might be in a better situation to see the significance of these weapons for Europe as a whole.

Stepwise overall progress

Nuclear weapons have been 30 years in Europe. They have too prominent a role in the military postures to be eliminated by one step. Yet, they are not something which cannot be eliminated at all, although many political and military leaders seem to believe so. There exist much better reasons for impelling the nuclear weapons into a domestic political trouble than there is for the peaceful nuclear power.

For instance the following types of steps probably could be achieved without risking the present stability in Europe. The proposed steps should not be understood as a rigid schedule:

---

† The WTO submitted in June 1976 figures for its ground and air forces: 805 000 and 160 000, respectively, totalling 965 000, 1 January 1976. In December 1976 NATO submitted figures which show that its ground forces (without the French forces within the reduction area, 60 000 men) were 731 000, air forces around 190 000 totalling 921 000, or 981 000 with the French divisions which are located in the reduction area. These figures, too, refer to 1 January 1976 [30].

1. After an agreement on troop reductions has been achieved in Vienna, the framework of the SALT or Vienna negotiations is widened or a new negotiation forum as described above is formed.

2. Doctrines are unilaterally changed to 'non-first-use' of nuclear weapons. Because of psychological reasons such a change may be more easily achievable unilaterally than through a formal agreement. Some improvement of conventional defence may be deemed necessary and is possible, for instance, by applying new conventional technology [31].

3. A narrow non-nuclear weapon zone – let us say 80–100 km broad on both sides of the east–west border – could be useful as an initial step, as a confidence-building measure. It would not endanger military security in any serious way but it would signal to the military leaders the changing course the military preparations have to take [32].

4. The final goal must be complete elimination of the threat of the use of nuclear weapons in Europe even if a conflict were to break out. This could only be achieved through negotiations on a proper negotiating forum which includes all factors relevant to the military balance in Europe.

Any schedule for disarmament adopted must be flexible enough to allow changes in procedure if a stalemate is threatening.

There seems little doubt that this is the right direction into which the development should be going and even may be going. But the existing political *détente* is frail and can easily be disturbed. Therefore, time should not be wasted but measures be taken now to direct Europe to a new course to a non-nuclear peace.

References

1. Conference on Security and Co-operation in Europe. Final Act. 1 August 1975. *Survival*, Vol. 17, No. 6, (1975), pp. 295–301.
2. SIPRI, *Force Reductions in Europe* (Almqvist & Wiksell, Stockholm, 1974).
3. SIPRI, *World Armaments and Disarmament, SIPRI Yearbook 1974* (Almqvist & Wiksell, 1974), pp. 29–47.
4. *Atlantic News No. 788*, (Bruxelles, 19 December 1975).
5. MBFR-Gegenvorschläge des Warschauerpakts, *Neue Zürcher Zeitung*, (29.2. and 11.3.1976).
6. W. De Vos Van Steenwijk, Press briefing (8 April 1976).
7: *Arms Control Today*, Vol. 4, No. 4, (April 1974), p. 4.
8. Gut, J., *Allgemeine Schweizerische Militär Zeitschrift 1976*, No. 6, 273.
9. Glasstone, S. (ed.), *The Effects of Nuclear Weapons*, (US AEC 1962).
10. Warshawsky, A. S., Radiation battlefield casualties – credible! *Military Review*, Vol. 56, No. 5, (1976), pp. 3–10.
11. *The Military Balance 1976–1977* (The International Institute for Strategic Studies, London, 1976).
12. Schlesinger, J. R., *The Theater Nuclear Force Posture in Europe*, (April 1975).
13. *The Military Balance, 1977–1978* (The International Institute for Strategic Studies, London, 1977), p. 78.
14. *The Defense Monitor*, Vol. 4, No. 2, (February 1975).

15. United States/Soviet Military Balance; a frame of reference for Congress, *94th Congress 2nd session*, (Committee Print, Washington DC, January 1976).

16. Ericson, John, Soviet force levels and force reductions, *Armies and Weapons*, (1976), pp. 4–5.

17. Ericson, John, Soviet military capabilities in Europe, *Military Review*, Vol. 56, No. 1, (1976), pp. 58–67.

18. Beecher, W., *International Herald Tribune*, (10 September 1976).

19. Rumsfeld, Donald, H., *Annual Defense Department Report FY 1977*, (Washington, DC), (27 January 1976), p. 102.

20. General Giller's testimony, *Military Applications of Nuclear Technology*, Hearing before the Subcommittee on Military Applications of the Joint Committee on Atomic Energy Congress of the United States, Ninety-third Congress, First Session on the Consideration of Military Applications of Nuclear Technology, (16 April 1973), Part I.

21. *Military Applications of Nuclear Technology*, 93rd Congress, (22 May and 29 June 1973), Part 2.

22. Sokolovskiy, V. D., *et al. Military Strategy: Soviet Doctrine and Concepts*, (McDonald & Jane's 1975, London).

23. Sidorenko, A. A., *The Offensive (A Soviet View)*. [Translated by US Air Force, US Government Printing Office, 1973.]

24. Savkin, V. Ye., *The Basic Principles of Operational Art and Tactics*, Moscow 1972. [Translated into English by the US Air Force, 1974.]

25. Novikov, Y. and Sverdlov, F., *Manoeuvre in Modern Land Warfare*, (Progress Publishers, Moscow 1972).

26. Shovkolovitch, A. K., Konasov, F. I. and Tkach, S. I., *Combat Action of a Motorized Rifle Battalion in a City*, (Voyenizdat, Moscow, 1971).

27. Miettinen, J. K., *Nuclear Miniweapons and Low-Yield Nuclear Weapons which Use Reactorgrade Plutonium: Their Effect on the Durability of the NPT, in Nuclear Proliferation Problems*, (SIPRI, Stockholm 1973), pp. 119–126.

28. Borba, 7 December 1975. *Yugoslavia and Nuclear Weapons, Survival*, Vol. 18, No. 3, (1976), pp. 116–117.

29. SIPRI, *The Law of War and Dubious Weapons*, (Stockholm, Almqvist & Wiksell, 1976), 228 pp.

30. *Atlantic News*, No. 888, (22 December 1976).

31. Miettinen, J. K., *Can Conventional New Technologies and New Tactics Replace Tactical Nuclear Weapons in Europe?* ISODARCO, 6th Course, (Nemi, Rome, June–July 1976). Published as *Arms Control and Technological Innovation*. David Carlton and Carlo Schaenf (Croom Helm, London, 1977), pp. 52–69.

32. Miettinen, J. K., Tactical nuclear weapons and the doctrines governing their usage are changing from deterrent to real battlefield role, *26th Pugwash Conference, Proceedings*, (London 1977).

8. Mini-nukes and non-aligned defence. The case of Sweden

Jan Prawitz†

I. Introduction

This chapter explores the consequences of the discussion in the years 1972–3 in Sweden on the possible introduction of 'mini-nukes' in the European theatre, and reviews the discussion and subsequent intervention at the Geneva Disarmament Conference (CCD).

The possible introduction of mini-nukes into the arsenals of nuclear weapon states as suggested some years ago would have important implications for the defence situation of a number of countries. In the first place this would be true for states which are members of the two military alliances in Europe. It was in the European context that the role of these weapons was first discussed. But it would also be true for the non-aligned countries, both in Europe and in other parts of the world.

The defence situation varies greatly between different non-aligned countries. An analysis of the impact of mini-nuke deployment by the nuclear weapon powers on the defence of non-aligned states must, therefore, be very general. The military power gap would certainly widen between on the one hand the nuclear weapon states and their allies and on the other individual non-aligned states, supposed to be non-nuclear weapon states. But what would this in turn mean in terms of stability and security in different parts of the world?

Factors like geographical size, population, and military strength of a country, its proximity to nuclear weapon power territory or to theatres where nuclear weapons are stationed or to the oceans, military commitments of neighbouring states etc., do influence the effect of a possible introduction of mini-nukes. Therefore, if any detail is required, separate analysis has to be carried out for each country.

The analysis in this chapter concentrates on Sweden, a middle-sized non-aligned country in northern Europe with a small population. It has a non-nuclear national defence tailored to provide sub-regional stability to the theatre military balance between NATO and the WTO. Its strategic situation is dominated by the two blocs and their military forces, which also include theatre nuclear forces.

---

† The views and opinions expressed are those of the author and do not imply the expression of any position on the part of Sweden's Ministry of Defence.

Some other non-aligned countries in Europe have a partly similar position, i.e. Austria, Finland, Switzerland and Yugoslavia, while there are still others with a quite different situation, e.g. Ireland, Liechtenstein, Malta and Spain. Outside Europe, a variety of other defence situations exist.

Neutrality and defence policy

Sweden's defence policy is based on several assumptions. One is that the European theatre for the foreseeable future will be dominated by the two major blocs having partially contradicting interests and generally balancing military forces. Sweden's political power is limited and cannot influence this situation. Sweden is further assumed not in itself to be of any substantial value to either bloc, but access to its territory or parts of it may in times of crisis be of value to one of the blocs in supporting its actions against the other. Furthermore, a potential invader will be able to mount only marginal forces against Sweden as he would primarily be engaged with his main adversary. An isolated attack on Sweden with the full forces of a great power is assumed to be unlikely.

To achieve its basic security ambition, which is the sovereign survival as a national state, Sweden has declared itself a non-aligned state intending to stay neutral to any military conflict observing the rules of the 1907 Hague Conventions. This policy is supported by a national defence intended to be strong enough to deter an attack from any of the two blocs using the marginal forces they may be able to allocate against Sweden [1].

The role of nuclear threat

Recognizing the express desire of both blocs to avoid the use of nuclear weapons by all available means, a military conflict in Europe, should such a conflict break out, would most probably begin as a conventional war and not be escalated by the use of nuclear weapons until all other possibilities, both political and military, had been exhausted. Were Sweden to be attacked within the context of a war in Europe, the attack would thus probably be a conventional attack. Only if the main conflict had already escalated into nuclear war or if such escalation were imminent, would this assumption not apply.

Should Sweden be drawn into a war that had already escalated to the nuclear level, it might also be subject to a major or minor nuclear attack. In such a situation, however, the solution to the problem of ensuring the survival of the nation and its population would not be military resistance but rather the opposite: to seek the termination of military fighting as soon as possible. The reason is that it would be impossible, both technically and economically, to organize sufficient protection against the effects of nuclear war to secure both effective military defence and the survival of the population. Limited measures of protection are possible, however, that could prevent immediate collapse following a few nuclear explosions and provide time for negotiations.

The non-nuclear status

Sweden signed the Non-Proliferation Treaty in 1968. Before that there was a debate for more than a decade on whether the Swedish military defence should include also a tactical nuclear force. The conclusion was negative, considering the dominance of the two blocs in Europe, their doctrine of a high nuclear threshold and also the limited defence expenditure that would be politically realistic in peacetime.

Should Sweden allocate a large proportion of its defence budget to building up a tactical nuclear force and thus spend considerably less than at present on its conventional defence, such a reduced conventional force might not be able to resist a conventional attack – the most likely contingency. Sweden would then in case of conventional attack be faced with a choice between surrender or escalating the conflict. The latter case would be the same as an implicit invitation to nuclear counter-attack in contradiction to the basic security ambition to survive. An adequate conventional defence would thus be necessary anyway and a nuclear force in addition would be beyond Sweden's economic means.

The high nuclear threshold, i.e. reluctance to use nuclear weapons, was thus a strong argument in favour of non-proliferation, a conclusion that would apply to several other non-aligned states as well [2].

II. The mini-nuke concept

The term mini-nuke has been used in different contexts and with different meanings from time to time. The presentation of the general idea of mini-nukes provoked some debate in Europe. In the case of Sweden these considerations resulted in a few official statements and an intervention at the Geneva Disarmament Conference.

A number of official statements by other governments, articles in professional journals and other documents played a role in stirring the debate in Sweden. A few are listed in the references, official statements in refs [3–6], others in refs [7–10]. Several more could be mentioned.

For the purpose of this paper the concept of mini-nukes means nuclear weapons in the sub-kiloton range down to yields comparable to the largest conventional charges. The effects of such weapons should be generally limited to the military targets at which they are aimed thus reducing the collateral damage associated with ordinary nuclear explosions. This requires precise guidance that will permit them to hit their targets closely, which presupposes that the intelligence support is similarly accurate in identifying and locating these targets. Finally, their use must not necessarily be decided at the political top level but might occasionally be left at the discretion of commanders in the field. This was indeed the definition behind the Swedish intervention in the CCD.

The use of present theatre nuclear weapons, as defined by the current stockpiles in Europe, would have nearly the same consequences as a strategic attack for the civilians living among the military targets. Such use would also mean a large risk for escalation to total nuclear war. They have, therefore, been considered almost as unusable as the strategic weapons.

Advocates of mini-nukes – as defined above – have pointed out that not only would they be useful on the battlefield from the beginning of a military conflict because of their limited side-effects on the civilian society, but they would not imply the same risk for escalation to higher levels of nuclear war. They would also provide cheaper fire-power than conventional weapons.

The crucial element in this mini-nuke concept is not the technology or economy, but the doctrine. It was argued that a new threshold or 'firebreak' could possibly be introduced between use of mini-nukes on the one hand and use of ordinary tactical and other nuclear weapons on the other. But the contrary view was also argued that the present clear distinction between conventional and nuclear warfare in general would be totally blurred. These are in fact the crucial problems.

III. Mini-nukes and non-aligned defence

The consequences for non-aligned countries such as Sweden if mini-nukes were to be deployed by the blocs in Europe would be negative, as would any drastic increase of the general firepower of the large armies. The marginal forces that could be allocated for an attack against Sweden could, if equipped with mini-nukes, easily be made strong enough to overrun the conventional forces of the defender.

Obviously Sweden's defence doctrine would be obsolete the moment mini-nukes and a new doctrine were introduced by the blocs on the European theatre. In particular, Sweden's reasons for abstaining from nuclear weapons of its own would be weakened. Such a development would go to the heart of Sweden's security interests and might require a re-examination of both security ambitions, defence policy and the role of neutrality.

Naturally, such an issue triggered the interest of the defence planners and strategic analysts. The first official reaction from the Swedish government was included in a speech in February 1973 by the Under-Secretary of State for Defence, Mr Anders Thunborg. Having referred to the arguments for and against the introduction of mini-nukes, he concluded that such weapons, if deployed, would make up a new element fitting into the escalation ladder and that the present nuclear threshold would remain high [11].

Comments on the possible deployment of mini-nukes and concern about related security implications were also expressed occasionally in the Parliament the following year [12].

Mini-nukes were also the subject of an intervention by Sweden's Minister of Disarmament, Mrs Alva Myrdal, at the Geneva Disarmament Conference in August 1973, where she expressed concern about the possible blurring of the distinction between nuclear and conventional warfare and put five questions to the representatives of the nuclear powers [13]. (The full text of her intervention and the replies are included as an Appendix to this chapter.)

In their replies the representatives of the three nuclear powers represented in the CCD insisted that no blurring of the distinction between nuclear and conventional warfare was under discussion. In addition all three nuclear powers interpreted two important arms control documents, the US–USSR Agreement on the Prevention of Nuclear War (1973) and the Security Council Resolution 255 (1968), as making no difference between mini-nukes and other nuclear weapons.

The role of tactical nuclear weapons in Europe has long been under review. It seems impossible to forecast in any detail what changes in hardware and doctrine will be decided in the years to come. Still, it seems safe to assume that whatever low-yield charges are introduced in the next decade, the nuclear powers would keep the control of all nuclear weapons at the top of government and the use of such weapons would not be delegated to field commanders on the first days of conflict. On the contrary they would probably think twice before using them, i.e. the threshold will remain.

This assessment is supported by the expressed policy of three nuclear powers and their interpretation of current international law. It is also true, as a study of the collateral damage on the civilian society following extensive use of mini-nukes will reveal, that such effects will be quite considerable also as regards radioactive contamination.

The mini-nukes are often called 'small, clean nuclear weapons'. But to say that a mini-explosion is clean – i.e. emits little radioactivity or none at all – is to express a qualified truth. To be sure, a charge of 0.01 kt – 2000 times weaker than the Hiroshima bomb – does produce 2000 times less radioactive debris. But the remaining radioactivity of a mini-charge is usually larger per kiloton than that of charges of higher yield. Following either a ground or an air explosion, the radioactive fireball rises to an altitude of only one kilometre or so. The radioactivity, which presumably also contains substantial quantities of unconsumed plutonium, will fall down immediately afterwards in the prevailing wind direction; faster and more concentrated, and thus more intense per kiloton and covered surface than corresponding contaminations from more powerful charges. When many mini-nuclear weapons are deployed against targets over a bigger area, the aggregate radioactive contamination can become so intense and widespread as to leave no room for ignoring safety considerations [14].

In addition, the continued development of precision guidance of munition to an assured head-on hit in all cases, implies that most military targets could be knocked out with a conventional charge smaller in yield than the mini-nukes. The need for the latter would thus decrease.

IV. Speculation on security

Although the conventionalization of subkiloton charges would probably not materialize, it might still be interesting, for the purpose of academic speculation, to consider what the introduction of mini-nukes in Europe and the related change of doctrine would imply for the defence of non-aligned states.

Assuming the same basic security ambition, i.e. maintenance of the national sovereignty of the country, a number of options are theoretically conceivable. They are here summarized into three main categories.

One could be to improve and strengthen the conventional defence as far as is financially feasible. At the same time the military element is de-emphasized in favour of political ones in the national security effort. This option might include seeking big power guarantees for the neutral status.

Another category of options could include a quantum increase of the defence expenditures in order to buy a limited nuclear force of low-yield weapons in addition to the conventional forces with the purpose of keeping up with the force level of potential adversaries. This option could be related to the ideas expressed by Dr R. G. Shreffler in Chapter 10 of this book. It does, however, imply withdrawal from the Non-Proliferation Treaty.

A third category of options could be to keep the conventional force but to seek the support of nuclear force from a nuclear weapon power, i.e. to give up non-alignment. In turn, this may dramatically change the strategic situation at the division line between the blocs in Europe and would very probably increase tension and confrontation.

In the case of Sweden, all the three options mentioned would mean very fundamental changes in its traditional defence doctrine. From the point of view of governmental decision making they would seem impossible or at least very difficult. However, these examples demonstrate the dimension of the question for non-aligned countries.

V. Concluding remarks

The introduction of mini-nukes somewhere between ordinary nuclear weapons and conventional ones in the arsenals of the two blocs in Europe would have drastic effects on the defence situation of some of the non-aligned countries. There might also be a weakening of the Non-Proliferation régime.

It has been said that the explosion of a nuclear device by India in 1974 set a bad example for other countries on the threshold of the atomic club. This might be true in the third world. In Europe, however, the risk that one or two countries on overseas continents fabricate a few nuclear devices of their own – bad enough – is not the immediate problem. There are the thousands of tactical nuclear weapons now deployed in Europe and ready for immediate use that constitute the immediate problem. Changes in the composition of the theatre nuclear forces or the doctrines for their use is a much more important factor than the Indian bomb, when European defence planners assess the implementation and the role of the Non-Proliferation Treaty (NPT).

The possible introduction of mini-nukes, as they are defined in this paper, on the European continent would certainly be a measure of sufficient magnitude to justify a reassessment of the role of the NPT. This does not mean, however, that such a reassessment would necessarily result in withdrawal from the NPT and subsequent acquisition of nuclear weapons. The question of reassessment has been raised, however.

Beside vertical proliferation by the nuclear weapon powers present in the European theatre, proliferation risks would include both independent acquisition of nuclear weapons by bloc members or non-aligned states and transfers of nuclear weapons between allies.

In her CCD intervention (see Appendix), Mrs Myrdal expressed the hope that the positions of the nuclear powers on the mini-nuke issue would 'be so reassuring that no State party to the NPT will interpret the situation as an extraordinary event which has 'jeopardized the supreme interest of its country' and act accordingly'. Her quotation was of course from the withdrawal clause (Art. X:1) of the NPT.

In his comment to the final declaration of the NPT Review Conference, the delegate of Yugoslavia expressing deep concern that the conference could not agree on any recommendation to withdraw tactical nuclear weapons from the territories of non-nuclear NPT parties, said that his country would be prepared to 're-examine its attitude towards the Treaty and draw corresponding conclusions' [15].

It is true that the two military alliances overwhelmingly dominate the strategic situation in Europe. But it is also true that the very existence of the five 'central neutrals', i.e. Finland, Sweden, Austria, Switzerland and Yugoslavia, separating the territories of the two blocs along the major part of their division line, is in itself a positively stabilizing factor in Europe. These countries could continue to promote stability and *détente* by staying non-aligned and non-nuclear. Appreciation of this role of theirs and of their security interests, including reduction of potential nuclear threats, would thus be in the interest also of the blocs.

The pressure for proliferation that an introduction of mini-nukes may cause, could lead to the substitution of the present stability for nuclear chaos. The main triggering factor would be the change of doctrine, i.e. the conventionalization of the weapons, not the introduction of new hardware alone. But as a massive deployment of sub-kiloton warheads would provide the physical basis for an overnight change of doctrine, such deployment without change of doctrine would still imply some risk of proliferation.

It would be beyond the scope of this chapter to continue the analysis into proposals for a strategy for peace and arms control in Europe. The purpose has been the limited task of reflecting a discussion among non-aligned defence planners caused by the mini-nuke suggestion a couple of years ago.

Postscript. A similar discussion took place on the concept of the neutron bomb in the summer of 1977. Although mini-nukes and neutron bombs differ considerably in terms of design and explosion yield, their strategic and doctrinal implications for a non-aligned state such as Sweden would be closely related. The same conclusions could therefore be drawn on the neutron bomb issue.

References

1. Sweden's national security and defence doctrines are published in parliamentary documents in Swedish SOU 1976:5, SOU 1977:1, Prop 1976/77:74, FöU 1976/77:13, parliamentary debate 26 May 1977. A recent book on the matter is *Swedish Security Policy and Total Defence*. Ministry of Defence, SSLP 1977:1.
2. A review of the early debate on a possible independent nuclear strike force for Sweden and the defence problems posed by various nuclear attack contingencies has been published in Holst, J. J. (ed.), *Security, Order and the Bomb*, (Universitetsförlaget, Oslo, 1972), pp. 61–73. (Prawitz, J., Sweden – A Non-Nuclear Weapon State.)
3. A statement by the US Secretary of Defense Melvin R. Laird on 'smaller, cleaner tactical nuclear weapons and new doctrines for their use' referred to by Beecher, W., US considers deployment of smaller A-arms in Europe. *International Herald Tribune* (17 April 1972).
4. Testimony of General Edvard B. Giller of the US AEC. Military Applications of Nuclear Technology. Hearing before the Subcommittee on Military Applications of the Joint Committee on Atomic Energy, Congress of the United States, (16 April 1973), Part I.
5. Testimony of General Andrew J. Goodpaster, Supreme Allied Commander, Europe. Same Hearings as in Ref. 4, (29 June 1973), Part II.
6. Statement by Ian Gilmour, British Minister of State for Defence, in Parliament, (10 May 1973). Verbatim Service 076/73.
7. Beecher, W., Over the threshold, *Army*, (July 1972), p. 17.
8. Lyons, W. C., *On Deterrence in Depth*. Mimeographed. (November 1972). This is the version that played a role in Sweden.
9. Bennett, W. S., Sandoval, R. R. and Schreffler, R. G., A credible nuclear-emphasis defence for NATO, *Orbis*, Vol. XVII, No. 2, (1973).
10. Douglas Home, C., Miniature nuclear weapons invented by Pentagon for use on battlefield. *The Times* 7 May 1973. This article reflects hesitation on mini-nuke deployment in Europe.
11. Thunborg, A., *Evolution of Doctrines and the Economics of Defence*. Swedish Institute 1973. Mr Thunborg's conclusion was: 'Naturally, one cannot overlook the possibility that the US, or both the US and the Soviet Union, will acquire mini-nuclear

weapons. In that case it lies readier to hand to see these weapons as a further extension of the escalation potentials and as a strengthening of the deterrent. This means that the political decision to deploy nuclear weapons will continue to be fraught with the same risks and that the nuclear threshold will remain high. A technical development of these weapons is now a bit under way. The fact that they are being analysed by the great powers also makes it incumbent on us to discuss them in doctrinal terms and from the protective aspect'.

12. Parliamentary documents (in Swedish) Prop. 1974:1. Bil. 6 p. 11; Foreign Policy debate, (20 March 1974); and others.
13. Mrs Myrdal has commented on mini-nukes and her own intervention in a recent monograph of hers. Myrdal, A., *The Game of Disarmament*, (Pantheon, New York, 1976), pp. 45–8.
14. The technological background for the discussion of the mini-nuke issue within the Swedish Government was prepared by the Atomic Department of the Swedish Defence Research Institute.
15. Document NPT/CONF/35/I, Annex II, p. 33.

Appendix

The issue of mini-nukes was introduced in the Geneva Disarmament Conference in August 1973 by the Swedish Minister of Disarmament, *Mrs Alva Myrdal*, who put five questions to the nuclear powers (CCD/PV. 620 pp. 14–15, 9.8.73):

Since the very entering into force of NPT there has been widespread recognition that it was not likely to survive because of its inherently discriminatory nature, if the Super Powers did not take 'effective measures relating to cessation of the nuclear arms race at an early date'. The SALT agreements reached in Moscow last year and the intentions expressed in Washington this year have rightly been hailed as promising steps in the right direction. Simultaneously, however, developments seem to be under way, threatening to render the NPT even more discriminatory to the disadvantage of the non-nuclear-weapon States. I am referring to news items that the major nuclear-weapon States are about to launch a new generation of *tactical* nuclear weapons systems, the so-called mini-nukes. Such a development would drastically aggravate the nuclear threat against non-nuclear-weapon States everywhere.

There has for a number of years been a common understanding that the effects of a strategic nuclear war would be such that no one could contemplate triggering such a war. The SALT I agreements in a sense codified this understanding. A similar understanding has developed as regards the tactical nuclear weapons. One reason for this is that the consequences of a nuclear campaign using tactical weapons for the civilian population living among the military targets would be similar to the effects of a strategic nuclear attack. More important to the civilian population of the leading nuclear-weapon Powers would be the formidable risk of an escalation to a higher level of nuclear exchange.

Therefore, it is so disturbing when we learn that ongoing research and development might lead to a new generation of tactical nuclear weapons with yields in the subkiloton range, overlapping the yields of the most powerful conventional charges, with extreme delivery precision, and with extra-accurate intelligence support. These weapons systems are by their proponents said to be not only usable on the battlefield but also preferable as providing cheaper fire-power than conventional systems.

Most important is, however, that an introduction of such mini-nuclear weapons would blur the present distinction between conventional and nuclear weapons. We are strongly of the view that an absolute 'firebreak' must be kept up between nuclear and conventional war.

Obviously, the introduction of mini-nukes and a decline of the nuclear threshold would create enormous proliferation risks. The main purpose of the NPT – to reduce risks for nuclear war – would be countered. This would occur at a time when in many countries a growing nuclear industry will produce considerable stockpiles of excess plutonium. Military arguments for acquiring nuclear weapons may again come to make themselves heard, albeit with different strength from nation to nation.

Without any doubt, such a new development in regard to tactical nuclear weapons would affect the very premises on which adherence to the NPT is based. In the view of the non-nuclear-weapon Powers, it is the tactical nuclear threat, rather than the one pertaining to strategic nuclear weapons, that today causes anxiety on their part. It is the option to produce ordinary and tactical nuclear weapons, if only of the 'old-fashioned' Hiroshima-size, which is primarily 'sacrificed' by adhering to the NPT.

There are enormous risks involved in unsettling the status quo between nuclear- and non-nuclear-weapon Powers by introducing changes in the tactical nuclear weapons capabilities. If the nuclear-weapon Powers were to enter upon a new race to 'improve the usefulness' of tactical nuclear weapons, a fundamentally new situation would be developing above our heads. It would affect possible war theatres around the globe, and could certainly not be restricted to scenarios for possible confrontations between the Super Powers themselves.

Fortunately, responsible statesmen in various countries have expressed themselves in favour of a continued high nuclear threshold and of regarding all nuclear weapons as qualitatively different from conventional ones.

As it would take a much shorter time to change the doctrines for the use of these new systems than to develop and procure them, my delegation considers it essential that the development of mini-nukes be stopped now.

This is such a serious matter that I must implore the delegations of nuclear weapons Powers, and most directly our two Co-Chairmen, to reply rapidly and fully to the following questions:

> 1. Is it true that a new generation of tactical nuclear precision weapons with subkiloton yields are being developed and tested?
> 2. Are plans for the development of such systems a reason for further testing and thus an obstacle to the conclusion of a comprehensive test ban?
> 3. Is it true that preparations are being made for the early deployment of such systems?
> 4. Does the recent agreement between the United States and the USSR on the prevention of nuclear war refer also to wars where no nuclear weapons are used except such mini-nukes?
> 5. Does Security Council resolution 255 (1968) in the interpretation of the Co-Chairmen refer also to nuclear aggression in which no nuclear weapons are used except such mini-nukes?

Complete clarity on these points of major concern must be given. I sincerely hope that the answers can be so reassuring that no State party to the NPT will interpret the situation as an extra-ordinary event which has 'jeopardized the supreme interests of its country', and act accordingly. This is, of course, quoting the withdrawal clause of the NPT. On the contrary, I hope that the answers will remove all fears. Efforts must be encouraged to stop a threatening tactical nuclear weapons arms race by means of appropriate agreement here in this Committee, at SALT II, and at the 1975 NPT review conference, when amendments could take care of dim or disputed points.

First to answer was the United Kingdom delegate, *Mr H. C. Hainworth* (CCD/PV. 625 pp. 18–19, 28 August 1973):

I come now to that part of Mrs Myrdal's speech in which she posed five precise questions. Perhaps it would be for the convenience of the Committee if I recall that the first three of these questions were worded as follows:
    1. Is it true that a new generation of tactical nuclear precision weapons with subkiloton yields are being developed and tested?
    2. Are plans for the development of such systems a reason for further testing and thus an obstacle to the conclusion of a comprehensive test ban?
    3. Is it true that preparations are being made for the early deployment of such systems?
Current press and other public interest in the question of the development of smaller-yield tactical nuclear weapons was rekindled, I believe, by an article in *The Times* (London, 7 May 1973). This article was subsequently referred to in other newspapers and gave rise to a number of articles commenting on it. There is, however, nothing new in the idea of 'mini-nukes', as they have become known; the feasibility of relatively small-yield tactical nuclear weapons has been recognized for some time, and it would doubtless be possible to link these with the precision guidance techniques which have already been applied to conventional weapon systems.

The production and deployment of a new generation of tactical weapons could of course raise very important political issues, some of which have been mentioned by Mrs Myrdal. For their part, the British Government have always been conscious, and remain very conscious, of the importance in political as well as military terms of the distinction between nuclear and conventional weapons and of recognizing this in any plans relating to their use. Mrs Myrdal's questions were addressed to the representatives in this Committee of the three nuclear Powers. Clearly I can speak only for the British Government. British Ministers have already made it clear in Parliament that no proposals for the deployment of such weapons within Europe have so far been made by the United States to their NATO allies. As far as the United Kingdom is concerned, no decision has been taken to develop, test, or deploy a new generation of small-yield tactical nuclear weapons. From our point of view this, I believe, answers the first three of Mrs Myrdal's questions.

However, with regard to question two about the development of such systems becoming an obstacle to the conclusion of a comprehensive test ban, I should myself like to ask a question of the Swedish delegation. It must be obvious from all that the Swedish and other seismological experts have said on this subject, particularly at our informal meetings in July, that the testing of such small-yield devices underground could not be detected or identified seismologically from outside the territory of even a medium-sized State. Yet the Swedish statement of 9 August suggested that the development of such miniaturized nuclear weapons might be a matter of such seriousness and concern that a State might feel that its supreme interests had been jeopardized. My question is: how would the Swedish Government, or indeed any other government, establish the fact that such weapons had been developed, tested, and deployed? Surely this is exactly the sort of issue which shows just how important it is to ensure that a comprehensive test ban should be adequately and effectively verified.

Mrs Myrdal's fourth question, which concerned the agreement between the United States and the Soviet Union on the prevention of nuclear war, was not of course addressed to the United Kingdom delegation.

The fifth of Mrs Myrdal's questions read as follows:

> 'Does Security Council resolution 255 (1968) in the interpretation of the Co-Chairmen refer also to nuclear aggression in which no nuclear weapons are used except such mini-nukes?'

Although it was not specifically directed to my delegation, I consider it right and proper that I should reply to it today. In the view of the British Government, Security Council resolution 255 (which was, you will recall, tabled jointly by the delegations of the United Kingdom, the United States, and the USSR on 12 June 1968) refers to nuclear aggression of any type whatsoever; and the statement made by the United Kingdom representative at the 1430th meeting of the Security Council is to be read in that sense.

In May 1974, the Swedish delegate, *Mrs Inga Thorsson*, reminded the two other nuclear powers of Mrs Myrdal's questions (CCD/PV. 633 pp. 9–10, 9 May 1974):

> . . . A CTB is not only connected with the development of strategic nuclear weapons. It is also important to underline the danger that an introduction of the so-called mini-nukes would entail. Such new tactical nuclear weapons would lower the nuclear threshold and could also lead to increased proliferation risks. Mrs Myrdal on 9 August last year (CCD/PV.620) put five questions concerning mini-nukes to the delegations of the United States and the Soviet Union. We attach the utmost importance to these matters and still hope for a reply to those questions.

Two weeks later the US ambassador, *J. Martin*, addressed the CCD with a carefully prepared statement on the question of mini-nukes (CCD/PV. 638 pp. 27–29, 23 May 1974):

> I wish to turn now to a subject introduced last year by the former leader of the Swedish delegation, Mrs Myrdal, and raised again by Mrs Myrdal's successor, Mrs Thorsson, at our 633rd plenary meeting. In her statement on 9 August last, the Swedish representative referred to newspaper accounts of the development of a 'new generation' of tactical nuclear weapons, so-called 'mini-nukes', which, in her words, 'would blur the present distinction between conventional and nuclear weapons'. Such a development, she said, would drastically aggravate the nuclear threat against non-nuclear weapon States and 'affect the very premises on which adherence to the NPT is based' (CCD/PV.620, p. 14). I should like to explain why such fears are wholly unwarranted.
>
> The term 'mini-nukes' is misleading in two important respects. First, the coinage of this new catchword in itself conveys the false impression that we are talking about a radically new and futuristic family of weapons. Secondly, the diminutive element of the term 'mini-nuke' falsely suggests some miniature nuclear device which can be handled and used in the same manner as conventional weapons. I would like to correct both of these misimpressions.
>
> We are not now, nor have we been in recent years, at the brink of some qualitative breakthrough in tactical nuclear weapons development. Instead, as

Secretary of Defense Schlesinger has stated, we have been engaged over many years in a gradual process of moderately upgrading our tactical nuclear stockpile. There is nothing new about the existence of tactical nuclear weapons of very low explosive yields. Indeed, it is public knowledge that nuclear weapons with explosive yields measured in the sub-kiloton range were introduced in the United States inventory many years ago. No decisions have been made to produce or deploy any new systems.

However, the main issue raised by the Swedish representative is not whether or when any particular improvements in the tactical nuclear stockpile have taken place, but rather how any such improvements would affect our objective of preventing the outbreak of war, particularly nuclear war. In deciding questions of possible modifications in the United States tactical nuclear weapons stockpile, we are governed by the objective of reducing the risk of nuclear conflict. That is to say, any improvements in our tactical nuclear weapons stockpile must make deterrence more effective and thereby reduce the likelihood of nuclear warfare of any kind.

In response to speculation that further development of low-yield tactical nuclear weapons would blur the present distinction between conventional and nuclear weapons, I wish to state categorically that the United States Government has no intention whatever to treat such tactical systems as interchangeable with conventional arms. We fully appreciate that the distinction, or 'firebreak', between nuclear and non-nuclear arms is a major factor in preventing nuclear warfare, and we will not act to erode this distinction. The very special command-and-control and safety arrangements that apply to nuclear weapons in general have of course always applied to small-yield nuclear weapons as well.

I think that, in making these comments, I have answered two of the questions posed last August by the former leader of the Swedish delegation, specifically her questions dealing with the development and deployment of sub-kiloton nuclear weapons. With regard to her question whether development of such tactical weapons would be an obstacle to a comprehensive test ban, members of the committee will recall that, in response to a similar question by Sweden in 1972, I stated that the United States stands ready to give up the advantages derived from nuclear weapon testing if we can be assured that others are abiding by the same restrictions (CCD/PV.580). That answer applies to the testing of any nuclear weapons, including tactical ones.

Finally, Mrs Myrdal asked last August whether the Agreement between the United States and the USSR on the Prevention of Nuclear War would apply to wars in which only 'mini-nukes' were used, and also whether the United States interprets Security Council Resolution 255 (1968), concerning security assurances, as applying to nuclear aggression in which only 'mini-nukes' were used. I have explained that my Government will continue to treat all tactical nuclear weapons in the same manner. Therefore, I can answer both of these questions with an unequivocal 'yes'.

Mrs Thorsson later expressed her appreciation for the two replies and again invited the representative of the USSR to answer the questions of her predecessor (CCD/PV. 647 pp. 7–8, 30 July 1974):

One particular aspect of the recent development of nuclear weaponry which has caused concern to my Government is the interest shown in making and deploying nuclear mini-weapons, i.e. nuclear weapons with very small yields, close to or even overlapping the yield range of conventional weapons. Last year my predecessor Mrs Myrdal posed a number of questions to the nuclear-weapon Powers in the CCD about the situation as regards those weapons (CCD/PV.620, pages 14, 15). I am indeed very pleased to note that two of the delegations concerned have now replied to those questions.

I first wish to thank the representative of the United Kingdom, Mr Hainworth, for his statement (CCD/PV.625) last year that the United Kingdom has taken no decision 'to develop, test, or deploy a new generation of small yield tactical nuclear weapons', and that

> 'in the view of the British Government, Security Council resolution 255 refers to nuclear aggression of any type whatsoever, and the statement made by the United Kingdom representative at the 1430th meeting of the Security Council is to be read in that sense.'

I also want to thank the representative of the United States, Mr Martin, for the comprehensive replies he gave on 23 May. We are particularly happy to note his explicit statement

> 'that the United States Government has no intention whatsoever to treat such tactical systems as interchangeable with conventional arms. We fully appreciate that the distinction, or 'firebreak', between nuclear and non-nuclear arms is a major factor in preventing nuclear warfare, and we will not act to erode this distinction' (CCD/PV. 638, page 28).

My delegation also attaches the utmost importance to Mr Martin's unequivocal 'Yes' (*ibid.*, page 29) to our questions whether the agreement between the United States and the USSR on the prevention of nuclear war would apply to wars in which only 'mini-nukes' were used, and also whether the United States interprets Security Council resolution 255 (1968) concerning security assurances as applying to nuclear aggression in which only 'mini-nukes' were used (CCD/PV.620, page 15). We think that the position taken on the issue of nuclear mini-weapons by the Governments of the United Kingdom and the United States is reassuring. It should remove one of the potential dangers to the NPT regime, something which is badly needed in these days. We hope that a similar statement will soon be forthcoming from the representative of the USSR. This would enable the CCD to close the issue.

The Soviet representative, Mr A. Roshchin, replied later in the same summer (CCD/PV. 650 p. 13, 8 August 1974):

I should like to take this opportunity to touch on the question of 'mini-nukes', that is miniature nuclear weapons, raised last August by Mrs Myrdal, the representative of Sweden, and again in the statement by Mrs Thorsson. In the last few decades nuclear weapons have received special treatment in a number of international instruments. So far there has been no need for distinctions among individual types of tactical nuclear weapons. We consider that they are still unnecessary. It seems to us that the attempts to equate certain types of these weapons with conventional armaments are dangerous in that they may conceal a search for loopholes to circumvent the prohibitions on nuclear weapons imposed by a number of international treaties and agreements, such as the Non-Proliferation Treaty. We believe that there is no justification for distinguishing among individual types of tactical nuclear weapons, or for trying to equate some types of those weapons with conventional armaments.

In accordance with this position, I should like to state that the Soviet Union's obligations under resolution 255 (1968) of June 1968 adopted by the United Nations Security Council in connexion with the Non-Proliferation Treaty, and under the Soviet–United States agreement of 22 June 1973 on the prevention of nuclear war, cover all types of nuclear weapons whatever their power.

In expressing her thanks to the Soviet ambassador, the Swedish delegate closed the debate on mini-nukes in the CCD at the following meeting (CCD/PV. 651 pp. 10–11, 13 August 1974):

Finally, Mr Chairman, I want to express the thanks of the Swedish delegation to you in your capacity of representative of the Soviet Union, for your reply to Mrs Myrdal's question on nuclear mini-weapons at our last meeting. We are particularly happy to note your statements that 'there is no justification for distinguishing among individual types of tactical nuclear weapons, or for trying to equate certain types of those weapons with conventional armaments'; and that the Soviet Union's obligations under Security Council resolution 255 (1968), and under the Soviet–United States Agreement on the Prevention of Nuclear War, 'cover all types of nuclear weapons whatever their power' (*ibid.*, page 13). I share your view that equating nuclear mini-weapons with conventional arms would, among other things, provide a danger to the NPT. Your statement and the previous statements of the representatives of the United Kingdom and the United States on this issue are, taken together, reassuring. The Swedish delegation welcomes the fact that, although the issue of nuclear mini-weapons obviously will remain under constant review, it can under the present circumstances be considered closed in this Committee.

By this exchange of views, the debate on mini-nukes in CCD closed.

9. Tactical nuclear weapons and European security

H. Afheldt

I. Introduction

For more than twenty years, nuclear weapons have been present in Europe as operational aids to military aims (tactical nuclear weapons). During this time, Central Europe has been threatened with annihilation by these weapons should deterrence fail. Every attempt to withdraw these instruments from European territory has foundered. Why?

In NATO's concept of deterrence and defence, tactical nuclear weapons play an important role, particularly for deterrence, and consequently cannot be abandoned. NATO's most important reason for keeping tactical nuclear weapons in Europe is the superiority of the WTO in 'conventional weapons'. The tactical nuclear weapons of the WTO are justified by the existence of NATO's tactical nuclear weapons. All attempts to induce the WTO to reduce this conventional superiority (MBFR) have been just as fruitless as attempts to withdraw the tactical nuclear weapons from Europe.

Manifestly, it is unrealistic to want to eliminate military resources to which are given important functions in the security policy of the participating states.

Whatever function the superiority in conventional striking power of the WTO, maintained with great expenditure, still has today, a review of this armaments policy needs to be made, because it is *one* key to the elimination of tactical nuclear weapons in Europe. But this is a matter for review by the states comprising the WTO and their citizens. It is to be hoped that this review will soon be undertaken.

A corresponding review of the security policy of NATO is equally essential. Whether tactical nuclear weapons in the 1990s will serve or hinder this security policy will finally decide if NATO maintains, expands or reduces these arsenals. The nature of NATO's military preparations will then influence the interest of the WTO in the maintenance of conventional superiority and the placing of tactical nuclear weapons in Europe.

II. Arguments for a review of NATO policy

In addition to the points just made, there are other objective grounds for a review of the security policy of NATO.

The present strategy of NATO, the 'flexible response', was already formulated in its basic outlines at the beginning of the 1960s, when the strategic nuclear capacity of the USA was still far superior to that of the USSR.

By the end of the 1960s, the USA's phase of strategic nuclear superiority was finished. The real parity between the strategic striking powers of the USA and the USSR made a strategic strike by the USA against the USSR suicidal. Consequently the credibility of the uppermost level of the threats of escalation in a flexible response strategy was lost, and with the disappearance of this ultimate degree of threat [2], the strategy itself became unconvincing.

Consequences of strategic parity reveal themselves also in arms-control policy. SALT I, the first agreement on the numerical limitation of strategic nuclear weapons, manifested in 1972 the intention of the two great powers to safeguard their survival by arms control, above all excluding the danger of one side's having superiority in strategic weapons (strategic instability).

Yet the foreseeable [3] consequence of SALT I was not stability of strategic armaments but a qualitative arms race. This took the form of increase in accuracy of delivery of the missiles and increase in the effect of a missile by multiplying the number of its warheads (MIRV and MARV). This qualitative arms race was carried out so far by both sides that nowadays it is surmised in some quarters of the USA, that the strategic equilibrium has been or will be lost to the advantage of the USSR.

At the same time, stocks of armaments on other levels, conventional and tactical nuclear, have on both sides been increasing.

The growing danger of strategic instability and continuance of the armaments race also on sub-strategic levels shows that the most important objectives of the arms-control policy were not realized.

A new formulation of the security policy of all states involved is therefore necessary. A new strategy must be developed, bearing in mind from the very start the reasons for possession of military means in the nuclear age: to keep peace in the long run, and in the short run to avoid escalation in a crisis and in a war, should deterrence fail nevertheless.

The first step towards fresh formulation of NATO's security policy in Europe is to determine criteria under which this policy must be seen. If these criteria are established, a second move can be made by way of example to sketch a new policy for security of peace with military resources, which will satisfy the above-mentioned criteria.

III. Criteria for peace policy with military means

3.1. When do military options serve the political purpose?

3.1.1. *Results of the recent US discussion on strategy.* In attempting to describe the US strategy one encounters a series of problems. The State Department, the Pentagon, Chiefs of Staff and individual military forces, Congress and President do not pursue consistent interests in all points. Any assessment of the US strategy is therefore a compromise; every formulation an act of diplomacy. An official US document [4] can be used to assess briefly the result of American strategy discussion up to the end of Ford's presidency, in order to obtain the foundations for a new security policy in Europe, but it is necessary to remain conscious of the range of variation in the individual interests thus brought together.

In *Verteidigung und Frieden* [5], eight 'norms' have been assembled to describe the US world power policy and military policy of the Ford era. For the purposes of this study, the most important of these are:

Norm A

Military strength is the basis for continuing the role of USA leadership in the world.

Norm C

Deterrence is no substitute for defence. Rational and feasible military options are the pre-condition of every credible deterrence.

Norm E

1. A greater imbalance to the disadvantage of the USA must not develop, either in total weight of strategic striking power, or in single factors such as total payload of missiles, accuracy of delivery, payload in relation to total weight, reliability or other factors which are of significance for the balance either of military efficiency or even the way the non-nuclear nations perceive the balance between the two great powers.

2. The capacity of the USSR for retaliation shall not be called into question. The capacity for retaliation by the USA is to be safeguarded.

3. The USA must be prepared and able to make limited strategic nuclear strikes (flexible strategic response, facility for coercive action).

Norm F

Tactical nuclear weapons in Europe are needed for deterrence and for possible military use. They should:
1. deter the use of tactical nuclear weapons by the WTO;
2. ensure flexible deterrence over the whole range of possible threats;
3. offer nuclear military options below the strategic level;
4. help to smash large-scale conventional attacks.

They furthermore serve the political purpose of demonstrating to the European allies the USA's commitment.

Norm G

1. Non-nuclear deterrence (deterrence with conventional means) is possible and necessary in Europe.

2. Conventional deterrence from the most important American strongpoint, Europe, demands strong and flexible forces, which defend the territories

of the NATO states, and can, consequently, also successfully attack the territories of the WTO states. Ability to attack in Central Europe is also necessary in order to compensate for weaknesses in NATO's flanks.

3. NATO must be able to conduct a non-nuclear war in Europe for an unlimited time.

Norm H

Sea-striking power, especially striking power for anti-submarine warfare must be so consolidated that it can ensure the maintenance of supplies across the Atlantic during a prolonged war in Europe.

3.1.2. *Doctrines from the recent US discussion on strategy.* The most important norm of the new US strategy is Norm C:

> 'Deterrence is no substitute for defence. Rational and feasible military options are the pre-condition of every credible deterrence'.

The only rationally feasible options (from the American viewpoint) are those which the USA retains reliably under control, so that *on* and *after* initiating this option, no unacceptable damage results for the USA.

If a conflict escalates to a major exchange of strategic strikes, then the damage is unacceptable for the USA. All rational US planning must therefore guarantee that escalation into this range is excluded (cf. Norm E.2).

It follows therefore that all options on military operations in Europe must be securely protected against escalation in the strategic range.

The simplest solution to the task defined according to these premises is given in the alternatives provided by the new US plans for Europe:

1. Establishment of a conventional defence for Europe, if need be until the enemy is exhausted, hence for unlimited time (Norm G.1 and 3).

2. Preparation of a limited tactical nuclear war (theatre nuclear war) in Europe for the case where either the conventional defence collapses (Norm F.4) or the enemy for his part attempts to destroy (Norm F.3) or could be tempted to destroy the strong conventional defence by means of nuclear weapons. In this strategy, nuclear weapons cannot therefore be abandoned. Hence, an extension of these arsenals by both sides would be expected in the 1980s.

These facts and their theoretical foundations can be used to derive criteria for a security policy which is rational not only for the USA but also for her European allies. With the proposition that 'rational and feasible military options are the pre-condition of every credible deterrence', the US security policy has to a great extent abandoned deterrence by incalculable risk, which hitherto governed NATO's security policy. The extension of rational feasible options allows war to be conducted in case of need, even between the great powers. Therefore neither the West nor the East Europeans can restrict themselves to military planning solely for 'pure deterrence', i.e. prepare for peace and crisis. They must also have a rational plan for the situation in which war comes in spite of deterrence. Planning military preparations only up to the crisis, only for the prevention of war, would nowadays be equivalent in the case of nuclear reactors of putting aside the question of possible mishaps to reactors on the grounds that mishaps are not the concern of reactors – reactors serve the purpose of supplying current [6].

So the necessity for rational planning, even for the situation where deterrence fails, follows from the possibility of war. The fact that deterrence itself is not credible if the options with which one threatens are fundamentally violating the particular interest of the one who is threatening, is a further cogent argument for rational planning for emergency [7].

From the above we derive

> **Criterion 1.** *Every (not only the US) policy of security based on military means must provide options that are rationally feasible if war comes in spite of deterrence.*

This leads directly to

> **Criterion 2.** *Only those options are rational which can guarantee that no unacceptable damage happens to Central Europe if war comes in spite of deterrence.*

> **Criterion 3.** *Damage will be unacceptable to all participants if war escalates into heavy strategic strikes. The certain exclusion of strategic nuclear war is therefore equally of European as well as American interest – and also Russian.*

Apart from this exclusion of escalation in the strategic range if deterrence fails, the recent American concepts of defence for Europe do not, from the viewpoint of the European allies, comply with the criteria developed in the American discussion. Neither a long-lasting conventional, nor a mixed conventional – tactical nuclear defence of Europe is rational for the European allies in the event of war, especially for the potential battlefield of the FRG. The destructions by these forms of war exceed what is tolerable for the industrial nations of Central Europe: they destroy that which should be defended.

The strengthening of Central Europe, or even the use of the FRG as a 'strong-point' [8] in the event of conflicts outside NATO territories, attracts (should deterrence fail) the danger of a war in the most densely populated area (Federal Republic of Germany, German Democratic Republic, Czechoslovak SSR), an area where no rational military option is available which will preserve the social and economic structure. The West European countries are industrial societies. An industrial structure, depending on export and growth-orientated in a densely populated country is directly opposed to the options for conducting a war.

> **Criterion 4.** *The ability of modern industrial societies (e.g., West and Central European countries) to tolerate damage is very limited [9].*

The requirement to minimize damage caused by an option drastically restricts the range of military resources that can be used; because with the destruction of each militarily important objective collateral damage is caused. How much depends on the disposition of the military target and on the weapons used for its destruction. The weapons used by an enemy will in turn depend on what weapons are available to him and how much resistance the objective can offer. The more destructive the means an enemy must use in order to overcome defences, the greater will be the collateral damage. As qualitatively unlimited means of destruction are available to the WTO in greater quantity, the more targets the NATO defences offer, the greater will be the damage caused in attempts to overcome them.

This phenomenon, which upsets the traditional rules of war according to Clausewitz, is not new. It was first discovered and discussed when there was threat of destruction to the great powers on the nuclear-strategic level. A principle was developed for the strategic level, opposing unlimited destructive power, which is therefore a criterion for a policy of peace with military means.

Criterion 5. *If an enemy has weapons which can cause unacceptable damage, and which cannot be intercepted, then rational strategy requires care to be taken that the enemy does not find a meaningful or even decisive military use for these weapons. (Principle to construct no targets whose destruction is militarily rewarding [10].)*

To permit, for the military security of Western Europe, only such weapon-systems and strategies as do not lead to the construction of militarily important and obvious targets has extraordinarily far-reaching consequences.

Every front defence line constitutes a target, whether it be fixed (Maginot Line), or mobile (NATO's forward-defence). Mobile forward-defence requires the sealing-off of intrusions by counter-attacks from previously disposed strong formations. The White Book for 1975–76 of the Federal Government describes this matter as follows:

'The defence against a numerically superior enemy with modern equipment must be mobile. The troops must be able to form strong points rapidly on the battlefield and make counter-attacks. Tanks and fighter bombers are of vital importance in this connection.

According to the NATO concept of forward defence, the integrity of the territory of the member states is to be preserved and territorial losses restored. . . . This entails the ability to carry out counter-attacks. To this end the Bundeswehr must be equipped and trained. Tactical attack is a component of strategic defence' [11].

Thus mobile forward defence develops an enormous pattern of militarily important targets on both sides, which really forces destruction of the FRG. The protective effect, which the forward defence should have above all for the FRG, becomes the opposite.

We then have the (negative)

Criterion 6. *Defence at the borders of the FRG (forward defence) can have no rational feasible military function for NATO, because it offers the WTO a multiplicity of targets, the destruction of which is worthwhile militarily and hence leads to the destruction of Central Europe.*

The fact that forward defence (i.e. defence of the borders) no longer serves the intended political purpose contradicts conventional thinking. Yet a glance at the history of military means shows that defence of borders represents the exception rather than the rule. Thus defence of the Great Wall of China or the *Limes* of the Roman Empire, or the defence of a front, which in any case protected parts of the national territory (e.g. France in World War I) was possible only under quite specific technical, social and geostrategic conditions. When Japan capitulated after the dropping of two atomic bombs, the end of meaningful defence of borders had really been demonstrated.

In addition, defence of borders is successful only if it prevents the enemy having access to the defended territory for however long the attack may last. Yet 'a conflict lasting any great length of time on the territory of the Federal Republic of Germany . . . would finally destroy the substance of that which should be defended' [12]. The special structure and the specific geostrategic position of the FRG makes lengthy conflicts particularly inexpedient and irrational for the FRG. Yet it is a general feature of modern wars that extended conflicts cannot serve the political aim.

Considering the consequences of warfare in World War I, German Army losses up to the time of freezing of the fronts in November 1914 amounted to 8 per cent of the dead and 13 per cent of the wounded; 92 per cent of the dead and 87 per cent of the wounded were sacrificed only after freezing of the fronts. The losses of the allies were similar [14]. Likewise, two-thirds of the losses in the Korean War were sustained by the belligerents after commencement of the two-year negotiations, most of the losses even occurring after the demarcation line, which was subsequently accepted, had been negotiated [15]. The military efforts of both sides procured this 'gain' in time.

Thus we find that

the true catastrophe of modern wars has usually been not the 'showdown', but the military efforts following the show-down, which made it possible to extend the time for negotiations (often for years).

Therefore the political purpose of military options for the defence of Western Europe cannot be the *greatest possible* gain in time. The purpose of employing military means is rather the classical aim of defence, namely, 'To gain time until conditions are more favourable' [16], i.e. until the other side can be made to desist from his aggression. That means, however, until he makes the *political* decision to abandon the attack, and this in turn means, until a politically successful conclusion to negotiations is achieved. It no longer follows that the best options are options for defence for a very long time. This conclusion would be logical only if the time period for the completion of negotiations were a fixed quantity. Yet that is not the case.

Thus it can be shown in the case of the World War I that from the end of 1914 until September 1918 no serious attempt was made by either side to press negotiations to a successful conclusion, because the military situation on both sides permitted waiting and hope [17]. The same is valid for the Korean War and the Viet-Nam Wars of France and the USA.

On the other hand, when there has been only very limited time available, either because one of the two sides had no means of preventing defeat, or because conduct of the war threatened to become a catastrophe for both sides, then agreements have been successfully concluded within days or even within hours. This has sometimes been in the form of capitulation by the weaker (e.g. Denmark in 1940), if capitulation alone satisfies the interests involved, or in the form of an agreement which restored the *status quo* if both belligerents feared the start or continuation of the conflict (the Cuba crisis of 1962). The interests of third parties, too, in particular of the great powers, have compelled the conclusion of truces within hours or days, when the military position inclined to a result which was inconsistent with their interests. This was the case, for example, for the conclusion of the British–French–Israeli Suez Intervention in 1956, for the Seven-Day War in the Near East in 1967 and for the Yom-Kippur War in 1973. Thus historical experience proves that

> *The time that elapses up to successful conclusion of negotiations is a function of the military situation.*

However, how can an enemy be quickly forced to restore the *status quo*, and without great havoc? It is impossible to consider this question separately from politics. Even in the 'primitive' way of conducting war – war to knock out the enemy's military equipment, 'murder war', as T. E. Lawrence called the war of Foch and Ludendorff – it was in the end a politically fatal error not to make the political conditions *after* conclusion of the war the criterion for conducting the war, but making 'victory first' the military aim. Previously it had been possible to develop a military strategy for this form of war, which would serve this short-sighted aim. If, however, a conclusive war employing all available military means is no longer a rational instrument, the military strategy adopted must be so arranged that it leads to *political* breaking off of the conflict after a short time; then this is a political war *par excellence*.

In this case the military aim can no longer be to knock out the enemy's military equipment, but rather it must become a direct lead up to the political aim. If the political aim is to protect the freedom and independence of the FRG or some other European NATO state, one must ask 'Upon what conditions would it depend?' Whether the USSR abandons the attempt, for example, to conquer or to blackmail the FRG? Obviously this depends on the intention of the Soviet government. The object of employing the military resources of NATO is therefore to influence such an intention.

The first task of NATO's forces is therefore to make possession of the FRG, or part of it, as costly as possible for the WTO. So costly, in the ideal case, that the disadvantages of continuing the aggression are greater than the advantages for the WTO, so that in consideration of its own interests it decides to stop the aggression.

However, NATO's military resources in Europe are subject to the restrictive criteria developed above. Therefore one cannot assume that with such limited resources it will be possible under all conceivable conditions to make the disadvantages for the enemy outweigh the advantages.

Instead, the military resources which must exert a decisive check on the intentions of the enemy are those comprised in the power of the USA. Under the particular conditions in Central Europe, a decision on the extension of the spheres of influence of the great powers (e.g. incorporation of West Berlin or the FRG in the Eastern bloc) is therefore in the end not dependent on the battle of the 'proxies', i.e. on the contest of the FRG with the GDR, or of the WTO with NATO, but on the 'struggle of intentions' (Clausewitz) between the USA and the USSR.

The battle of the proxies can therefore only be critical for the final outcome of the struggle of intentions of the great powers if it creates a *fait accompli* before this struggle of intentions is established in all seriousness.

Hence the following criterion is valid:

Criterion 7. *If the USSR wants to prevent the start of a struggle of intentions between the great powers, it must try to create a* fait accompli *quickly.*
The converse conclusion follows:
If the European NATO-states want to gain use of American power in the decisive struggle of intentions, they must prevent a fait accompli. *This is the root of the concept – to buy time.*

This political aim does not contradict the requirement for the rational options to enforce a decision *quickly*. A war which lasts weeks or months *without* decision cannot fulfil any political function. The old function of war, to make the relation of power clear, is not served, since the balance of power between the final decisive powers is obvious. The recent function of testing the *resoluteness of the intentions of the great powers*, the will to risk using their ultimate means of power in order to bring about a political decision, cannot be fulfilled by an extended war, because nothing is decided by conducting an indecisive war in Europe for a long time. On the contrary, so long as both great powers 'remain secure' – and that is their paramount common interest – such a war only puts off the political decision between them. It proves only that neither of the opponents has *so far* decided upon capitulation. Such an undecided limited war (from the standpoint of the great powers) therefore serves simply to postpone the political function of the test of intentions.

The period of irresolution is therefore non-functional, destructive, and should be ended as quickly as possible. This leads to

Criterion 8. *The options available to NATO military means must on the one hand prevent a* fait accompli. *Yet on the other hand they must serve that aim for a limited period only, because the time of the negotiations is a function of the ability of these means to hold out, and it is one of the requirements for rationality of the options to limit the negotiations to a short space of time.*

What then should the decisive test of the intentions of the great powers be like? Such a struggle between great powers can no longer ignore the existence of the strategic nuclear arms on both sides. These means of exercising power are part of reality, and to exploit them to political advantage is therefore in the interest of the great powers.

That leads to the question: 'What criteria must nuclear options satisfy in order to be useful in turning strategic nuclear power to political benefit?'

In developing the criteria for such use of nuclear means of exercising power we are helped by the findings of the recent American discussion on strategy [18]. In *Verteidigung und Frieden* [19] it was shown that the fundamental principle of intimidation may be formulated thus:

Criterion 9. *Fundamental principles of coercion.*
1. *The ultimate target for military action is the government of the enemy.*
2. *The action serves exclusively the purpose of influencing the intentions of this government in such a way that the enemy neither realizes the technically possible action of his own military destructive resources nor can threaten with them credibly.*
3. *The military action must threaten the enemy with disadvantages unacceptable to his government.*
4. *Dissolution of sovereignty in particular is unacceptable for this government: total destruction of the territory and death of a great part of the population by a nuclear attack on the populated areas is admittedly a conceivable means of causing unacceptable damage for the government, but by far not the most appropriate. Therefore, such a murdering attack cannot be justified* [20].

It was further shown [19] that

Criterion 10. *The guiding criterion for politically determined application or threat of nuclear weapon strikes, is: to keep the greatest possible reserves in weapons and to spare sufficient important targets that in the end it is always possible to threaten the enemy's government with further unacceptable damage. This guideline for action is directly opposed to the guideline for military action according to Clausewitz. The use of nuclear weapons for military purposes cannot therefore be an optimum political strategy.*

Nothing reveals more clearly the distinction between the new strategy on the strategic level and the classical rules of Clausewitz than the roles of reserves of troops and weapons. According to Clausewitz 'A reserve has two purposes, which allow of clear distinction, namely, in the first place prolongation and renewal of the conflict; and secondly, use in the event of unforeseeable situations. The first purpose assumes the use of successive applications of force and therefore has no place in strategy' [21].

The use of nuclear weapons for military purposes jeopardizes the political aim. Military requirements demand applications of nuclear weapons on a scale determined by the attacker and his forces at a level much higher than that suitable for political purposes.† To give nuclear weapons a classical military role in this way destroys their political character, and makes them instruments of uncontrollable consequences. These consequences may include the destruction of Western and Eastern Europe [22]. Thus it can be concluded that:

> **Criterion 11.** *A policy of security which ascribes a* military *role to nuclear weapons is irrational.*

3.2. What criteria must military options satisfy in order to stabilize the politico-military relationship between the blocs (arms-control criteria)?

The concept of arms control is understood in various ways. The methods and objectives of arms-control policy vary. But the most exact description of the highest aim of arms-control policy would be 'stability' [23]. This politico-military stability in peace, crisis and war presents further requirements in the preparation of military options, that every policy of security must satisfy.

3.2.1. *Stability in peace.* From the general term stability it is possible to derive far-reaching starting points for a policy of peace with military means. Stability in *peace* comprises political and military stability. Political stability means in this connection *at least* to abandon reciprocally the attempt to compel the other side to capitulate ('peaceful co-existence'). The attempt to build up armaments that could compel this capitulation is therefore inconsistent with political stability and moreover inevitably leads to the building up of armaments by the opposite side. The consequence is an arms race, which inevitably produces *military* instabilities, because in the course of such a reciprocal strategic arms race phases arise in which a crisis for one – or even for both sides – it is advantageous to attack, instead of waiting for the first strike from the other side [24]. So for stability in peace we have

> **Criterion 12.** *Stability in peace demands that there be no build-up of armaments which could lead to the superiority of one side on the strategic level.*

The second requirement for stability of a new policy of security in *Europe* is to end the arms race at the level at which the *European* nations are arming,

---

† See, for example, the conditions for 'limited' application of nuclear weapons within the framework of the flexible response for the defence of Europe: the object here is to bring a Russian attack to a standstill by a massive shock by employing nuclear weapons.

and at the same time, to maintain the military means effective for defensive aims. This can be achieved only under certain conditions. Such conditions therefore become criteria for stability.

The classical policy of armaments does not achieve this aim. If one side assumes, or even only imagines, a build-up of armaments by an enemy, it will build up its own armaments, in order to maintain 'balance'. This build-up will in turn provide the other side with grounds to arm itself further.

How such a feedback can be avoided is shown by the theory of stable deterrence, which marks the beginning of arms-control thinking. In this theory, increased armaments give no military or political advantage. Therefore each side can maintain its armaments at a low level without endangering its own interests. The rationale of the reciprocal arms race is thus removed. This leads to

> **Criterion 13.** *In order to remove the rationale of the reciprocal arms race, the military options must be developed in such a way that one-sided limitations in armaments are possible for friend and foe without loss of position within the widest limits, because* more *armaments offer no politically relevant advantage, either qualitatively or quantitatively* [25].

3.2.2. Stability in crisis. Stabilization of the politico-military scene in possible crises against outbreak of war aims to make war improbable, if not (rationally) unthinkable.

Formulation of the criterion for stability in a crisis is likewise given by the theory of stable deterrence. The object is to remove the pressure of time to which every decision is subjected, when 'the advantage for the one who makes the first strike is great, or even decisive' (Schelling [26]). The theory of stable deterrence removes this pressure of time by eliminating all interest in prevention or pre-emption. This allows of deliberation without disadvantage in crises which otherwise would have broken out into war.

The condition for stability in a crisis may therefore be formulated thus:

> **Criterion 14.** *The position of neither side should be substantially improved by its being the first to go to war in a crisis (no prevention or pre-emption premium)* [27].

3.2.3. Stability in war. Stability in war means avoiding unintended [28] escalation to a more destructive level, should deterrence fail. Damage-limiting is another arms-control aspect of the same problem. The following criterion is valid here:

> **Criterion 15.** *If deterrence fails, one's own military means are rationally employable only if one's own strategy reliably guarantees that the desired solution of the conflict will be achieved on a level of destruction on which the war is still a rational means of politics for one's own side* [28].

On the one hand, with the present strategy of flexible response, the WTO can deprive NATO of liberty of action over the use or non-use of its nuclear weapons, either because it conducts conventional warfare in such a way that NATO cannot control it with its conventional defences, or because there are signs by which it may be recognized that the WTO threatens to set about a preventative attack on the nuclear tactical system stationed on NATO territory.

On the other hand the enemy can also get into a position – particularly with a conceivable build-up in the conventional armaments of NATO – in which he believes he is able to achieve his purposes only by a limited use of nuclear weapons (e.g., limited to the territory of the FRG).

The basic reason for fear of unintended escalation in the tactical nuclear range is in the former case that *NATO* has ascribed a military role to its tactical nuclear weapons, although it does not lie with NATO but with the WTO whether or not this function needs to be fulfilled.

In the second case, the basic reason for fear of unintended escalation is that the WTO is able to carry out its 'aims' if, and only if, it employs nuclear weapons.

The questions of whether nuclear escalation comes through the agency of the opposite side and whether one is compelled by the military situation 'to become nuclear' oneself is of highest importance. So in a crisis, and especially after outbreak of war, this military question will pre-occupy the political apparatus of decision up to the highest level, and thus to a considerable extent blocks the decisive political questions of the conflict just when they are most critical.

This card-house made up of reciprocal threats and military pressures can hardly reliably halt a war in Europe in case of emergency. Criterion 15, the wish to achieve stability against unintended escalation in war in this way, seems to be an illusion. Yet how then may such stability be achieved?

The way out lies only in *eliminating* the adverse *interest* in the *military* use of nuclear weapons. The meaning of that for NATO's strategy is

> **Criterion 16.** *The military purposes of NATO's policy of security in Europe must be achieved without use of nuclear weapons for military aims.*

However, the following criterion is necessary at the same time:

> **Criterion 17.** *An option for defence is rational from the standpoint of the Western European countries only if it sets up no military interest for the WTO to make use of nuclear weapons should deterrence fail. Otherwise the option violates the fundamental law not to provide an enemy with any militarily decisive options for such military means as would destroy these countries (Criterion 5).*

The WTO has then an extremely great interest in the use of nuclear weapons if it can smash NATO's defences with nuclear weapons, and only with nuclear weapons. Therefore defence strategies which can be knocked out by nuclear weapons and only by nuclear weapons should not be established.

3.2.4. *The aims of arms-control and the reality in Europe today.* If we evaluate the results of nearly twenty years' arms-control policy in respect of the aims and criteria of this policy, it is clear that the policy of arms-control has gone to pieces. The arms race could not be avoided. It was not stopped even on the most dangerous level for the great powers, the strategic level. On the contrary, a qualitative arms race succeeded the quantitative arms race. Neither was the situation on the strategic level felt to be stable, nor was military stability in a crisis achieved.

The latter is particularly important for Europe because the military premiums for a first strike have never been so high. The fear of a surprise attack by the WTO is an example of such a premium. The preventative nuclear strike against the tactical nuclear weapons of one or the other side in Europe is another example [29]. At the same time, however, war between both great powers will again appear to be a rational instrument of politics. Paradoxically, precisely the attempts to guarantee the arms control aim of 'damage-limiting' *for the great powers* lead to this result, because the more reliably war seems to be restricted to forms which the great powers can bear, the more easily can one of them decide to use military means in the case of a crisis. To offer this option is the declared aim of the preparation of graduated, limited military actions [30].

On the other hand, it is not lack of consideration towards the allies that is the cause of establishing such options for the great powers, but the incontestable realization that military options which cannot be realistically carried out without harming one's own interests do not deter.

As this failure of arms-control policy shows, it is easier to establish principles of arms-control than to carry them out in politico-military reality. The reason for this could naturally be that politics with military means inherently keeps any stabilization out of reach. If this is true, the build-up of armaments will always lead in the long run to instability, i.e., finally to war, whatever the intentions with which it was started [31].

The reason could, however, lie much less deep. It could be that previous attempts to stabilize the build-up of armaments have infringed important rules, which must be adhered to, if such a stabilization is to be a success. Such a difficult undertaking as stabilization of the build-up of armaments must be subject to its own rules, if it does have solutions.

How can these rules be ascertained? Comparison of unsuccessful with *successful* partial stabilizations might provide information.

The temporary stabilization of the strategic level, documented in SALT I is an example of a relatively successful attempt at arms-control. It was one characteristic of this stabilization that limitations to the build-up of armaments on one side were possible without loss of position (Criterion 13 above).

Because limitations to the build-up of armaments are possible without loss of position, they are not dependent on bilateral agreements.

Besides, the question of control of disarmament measures loses force precisely when such limitation is possible on one side without aggravating disadvantages. Otherwise, the question of control not only strains every agreement towards impossibility, but also destabilizes on the strategic level, since both sides will prepare for the worst possible case in the preparations of the other side [32].

This is shown by the fact that stabilization on the strategic level threatens to collapse today, precisely because there are such difficulties, for each side, in estimating the other's capabilities [33].

It was moreover typical for introducing stabilizing elements into the strategic level, that the aims of stropping the arms race and removal of every first-strike premium were present when planning began [34]. Purposeful military planning was established on the basis of these aims of stabilization. Attacks on the strategic arsenals of the enemy (counter force strategy) were banned. Retaliatory second strikes on the civil population were prepared [35]. On the basis of that strategy the necessary weapon-systems were built (intercontinental missiles in bunkers) and developed (ballistic missile submarines).

Thus we have as the criterion for successful arms control

> **Criterion 18.** *The precondition for successful arms control is that arms control comes at the beginning of military planning and is the basis of military strategy.*

If military planning existed in the first place and military resources were introduced for this planning, and if one *then* tried to invert the criteria of stabilization on the planning, failure would almost be certain; because all military planning that has not been developed on the basis of the above-described criteria for stability is almost immune to measures for arms control. No responsible military planner will ever be sure he has sufficient or even too much military means for all cases. If however military preparation is felt already to be inadequate, then all further limitations are seen as bringing about an intolerable situation – whether these limitations are on one side only or laid down in agreements. The hopeless attempts at disarmament since the days of the League of Nations prove this, as does the MBFR spectacle which is now ten years old.

The necessity of formulating afresh the policy of peace and its ancillary military strategies in Europe provides the opportunity for adopting a policy which gives arms control the necessary priority.

IV. Criteria for the policy of peace

Stabilization of the military resources on both sides is a prerequisite for a militarily secured policy of peace. If the criteria developed in the previous section cannot be fulfilled, then the military means are useless for a policy of peace. Yet even if these criteria are satisfied, the policy so outlined is not necessarily a policy of peace.

The first conflict could result from the fact that armed forces which cannot be made operational under any circumstances for any purpose whatsoever are useless [36]. A militarily secured policy of peace must therefore provide for the operation of military means at least in an extreme case of emergency. For that reason, peace cannot have the highest value in a militarily secured policy of peace. There must be higher-ranking values. So the decision to go to war remains a real possibility.

On the other hand, world peace is demanded by common sense [37]. Thus one arrives at

> **Criterion 19.** *A militarily secured policy of peace must develop means which simultaneously serve the long-term security of world peace equally with the short-term and the medium-term prevention of war.*

However than presupposes

> **Criterion 20.** *Military means for a policy of peace must be so organized that they either ensure peace in the long-term or the military means ancillary to the short-term and medium-term prevention of war must be so chosen that they can be replaced in the long-term by other means which ensure peace in the long-term. These other means may be again military means (armed peace) or exclusively non-military (peace with complete disarmament).*

The second conflict between a successful policy of arms-control and a policy for peace could manifest itself in the confrontation between the ability to conduct war and the prevention of war [38]. On the one hand, rational realizable military options are assumed for credible deterrence (see above, US Norm C). On the other hand if both opponents prepare limited military reactions, which they deem to be military or political instruments, then once more they make war an accepted instrument. However, this means that elimination of war (as a rational instrument of politics) by deterrence becomes impossible because the assumption of this kind of deterrence is precisely the *possibility* of conducting war.

Thus the question arises: Does not the establishment of rationally realizable military options for the worst case, i.e., if deterrence fails, contradict the fundamental principle of a policy for peace which is to make war as improbable as possible? Even if the purpose of developing such kinds of rational options were precisely to prevent war by credible deterrence (US Norm C)?

So we are forced to ask ourselves if the sequence

deterrence→rational options for military means→possibility of war

is really insoluble, or under what conditions it can be counteracted.

Once again, a glance at the first model of a militarily secured policy for peace, namely, the model of stable deterrence, is helpful. On the one hand strategic weapon systems having quite definite properties are required in this theory, which allow of a retaliatory strike (e.g. assured destruction) so as to effectively deter the enemy. On the other hand, in this model, use of the system in a crisis is not rational for either side. This contradiction no longer startles us as long as we think of its solution as well. What is irrational in the theory of stable deterrence is always the *first strike*, the attack, for it leads to an annihilating retaliatory strike. This second strike is *not irrational*, because the retaliating party has nothing more to lose [39].

Criterion 21.
1. As in the historical precedent of the strategy for stable deterrence, it should not be possible for either side to employ military means in the case of a crisis in a rational manner.
2. However, if one of the opponents has used his weapons to make an attack, this step should make it possible for the one who is attacked then to employ his military means in a rational (or at least not irrational) manner so that the attacker receives so much damage that his attack becomes irrational.
3. This possibility of such a second strike (Section 2) ensures the prerequisite of no rational one-sided use in the case of a crisis (Section 1).

The American military policy does not satisfy this criterion. Furthermore, it cannot satisfy it, because the laws to which this policy appears to be subjected favour the unilateral availability of military power (American Norm A). The one exception is to be found where the existence of the USA is at stake; in order to prevent a massive attack by the great powers one against another, the USA grants the USSR an *assured* capacity for the second strike (American Norm E.2); this capacity for a second strike by the USSR excludes by its existence the rational one-sided availability of the strategic capacity for the USA (policy of peace with military means on the level of imperilled existence).

The European states pursue no world power policy. In searching for a militarily secured policy of peace for European states, one is not subject to the compulsion of the (US) norms of world power policy, which demand the free availability of military means. However, ensuring the existence of individual states is just as much the object of the policy of the European states as of the USA. To exclude the unilateral availability of military means in Europe, where the use of these means endangers the existence of the European states, must therefore be just as much a principle for European defence policy, as it is today for the American. Europe's existence is endangered by atomic or a large conventional war in Europe as much as is the existence of the USA by a massive attack with nuclear weapons.

Criterion 22. *Policy of peace for the European nations demands that the military options available are not applicable unilaterally in a rational way, but only in reaction.*

The old principle of a militarily assured policy for peace, namely, to develop only *defensive* military equipment, is a special case of this general rule. The fact that, for all sub-strategic levels, unilaterally applicable options are sought by the *great powers* is certainly partly to be explained by their interests in power. However, this explanation does not go far enough; for example, the American norms demonstrate the USA's interest in securing the 'independence of allies'. Now, ensuring the independence of allies by great powers is certainly part of power politics. Securing the interests of allies can even become a mockery when it camouflages repression ('brotherly help' by the USSR for the Czechoslovak SSR in 1968). However, in all other cases – and in many cases of such repression – this help serves peace *also*; indeed the declared help and its credibility are often the sole guarantee of peace for a smaller nation. And in a world with military equipment of present-day structure, such help can only be furnished by unilaterally employable military means. And these can naturally also be used for attack [40]. But if these attacks by a great power – for example in Europe – encounter defence reactions which make these attacks irrational in the same way that mutual assured destruction makes a massive nuclear attack irrational, the sub-strategic level would be stabilized just as the strategic level has been stabilized.

If this could be achieved, the attempt to exploit deterrence on the strategic level for consolidation of the sub-strategic level of European commitments would be superfluous. Only a power system in which the great powers' own interests make them renounce such use of the strategic level allows stabilization for any length of time. For the attempt to exploit the strategic level for the guarantee of commitments (exploitable capability) leads, as experience has shown [41], in the first place to an arms race on the strategic level and ends by destabilizing the situation and to pressure for preventative action – without solving the insoluble problem of stabilizing the strategic level and yet using it to stabilize the sub-strategic commitments. Thus the following is valid:

> **Criterion 23.** *Peace policy on the strategic level requires stable militarily assured peace on all other levels (peace policy on the sub-strategic level).*

Renunciation by the European allies of unilaterally available military means, which is demanded by Criterion 22, does not therefore conflict with the interests of the great powers, particularly the interests of the USA; but is a prerequisite for the long-term security of these interests.

V. Models of a new security policy for Western Europe

Arising from the political aim to protect the freedom and independence of the Western Europe countries from military threat and political pressure, and the criteria valid for rational options in Central Europe, the following tasks are to be imposed on the plans for exercising power by the NATO alliance in the later 1980s and 1990s:

Task Complex I. *The* first task *for the military means of a European NATO partner is, within the framework of the criteria for their rational employment to make attack on Central Europe and occupation of NATO territories as costly as possible politically and militarily to the aggressor* (*credible deterrence of restricted range*).
The second task *is, within the same framework of criteria to prevent a* fait accompli *within hours or days, in order to give the USA time to bring its means of exercising power, especially its nuclear military resources, into play.*

Task Complex II. *The task of the* United States *is in the first place to keep in readiness such nuclear options as could be employed rationally for both the USA and its allies should deterrence fail, and which will lead in a rational manner to restoration of the* status quo.

From the criteria for a rational employment of nuclear weapons (Criteria 9–12, 15–17) it then follows:

First: Nuclear weapons cannot be employed for the *military* task of Complex I without destroying the rationality of the options (Criteria 10, 11, 16, 17) [42]. Consequently in the case of Task Complex I, it is necessary to develop military strategies which can be employed successfully without nuclear weapons and which also provide no military interest to the WTO for its part to employ nuclear weapons (Criteria 16, 17). A strategy which satisfies these conditions, and a model which fulfils the criteria for organizing and arming the striking powers of NATO in Europe will be outlined in the following section (5.1).

Second: US nuclear means for exercising power serve the tasks of Complex II. *Military* tasks cannot be given to these means (Criteria 10, 11, 16), because the necessary prerequisite for a limitation of the level of escalation (Criterion 15) is the fact that nuclear resources are employed exclusively in accordance with *political* criteria (Criteria 9, 10).

5.1. A model for rational options for prevention of the *fait accompli* and for establishing a credible deterrence of restricted range with non-nuclear military resources

A first model for fulfilling these tasks can be developed from the ideas put forward by the French Major (now Lieutenant-Colonel) Brossollet in his book *Essai sur la non-bataille* [44], which has appeared in German in the book *Verteidigung ohne Schlacht* by Spannocchi and Brossollet [45]. The discussion of Brossollet's concept in *Verteidigung und Frieden* [46] shows in what places Brossollet's ideas must be rejected, further developed or supplemented, in order to arrive at the first model put forward here.

Commandos, each consisting of twenty men, form the basis for the concept of this model. These commandos are equipped with anti-tank weapons (rockets), mines and light infantry weapons. Their technical equipment should be optimal for the specific purpose of repulsing the enemy's mobile heavy equipment (target-seeking missiles, semi-automatic and fully automatic means of destruction). This makes the term 'techno-commando' appropriate.

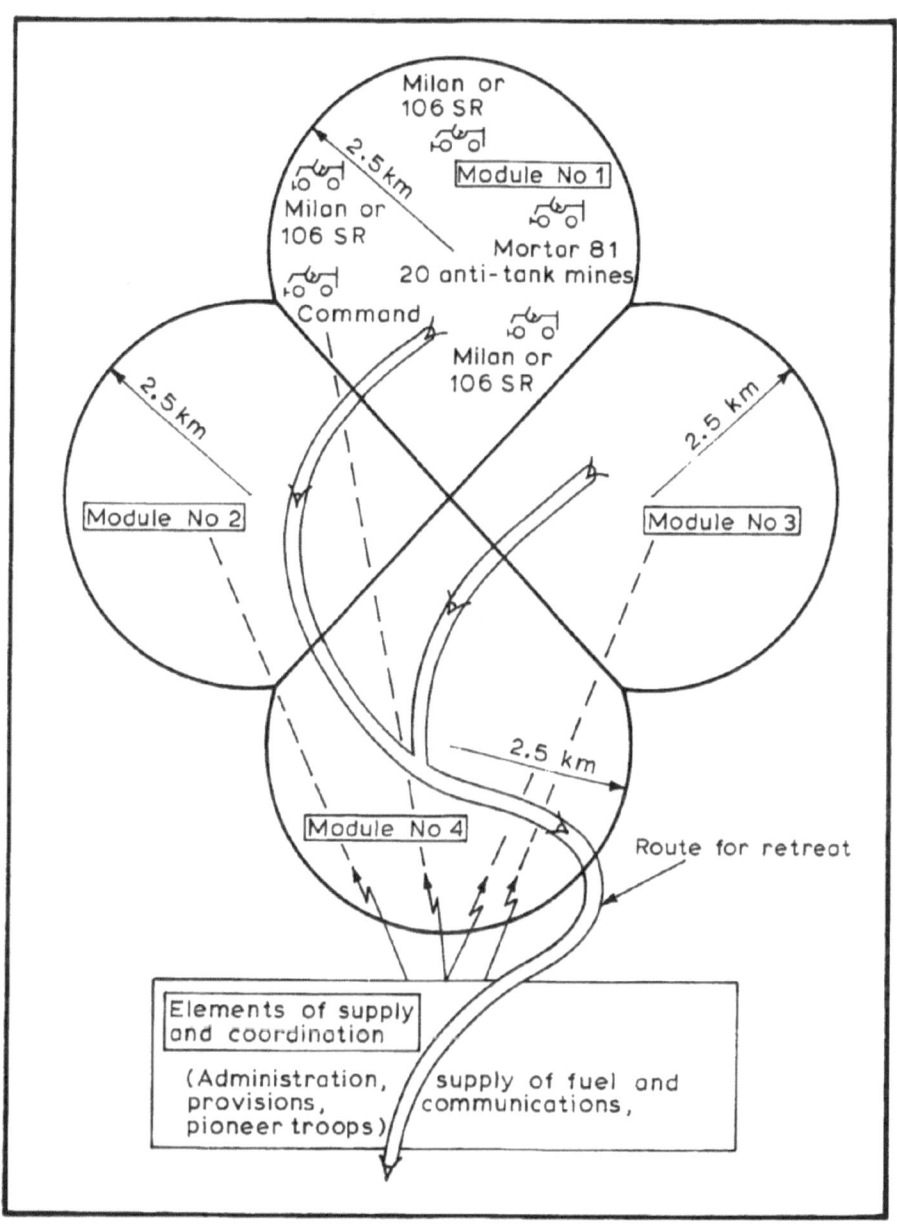

Figure 9.1. Light, stationary techno-commando [46]. 'Milan' is a precision-guided anti-tank missile (*Jane's Weapon Systems* 1977, p. 37), and serves as an example. For the later 1980s and 1990s, more advanced anti-tank weapons (third generation of precision-guided weapons) are being developed. (From: Spannocchi/Brossollet: *Verteidigung ohne Schlacht.* Munich, 1976, p. 164.)

In time of peace each commando will be already stationed and equipped in his region, a region of about 20 km². In the event of deterrence failing, the commando in *this* region will autonomously fulfil his task to hinder, or to destroy, as far as possible any hostile advance.

All the techno-commandos together constitute a network which covers at least the FRG completely. The density of the network can be adjusted to local conditions.

Each techno-commando has its communications division. The network of autonomous techno-commandos is thus able to provide a realistic survey of the situation at all times without great expenditure. In order to make an advance of tanks in corridors impossible, the stationary commandos of the network should be equipped also with some long-range target-seeking missiles in addition to light short-range anti-tank missiles (e.g., Milan). These missiles must have a range of at least 25 km in order to make attacks over a breadth of less than 50 km unattractive. On the other hand these weapons should not constitute any detectable target. Mini-cruise missiles and TGSMs [47] delivered from suitable carriers should fulfil these requirements. If each autonomous stationary techno-commando is allocated one missile-carrier unit, then each commando needs about six additional men [48].

Such a network of autonomous techno-commandos might or might not be supported by an *air force*. Furthermore, the air force might be employed for the political aim of deterrence.

If we consider the *effectiveness* of the network of autonomous techno-commandos as outlined, then the assumptions of Brossollet indicate that even with the *present* technical equipment (Milan) each techno-commando will destroy three enemy tanks [49] on average in his region, if it is attacked. For the whole territory of the FRG, approximately 10 000 such techno-commandos are required [50].

If it is correct, as Brossollet assumes, that one such light techno-commando can destroy on average three enemy tanks, then in order to invade the whole FRG, the enemy requires 30 000 tanks. However, at the present time there are only 20 000 tanks belonging to the WTO stationed in North and Central Europe.

If the network has the effectiveness which Brossollet ascribes to it on the basis of manoeuvres, then the *light* commandos alone could overcome the strength of the WTO in tanks, if the WTO should attempt to occupy the whole territory of the FRG.

The WTO tanks cannot therefore achieve the objective of conquering the FRG in a short time, thereby creating a *fait accompli*, which makes the intervention of the USA's nuclear resource improbable.

Can the WTO not overwhelm the network by intervening with its mass of infantry so that the network cannot achieve its task? The infantry of the WTO could be stopped by each individual techno-commando only for a limited time, so could overcome one commando region after another and mop up anti-tank weapons. However, such an inversion of the roles of tank and infantry takes away from the tank attack its decisive advantage – speed. Tank units which can advance only *behind* infantry in the fighting-line, cannot break through or even overwhelm the territory of the FRG either in hours or days; if one assumes even the improbable rate of 30 km per day for the advance [52], the attacking troops would be at the Rhine only after a week. A network of tank-breaking commandos which forces the tanks to let the infantry go *ahead* consequently prevents the *fait accompli*. Besides, in the 1990s the techno-commandos would be equipped with appropriate anti-personnel weapons with which to halt an infantry advance for a considerable time, or even make it impossible [52 a].

The network of autonomous techno-commandos could thus accomplish the assigned task, namely, to prevent a *fait accompli*, even with the weapons available today. This together with the circumstance of assured heavy losses by the hostile motorized striking forces simultaneously constitute a reliable, credible deterrence of limited range without nuclear weapons.

Even weapons already developed for a quite different kind of conventional war show a trend which will increasingly favour the use of options in accordance with this kind of model in the future – miniaturization and automation.

Miniaturization leads to the situation where it is possible to employ means of destruction composed of still smaller units which are still more difficult to locate and still more effective (over greater distance and more accurate in reaching the objective) against heavy material. Also options for defence according to this model make the best use of the trend towards further automation. Because this model renounces tanks and heavy mobile units, obstructions and automatic means of destruction which are effective against mechanized formations can be employed without danger to the defender's formations. Likewise, automatic missiles or unmanned aircraft can be used against tanks without endangering any defending tanks, for there are none. Semi-automatic or even fully automatic weapons are best employed for defence in depth, for an enemy making an advance continually encounters fresh automatic weapons, which have been prepared beforehand in peace, without exhausting the store of automatic weapons in one zone (the front) and without being able to put a weapons-system concentrated in one zone out of action by massive resources of destruction.

There is no doubt that a technically advanced country could make great advances in weaponry most appropriate to this kind of warfare if this were its specific aim.

If everything thus seems in favour of the efficiency of such a network of techno-commandos, then the question arises as to whether the *cost* is tolerable.

Table 9.1. The total requirement for the striking forces of a network of autonomous techno-commandos in the FRG

1. For the permanent light commandos 10 000 commandos of 20 men each	= 200 000 men
2. For air defence: 5 additional men for each permanent commando	= 50 000 men
3. For the long-range tank defence: 6 men for each commando	= 60 000 men
4. Liaison units (Reserves in peacetime)	= 30 000 men
Total to be provided in peacetime as striking forces in position	**340 000 men**

To the above must be added mobilization striking forces for combatting landings by aircraft, which forces will be summoned by the autonomous permanent commandos. It is difficult to give firm figures. As a starting point, cadres of five men each per commando should be provided for the leading positions of these reserve units. There will then be a further need of 50 000 men in peacetime.

At present, the strength of NATO's land striking forces is about 775 000 men on the territories of the FRG and the Benelux states [53]. This shows that it is possible to recruit manpower at the level necessary for this model, even if this first calculation is a serious underestimate. For a network of autonomous techno-commandos subject to the requirement not to create any objective which is worth being destroyed militarily, von der Heydte's comment on guerillas is particularly relevant: 'Guerilla war is not only . . . in itself a war of the weak, but also a war in which numerous weak groups can mean operative strength. . .' [54]. Because quantitative reinforcement makes guerillas into a worthwhile target for the attacker, it thus acts against the guerilla.

A *quantitative* strengthening of autonomous commandos beyond those needed to prevent the *fait accompli* would therefore bring no overall advantage. In that way, the arms-control criteria for 'stability in peace', which require the demolition of the feedback mechanisms that lead to an arms race, are satisfied (Criteria 12, 13).

As to the financial cost, calculations for the 1990s are at present almost impossible, but bearing in mind that the concept of such a network of techno-commandos does not demand heavy equipment (tanks, armoured vehicles, naval units for ensuring supplies etc.), it is nearly certain that the network could be built up with the resources available within the NATO alliance today. This would not necessarily be the case for equipping NATO with the next generation of heavy equipment of the type being used today.

5.2. Models for strategies of coercion for the nuclear resources of the USA [55]

How could US nuclear means of exercising power be applied, so that intervention

1. from the US standpoint is a rational option, and does not finally lead to irrational, more destructive levels of warfare;

2. from the standpoint of the Central European states complies with the criteria for rationality; and

3. nevertheless satisfies the requirement of inducing an aggressor finally to restore the *status quo*.

The second requirement defines most closely the range of conceivable options. If threats to use nuclear weapons should drive invading troops of the WTO from NATO territory, then nuclear weapons will have to play a decisive role in the conflict. No long-range highly destructive weapon can however be stored in Central Europe that cannot be located and preventatively destroyed. It is forbidden by Criterion 5 to furnish Central Europe with US nuclear weapons which form a target for a decisive strike by the nuclear weapons of the WTO.

For the NATO partner a nuclear option can therefore be rational only provided that the necessary nuclear weapons for this option are *not* stationed in Central Europe.

This is the exact reverse of the requirement to station US nuclear weapons on the territory of the FRG to which one comes if in place of rational political options for using military power, one puts deterrence by non-calculable risk.

On the other hand the nuclear weapons destined for the 'European theatre' should on no account stand in the USA, because that would increase the danger of drawing the territory of the USA into military action. This leads to the conclusion that the US nuclear weapons destined for the European theatre should not be stationed in either the USA or in Europe.

Consequently, no-man's land, i.e. the oceans, alone are appropriate for stationing [56]. Seeing that the USA possesses, for example, 41 submarines with 656 missiles and 4750 warheads [57], and furthermore that only a fraction of these explosive heads can be used for MAD, it is evident that the arsenals for politically determined options of threat and pressure for the European theatre are now ready for disposal. Indeed, they are already partly designated for NATO [58].

What should the targets for these nuclear weapons be in order to satisfy the political aim? Certainly the WTO troops invading NATO territory cannot be the target, primarily because any military application of nuclear weapons must be excluded on account of the danger of escalation (Criterion 11), but also because such a strike would destroy the country which was the combat area and which NATO intends to defend (Criteria 2, 15, 16).

The nature of the targets chosen is determined on the fundamental principles of rational deterrence and the need to restore the *status quo* in Central Europe. The fundamental principle of deterrence is given in Criteria 9 and 10 (see p. 275 above). Consequently the task which the threatened nuclear strikes should be given would be to threaten the Soviet government's ability to rule. However, in doing this, there is the danger of escalation into a general strategic exchange. A strategic nuclear war is, however, irrational for all participants (Criteria 1, 3). If the Soviet regime cannot be threatened without the danger of such escalation, Soviet domination can be treated as an object of threat only in regions outside the USSR. Seeing that in case of an attack by the WTO the starting point of the aggression would be Eastern Europe, and the East European states would support aggression in consequence of Soviet domination, NATO should, in this case only, consider threatening this Soviet domination over Eastern Europe. How could this be done?

The proposal made by Martin [59] contains a concept according to which, in case of an attack on Western Europe, separate nuclear strikes from a sea-based force would be used against Eastern Europe. Martin recommended a NATO covering force consisting of submarines carrying missiles. The object of this force should be to threaten in Eastern Europe about 200–300 km behind the front. Targets are aerodromes, headquarters, supply lines, reserve units etc.

If we seek a politically determined option for threat or pressure, then the targets threatened must take a different form in order to keep decisions free of military contingencies and their destructive consequences in Eastern Europe. What then should the covering force threaten, if the political aim is to put domination by the USSR in Eastern Europe in question *in the event* that it attacks Western Europe with the help of the WTO?

A similar question was put forward by M. A. Kaplan. His 'dissuasion strategy' [60] contains a series of elements, which could serve as model or foundation for a strategy of rational politically determined options for coercion [61], although his purpose does not coincide with the aims being considered in this section. Kaplan wants to develop a strategy in which the defensive battle can be won. The armies of the Eastern European states should according to this strategy be separated from the armies of the USSR, so that the fighting forces of the attacker become so reduced that the NATO troops can achieve a *defensive victory* [62]. For this purpose Kaplan wants to dissuade the Soviet Union's WTO partners

1. From participating in an attack upon the West.
2. From holding static positions on central front lines.
3. From interfering with NATO flanking manoeuvres.

'WTO members would not be asked to refuse permission to Soviet troops to use their territories for attack or to interfere with Soviet utilization of local supplies or logistic facilities' [63].

Any state belonging to the WTO which co-operates with the USSR in any of these 'forbidden' ways is threatened:

> '. . . after population warning to permit the local populations to avoid attack zones, NATO would begin with small attacks upon their economic infrastructures. If this warning were not heeded, and if the regimes did not refrain from the interdicted activities within a reasonable period of time, their economic infrastructures would be destroyed if necessary' [64].

The principle not to attack any target in Eastern Europe without ample warning to allow evacuation of the population should be a precondition for any use of nuclear weapons for the security policy of Western Europe.

Once again it is apparent how important it is that military victory on the battlefield is not an aim of the strategy in our model. Only after this has been renounced does it become possible to incorporate vigorous de-escalation factors: prevention or pre-emption brings no advantage (Criterion 14). On the contrary, every advance of the attacker from commando to commando in the network brings additional disadvantages, because his loss increases from commando to commando. These losses can mount up enough to endanger the Soviet domination over East Europe. At the same time with every town the attacker captures, he is losing an objective for counter-threats, until finally no such objectives are left, because the troops of the WTO screen the territory being invaded from Soviet nuclear strikes.

If, on the other hand, one followed the military aims given by Kaplan, one would be constrained to separate the WTO states from the USSR *before* NATO's defence collapsed. At the same time concentration of NATO troops would offer an important objective for Soviet nuclear strikes. So once again there would be pressure to undertake preventative military strikes, with the danger of atomic annihilation, first of Eastern Europe and then of Western Europe.

If one enquires about the effectiveness of threats to the Eastern European states in the concept proposed in this chapter, one must take into account that during an attack on the FRG, the network of autonomous techno-commandos would continually be reducing the Soviet tanks and mechanized formations. It is not ruled out that the Soviet military machine would become so decimated that the Soviet domination of Eastern Europe would be shattered. This can occur especially if the commandos of the network can concentrate on destruction of the *Soviet* striking forces, but spare these of states of the WTO which no longer wish to take an active part in the attack.

A nuclear threat of that kind against governments of the WTO whose troops continue to participate in aggression against the FRG will certainly be followed by a Soviet nuclear counter-threat. In consequence both parties to the conflict are to reopen the political dialogue.

VI. Does the policy of security described accomplish the tasks put forward?

A policy of security which is designed to fit the strategy outlined in the models and the military resources mentioned above, complies with the criteria developed in section 3. This is valid both for the criteria for rational use of military resources and for the criteria for stability in peace, crisis and war.

The employment of military resources remains rational because the destruction is limited. There are no targets for the enemy's nuclear weapons; no military use for one's own nuclear weapons.

The build-up of arms by NATO is as far as possible uncoupled from that of the WTO, and vice versa, for the task was not to prevent occupation of NATO territories but only to make it very difficult (limited deterrence effect). A task formulated in this way gives no reason to set very high 'minimum strengths' for fulfilment of the mandate even in the worst case. However, for the enemy also there is no temptation to build up his armaments so as to be able to overcome the NATO preparations. The more tanks he brings into the field, the further he may go in a certain time. However, no *fait accompli* is possible thereby, nor are his losses smaller – rather the contrary. Unilateral limitations to the build-up of armaments thus become possible without loss of military position.

Neither side improves its position by attacking in times of crisis. The techno-commandos cannot do this at all; they are not 'unilaterally available', but must wait until an enemy comes within reach. Surprise brings no decisive advantage for the enemy, for the network is present in peacetime, and strategic break-throughs by surprise attacks are not realizable in a network since there is no line or combat-zone to break through. Tactical nuclear weapons play no role in the concept, and could therefore at last be removed.

The military means so described stabilize the European field of conflict without destabilization of the strategic nuclear level. For a threat by the USA of heavy strategic strikes is not part of the strategy.

These military means finally can serve for peace in the long term. They threaten no one. They can also slowly be converted into non-military forms of resistance (civilian defence) as political security of peace increases.

The fact that such a policy of security is far superior to the present ones should deterrence one day fail is thus well proved. It is evident that this policy for security – unlike to the present ones – is compatible with arms control and policy for peace.

However, most people have become so accustomed to living with the present strategy of deterrence, which certainly in the event of war would lead to inconceivable losses and destruction, that they can only take comfort in the thought that precisely this 'inconceivableness' of war prevents war.

The premise which is at the root of the US concept of defence is 'Deterrence is no substitute for defence. Rational and feasible military options are the pre-condition of every credible deterrence' (US Norm C). Someone who does not accept this might say that a policy of security according to the model is weighted too heavily on the case where deterrence fails. It makes war still more probable on account of its ability to control it in case of a crisis and failure of deterrence.

Primarily, this view ignores the fact that, whatever deterrence is built up, war remains a real possibility. Secondly, it assumes that it is more unlikely for the USSR that the USA would really threaten and put into effect rational politically determined responses of coercion in emergency, than that the present militarily programmed tactical responses would be used; and that in the second case there would then result an escalation in the strategic sphere followed by unacceptable consequences for the USSR. The nub of the argument is therefore the question whether the USA would be more inclined to respond with its nuclear means of exercising power given the present or the recommended strategy.

In comparing the present role of tactical nuclear weapons in Europe with the role of nuclear weapons for politically determined options as given in the model, it is necessary to ask if the USSR should expect today's military responses as more probable than the political responses of the model, and what type of responses more strongly thwart its plans.

At present, tactical nuclear weapons in Europe have the following functions
1. to deter the use of tactical nuclear weapons by the WTO;
2. to ensure flexible deterrence over the whole range of possible threats;
3. to offer nuclear options of response below the strategic level;
4. to help to repulse conventional large-scale attacks.

They furthermore serve the political purpose of demonstrating to the European allies the USA's commitment.

The last point, to demonstrate to the European allies – and to the USSR – the commitment of the USA is an important function of American policy for security with military resources. Our models do not abandon this function. On one hand, the stationing of further American troops in the FRG is naturally provided for [65]. The optimal form would be to let American units advance to the borders of the WTO, not with the mandate to defend this border, but solely to delay the enemy's advance for some time, and to stamp the conflict as an American–Russian one.

Returning to the first three functions of US tactical nuclear weapons in Europe at the present time, the first task, to deter the use of tactical nuclear weapons by the WTO, is in the context of NATO's present strategy unquestionably an important one. The present NATO strategy for Europe gives an almost compelling military advantage to the WTO for use of its nuclear weapons. Consequently it becomes necessary to try to deter the leadership of the WTO by the threat of NATO's tactical nuclear weapons. In the model described above, the WTO is restrained from using nuclear weapons on NATO territory by ensuring that no militarily significant use is available to these weapons. This is clearly a more reliable way than threats of use of NATO nuclear weapons, which the WTO always has the option of destroying by massive preventative nuclear weapon strikes.

Coming now to the second and third tasks, the role of tactical nuclear weapons in fulfilling the task of ensuring flexible deterrence over the whole range of possible threats is to provide an intermediate stage between conventional warfare and strategic nuclear warfare [66], in order to make escalation to the latter a credible process. Now it is certain that the new US policy of defence provides for determined 'flexible' strategic nuclear options as the first stage of strategic response, because comprehensive strategic attacks violate the USA's own interests, and therefore are not credible [67]. Will these 'limited strategic options' become incredible if the 'preliminary stage' of tactical nuclear warfare is taken away?

That leads to the question:

Are tactical nuclear weapons then a *preliminary stage* of political responses, which is necessary for the credibility of political response?

The credibility of every nuclear threat depends on the consequences of carrying out that threat. Politically determined responses can be limited. Our examples, in which separate responses were threatened after forewarning to evacuate the population, serve to demonstrate this. It is otherwise with the *military tactical* use of nuclear weapons. In this case military requirements are combined with political calculation and demand, in certain circumstances, further responses with nuclear weapons in order to annihilate the enemy's reserves, or to cover deficiencies that have arisen through the enemy's nuclear replies. It is precisely this preoccupation with classical. military tasks which hides the political character of nuclear weapons in these cases and converts them into instruments with *uncontrollable* consequences. Such responses thus represent a much higher stage of escalation than politically determined limited options. They cannot therefore be regarded as *preliminary stages* of politically determined limited options. So they do not have to take precedence in order to make political responses credible.

Consequently, even in today's NATO strategy, the very first use of nuclear weapons by NATO will most probably be a political use: a nuclear demonstration.

Attachment to the habitual is certainly one explanation for continuing the flexible response. But another is the fourth task of tactical nuclear weapons in Europe: repulsing conventional large-scale attacks. Because if nuclear weapons really fulfil this task – although this is not at all certain [68] – then deterrence seems to the optimistic observer to be 'complete'. However, is the defence then really complete? Or must there be even today recourse to the USA's nuclear means of exercising power available behind tactical nuclear weapons, in order to achieve the political aim of restoring the integrity of the NATO territories? For what, on the most optimistic consideration, can be achieved at the present time by recourse to tactical nuclear weapons in the event of a large-scale conventional attack? Obviously, the most optimistic outcome would be that this conventional attack was halted at the point which it had reached when NATO found itself obliged to take up tactical nuclear weapons, and that the WTO did not continue the attack as a nuclear-conventional one [69].

This leads to the conclusion that restoration of the integrity of NATO territories cannot be achieved even now by a *military* use of the resources of NATO. Even now we have to rely upon political pressure to achieve this purpose [70]. So the means of exercising power which should serve for restoration of the integrity of the NATO territories, are the same that have been considered in the models here. If the USA wants to constrain USSR to release the conquered NATO territory in such a way, then it makes no difference whether one fifth, one half or two thirds of that territory is occupied. However, for Central Europe, it is the difference between life and death, whether it is already destroyed at that moment by a tactical nuclear war, as is likely according to present-day strategy, or not.

It is precisely at this weak point in present-day strategy, namely the necessity of forcing the USSR back with politico-military means, that certain recommendations intended to prevent the enemy from having initial success by nuclear means come into play. A prototype of these recommendations, in which the structure of the problem of defence and nuclear weapons becomes especially clear, is that of Bennett, Sandoval and Shreffler [71]. In the revised form submitted at the SIPRI Symposium in October 1976, this recommendation provides for defence of the FRG with nuclear weapons (4000 warheads) in a blockade-belt about 100 km deep at the border zone†. Authority to employ the nuclear weapons (mini-nukes) stationed there would de delegated to the commanding officer in peacetime [72].

Naturally, it is possible for the USSR to destroy this belt of mini-nukes with the heavier tactical nuclear weapons already available to it today. The USA must therefore, according to this concept, force the USSR not to act against this defence with larger tactical nuclear weapons. To that end the USA must make it credible that they will proceed to higher stages of nuclear escalation in the event of the USSR exceeding the limits of nuclear warfare established in

† The latest version of this proposal is presented in chapter 10.

the interest of NATO (mini-nukes in the defensive belt). So border defence by mini-nukes depends at least as much on the readiness of the USA to proceed to certain forms of escalation in the nuclear strategic range as does the concept proposed in section V of this chapter. But, in contrast to the proposed concept, nuclear border defence demands decisions to proceed to nuclear escalation under heavy time pressure to prevent any gain of NATO territory by an attacker. The danger that war would escalate, causing unacceptable damage at least for Central and Eastern Europe, would be high.

The necessity for developing flexible politically determined strategic nuclear options is thus manifest, not as a weakness of our concept, but as a basic condition for the utilization of military resources of a 'super'-power at the present time.

How could it be otherwise? The description of a power as a 'super'-power is linked to possession of this strategic–nuclear means of exercising power, and the USA is already developing such 'limited strategic options, not under the strategy recommended in this chapter but under the doctrine of flexible response (US Norm E.3).

VII. Conclusion

Any strategy may fail, if deterrence fails. But there is no doubt that it is time to abandon the present strategies which employ nuclear destruction in Eastern and Western Europe as a declared instrument for defence, if deterrence fails; these strategies with high premiums on striking first in case of a crisis and a strong feedback in armament levels lead to an arms race in peacetime.

The model presented here serves as an example for a strategy which avoids these shortcomings of present-day strategies. Such new methods and purely defensive (purely reactive) strategies can allow the limitation of armaments, the removal of tactical nuclear weapons, and at the same time reduce the probability of war and limit the war and the damage, should deterrence nevertheless fail.

References and Notes to Chapter 9

1. On the consequences of a tactical nuclear war in Europe see von Weizsäcker, C. F. (ed.), *Kriegsfolgen und Kriegsverhütung*, Hanser Verlag, (1970).
2. The problematical nature of the matter is complicated. It can be described only in a simplified way in a short account. For comprehensive information reference should be made to the author's book *Verteidigung und Frieden, Politik mit militärischen Mitteln*, Hanser Verlag, (1976). See Chapter 4 for the theory of intimidation with strategic weapons.

3. See, for example, H. Afheldt, *SALT und qualitatives Wettrüsten*, Wehrkunde, (1972), pp. 628 ff.

4. Report of Defense Minister Schlesinger to Congress 1976–77. This report is particularly suitable as a summary, because it contains relatively clear formulations. There are no significant deviations in the later report of the last Defense Minister of the Ford Administration (Rumsfeld).

5. Afheldt, H., *Verteidigung und Frieden*, Hanser Verlag, (1976), pp. 107 ff.

6. Feldmeyer in *Frankfurter Allgemeine Zeitung*, (24 January 1977).

7. Afheldt, *Verteidigung und Frieden*, Ch. 4.222 and p. 726.

8. The best known example is the use of NATO bases in the Federal Republic of Germany during the Yom Kippur War, in which the Federal Government at that time had to consider West German interest.

9. Result 55 in *Verteidigung und Frieden* [2].

10. Norm 9 in *Verteidigung und Frieden* [2].

11. *Verteidigungs-Weissbuch der Bundesrepublik* (1975–76), No. 152.

12. *Verteidigungs-Weissbuch der Bundesrepublik* (1975–76), No. 156.

13. In a different way the guerilla strategy of exhaustion *against* industrial nations, e.g., Mao Tse-tung's war against Japan and Ghiaps war against the USA in Viet Nam.

14. *Statisches Jahrbuch für das Deutsche Reich 1921–1922*, pp. 28–9. Up to the end of November 1914 the British losses amounted to 2.7 per cent of the total losses in dead and 2.2 per cent in wounded.

15. Iklé, *Every war must end*, pp. 55, 89.

16. Clausewitz, *Vom Kriege*, Book 8, p. 902.

17. Only the break-through of some divisions on the western front brought the German government, under pressure from the military leadership (Ludendorff), seriously to the negotiating table.

18. Afheldt, H., *Verteidigung und Frieden*, Chapter 4.2 with references to Burns, *Ethics and deterrence*, Adelphi Papers 69. Russett, *Survival*, (1974), p. 136 and Schlesinger, (Report 1976–77), pp. II-1 ff.

19. *Idem*, Chapter 4.3, pp. 85 ff.

20. Afheldt, H., *Verteidigung und Frieden*, Chapter 4.2.

21. Clausewitz, *Vom Kriege*, pp. 295, 298.

22. Afheldt, H., *Verteidigung und Frieden*, p. 273.

23. A comprehensive exposition of arms-control policy was first made in the *Daedalus-Arms-Control Special 1960*. (German: Strategie der Abrüstung, Bertelsmann, (1960). See there, for example, Schelling, pp. 198–190, or Wiesner, pp. 236–7.)

24. A demonstration model describing these phases of instability has been given in *Kriegsfolgen-Studie*. See Afheldt/Sonntag, *Kriegsfolgen und Kriegsverhütung*, pp. 303 ff. English: *Stability and Strategic Nuclear Arms*. World Law Fund occasional papers, (New York 1972).

25. See Result 17 and Norm 8 in Afheldt, H., *Verteidigung und Frieden*, pp. 56 and 59.

26. Quoted and discussed in more detail in Afheldt, H., *Verteidigung und Frieden*, p. 56.

27. Norm 7 in Afheldt, H., *Verteidigung und Frieden*, p. 59.

28. On the problematical nature of employing *intended* options of escalation for preventing war and stabilizing war, see *Verteidigung und Frieden*, pp. 131 ff., particularly Result 40, p. 134: If on both sides there exist military resources the use of which would surpass the level of power on which the war can still be regarded as a rational means of politics, then it follows from the premises of the strategy of flexible militarily determined reactions that this level of power will probably also be surpassed in the event of deterrence failing. Historical experience provides evidence for this conclusion.

28a. Afheldt, H., *Verteidigung und Frieden*, p. 134, Norm 11.

29. With regard to the first example see the book *L'Europe sans défense* by R. Close. With regard to the second example see, for example, 'Kongressbericht über die Verwundbarkeit der derzeitigen taktischen Nuklearwaffensysteme der NATO', *Frankfurter Allgemeine Zeitung* (10 January 1977).

30. Concerning this theory and its weaknesses see the discussion of the theories of Schlesinger and Wohlstetter in Afheldt, H., *Verteidigung und Frieden*, pp. 131 ff.

31. Considered in more detail in Afheldt, H., *Verteidigung und Frieden*, pp. 18 ff.

32. With regard to this problem see SIPRI, *Strategic Disarmament, Verification and National Security*, London: Taylor & Francis (1977).

33. Danger threatens strategic stability at once by improving the land-based systems to still more accurate systems, whereby the advantage of the one making the first strike is re-established. At what limit this advantage becomes real, cannot be reliably determined, because neither the number of war-heads (MIRV or MARV) nor the accuracy of the opponent's missiles can be reliably calculated.

   Furthermore there is the threat of danger likewise from the qualitative 'advances' in anti-submarine warfare which are hardly controllable.

   Finally, there is the threat of danger from cruise missiles, which are rockets of the type V 1, fitted with nuclear or conventional warheads and have the highest accuracy for reaching their objective. Their number and armament (nuclear or conventional) cannot be ascertained by the other side. For further information on this matter see Afheldt, H., *Verteidigung und Frieden*, Chapter 4.221 (pp. 77 ff.) and 6.322 (pp. 154 ff.).

34. Afheldt, H., *Verteidigung und Frieden*, Chapter 4.1, pp. 53 ff.

35. The problems of such planning are discussed in more detail in Afheldt, H., *Verteidigung und Frieden*, Chapter 4.21.

36. Considered in great detail in Afheldt, H., *Verteidigung und Frieden*, pp. 18 ff.

37. von Weizsäcker, C. F., *Wege in der Gefahr*, pp. 109 ff.

38. Afheldt, H., *Verteidigung und Frieden*, pp. 143 ff.

39. Afheldt, H., *Verteidigung und Frieden*, Chapter 4, p. 82 (in particular Result 24).

40. Former Defense Minister Schlesinger in his Report 1976–77, pp. III-8.

41. Afheldt, H., *Verteidigung und Frieden*, Chapter 4.22.

42. Such an employment is moreover prohibited by Criterion 16 for the Central European countries (destruction much too great by employing nuclear weapons).

43. From Brossollet, p. 71.

44. Brossollet, G., *Essai sur la non-bataille*, (Paris 1975).

45. Spannocchi-Brossollet, *Verteidigung ohne Schlacht*, Hanser, (1976). Teil 2: G. Brossollet, *Das Ende der Schlacht – Versuch über die 'Nicht-' Schlacht*.

46. Afheldt, H., *Verteidigung und Frieden*, Chapter 9.1, pp. 234 ff.

47. *Jane's Weapon Systems* (1977), p. 46. A terminally guided sub-missile (TGSM) is an anti-armour weapon to be used with any launch vehicle. Six or nine TGSM would be loaded into the missile's warhead. When released in the neighbourhood of an attacking force, each TGSM will automatically seek and destroy an enemy tank.

48. Six men when using a rocket-launcher. *Jane's Weapon Systems* (1974–75), p. 37.

49. See Afheldt, H., *Verteidigung und Frieden*, p. 246, and Spannocchi-Brossollet, *Verteidigung ohne Schlacht*, p. 184.

50. The territory of the German Federal Republic comprises about 248 000 km². Neglecting high-mountain regions and large towns, there remain about 200 000 km². Allowing about 20 km² to each techno-commando then 10 000 commandos are required.

51. *The Military Balance*, (1974–75), p. 99. This includes 'main battle tanks in operational service in peace time'. A similar number (22 300) is given in the *Weissbuch*, (1973–74),

p. 13. These figures do not include light tanks and reserve units.

52. Such a demand for the advance of front-line troops is certainly almost impossible to fulfil, even in case one starts from the assumption that the combat troops can be relieved each day on the advance route won by fighting.

The depth of the Federal Republic varies between about 200 km (Thuringen–Rhine) and 500 km (South Germany).

52a. SIPRI, *Anti-personnel Weapons*. London, Taylor & Francis (1978) (in the press).

53. Air and sea striking forces not included. *Weissbuch* (1973–74), p. 13.

54. Heydte, von der, *Der moderne Kleinkrieg*, p. 105.

55. This section (5.2) is mainly a slightly condensed version of Chapter IX.2 in Afheldt, H., *Verteidigung und Frieden*.

56. Theoretically one can imagine stationing out in space. However, that is forbidden by the treaty on the peaceful exploration and utilization of space (*Europa Archive*, 1967, p. D1–5).

57. *Military Balance*, (IISS London, 1976), p. 71. (The number for the warheads of the Poseidon missiles is apparently not quite reliable. SIPRI (1974) speaks, for example, of 10–14 heads of 40 kt.)

58. See *International Herald Tribune*, (19 June 1975): US bolsters NATO A-force with more Poseidon missiles. The modernization of nuclear potential for the European theatre of war decided upon at NATO's Spring Meeting 1976 also seems to provide for a considerable shift to the sea. See *International Herald Tribune*, (16 June 1976).

59. Martin, Lawrence, *Survival*, (1974), pp. 268 ff.

60. Kaplan, M. A., *The Rationale for NATO*. AEI Hoover Policy Studies 8, 1973.

61. The fundamental requirements Kaplan puts forward for his strategy resemble on the other hand some of the norms laid down above: 'It should be effective in war, credibly deterring to possible aggressors during peace, stable during crises, supportive of alliance cohesion, understandable both to professional soldiers and to professional politicians, consistent with moves toward détente, and not inconsistent with programs for arms reduction', p. 54.

62. Kaplan, M. A., *The Rationale for NATO*, pp. 58 ff.

63. Kaplan, M. A., *The Rationale for NATO*, p. 66.

64. Such a requirement seems to Kaplan rightly to be impossible of fulfilment in most cases, because it presents the Eastern bloc state with the alternative of being attacked with nuclear weapons either by NATO or by the WTO.

65. Afheldt, H., *Verteidigung und Frieden*, pp. 275–6.

66. For the role of intermediate stages in 'flexible reactions', see Afheldt, H., *Verteidigung und Frieden*, pp. 131 ff.

67. See American Norm V and Schlesinger, *Report 1976–1977*, pp. II-6. Quoted in Afheldt, H., *Verteidigung und Frieden*, p. 131.

68. See the discussion of this problem in Afheldt, H., *Verteidigung und Frieden*, pp. 182 ff.

69. Regarding this see the review by Record J., in *US Nuclear Weapons in Europe: Issues and Alternatives*, (Brookings, 1974), p. 45; and the discussion in Afheldt, H., *Verteidigung und Frieden*, pp. 78 ff. (183).

70. Steinhoff, in his book *Wohin treibt die NATO*, comments on this view as follows: 'One had become much too accustomed to ideas of war in which after the apocalyptic use of the total nuclear potential of NATO the 'Lieutenant sprang out of the grave, shook the dust from his sleeves and carried his soldiers along to the attack'' ', p. 64.

71. Bennett, Sandoval and Shreffler, *Orbis*, (1973), particularly pp. 464 ff. Limitation of the use of nuclear weapons on the territory of the Federal Republic of Germany is now provided for.

294

72. In order not to provoke destruction of the Federal Republic in depth, no large military objectives in the Federal Republic should remain – according to this concept (as well as in that recommended by us) – outside the blockade belt. There is no longer place for heavy mobile units either in this concept.

10. The new nuclear force

R. Shreffler

I. The new nuclear force

The purpose of this chapter is to describe a new military force that depends strongly upon nuclear weapons – particularly low yield nuclear weapons – and the influencing products of modern technology. Within reasonable economic constraints, the proposed force enhances deterrence and significantly reduces the need to contemplate the use of strategic forces. The proposed force would promote political stability because its non-strategic elements are purely defensive and non-provocative; its deployment would reduce concern over proliferation and arms races. In brief, such a force would provide a rational basis for a peaceful world.

The major element of the proposed force, a defensive nuclear force, is designed to defeat an aggressor as he penetrates the borders of the defended country. The essential component of this defensive nuclear force is a guided missile of 100 km range which is tipped with a subkiloton nuclear fisssion warhead. Other elements of this purely defensive force are designed to protect and support the missile-firing units as well as the country at large from subversive and airborne attack. The force is argued to be extremely effective in its primary task of defeating an aggressor while minimizing unwanted damage; it is quite invulnerable; it requires minimum logistic support; it is relatively cheap; it is adaptable to a wide variety of geographic and political constraints; it is devoid of such elements as battle tanks, air force installations, nuclear storage sites, and other massed forces which serve as essential and lucrative targets for pre-emptive nuclear strikes by aggressors. When associated with an independent punitive nuclear strategic force, and those other measures which counter political subversion, it would serve as the principal element of an optimum political military deterrent, the New Nuclear Force.

One might inquire why a force with such redeeming qualities has not been chosen as a pattern for modern forces. This question becomes the more appropriate when one considers present military forces which champion strategic elements of horrifying potential, and expensive anachronistic tactical elements which have only a modicum of military capability. This question is addressed in the following paragraphs.

Western force development

World War II ended leaving the allied powers blessed with victory and cursed with a prescription for future political/military action. The destruction of Hiroshima and Nagasaki served to present the United States Air Force with a demonstration of an ultimate weapon which served to further entrench their punitive strategic philosophy which has survived to the present.

It was only natural that the US Army would choose to give the tactical role of the nuclear weapon whatever attention it was to receive. To this end they made a number of commendable efforts. However, the Army was never able to assimilate the recommendations of these works. Today the efforts of the US Army, as well as the other services, to understand and incorporate nuclear weapons into US tactical force remain minuscule. In all fairness, it should be added that the inability to cope with this problem remains a principal frustration, an irritating thorn in their side.

It is much to the point to ask why organizations of unquestioned integrity, who share the responsibility for protecting their country, place such low priority on this tactical nuclear capability. Why do they choose to field an obsolete force which would be sorely pressed to either fight or protect itself in an engagement with its principal adversary, the Soviet Union? The answer rests in the fact that an optimized defensive nuclear force represents too gross a departure from the offensively oriented conventional force which had served so well in World War II. To this must be added a justified fear that the nuclear weapons around which the force would be structured would not be released by political authority when they were needed.

The US Army, in particular, was aware that a force dependent upon nuclear weapons would be much smaller than its World War II force, and totally different in both its structure and supporting equipment. The reorganization from the status quo to such a military posture, which was totally unsupported by any experience, represented too irreversible a step. Under the protection of the US strategic umbrella, and in the absence of an imminent tactical military confrontation, pressures were not sufficient to force serious consideration of such a drastic change. As a consequence, the US Army continues to consider tactical nuclear weapons as some sort of not-too-well defined extension of its current conventional capability.

The frustrations inherent in understanding the role of nuclear weapons have been shared by persons outside the US military organizations. Former Secretary of Defense Schlesinger acknowledged the inherent difficulties by noting that 'after 30 years we are still struggling to adapt our concepts of conflict and its deterrence to nuclear weapons that range in yield from subkiloton to the multimegaton' [2].

Alain C. Enthoven contributes the following comment:

> 'I am prepared to conclude that 20 years of efforts to find an acceptable doctrine for the use of nuclear weapons in the defense of Western

Europe have failed because one does not exist. The planned first use of nuclear weapons for the defense of Western Europe simply doesn't make sense. It amounts to saying, "we'll have to destroy this continent in order to save it" ' [3].

Of recent date there have been repeated efforts to come to grips with the problems associated with the tactical use of nuclear weapons. The subject received its widest advertisement on the political stage in what might be termed the 'mini-nuke debate' (see Appendix 1). Most papers [4–6] on the subject have been based upon the premise that a nuclear capability was quite secondary to the existing conventional force. These papers characteristically present force options including the existing force with its present nuclear weapons as one limit and with the nuclear weapon stockpile reduced and modified to include no nuclear capability at all as the other. The impression left with the reader is that the options presented pose a complete list, instead of perturbations of the existing force. There is a general conclusion that the basic war fighting element is our conventional force, and that the psychological value of the deployment of nuclear weapons far outweighs whatever military contributions the weapons may make to overall deterrence.

At the other extreme is the position that the force should be defensively oriented as described in this chapter; the essential force element is the low yield nuclear weapon [7–13].

There are intermediate positions. Some represent the first step in a departure from the present force to a defensive force configuration designed to accommodate conventional precision-guided munitions [14]. In this volume the preceding chapter describes a force which is an extreme limit of a defensive conventional force. It merits careful consideration. Unfortunately such forces in their implementation must surrender territory. Others [11, 15], hold positions close to those presented in this chapter; there is agreement that the force must be defensive in quality and depend upon the firepower of nuclear weapons. The differences largely rest upon the degree to which the authors choose to depart from the elements of the existing force.

Even a brief review of literature would be incomplete were it not to recognize the wise words of such authors as Bernard Brodie [16], who warned in 1963 of the dangers inherent in the present NATO force and the danger to the United States of expending its prestige in maintaining objectives which are peripheral at best.

There has been some recent encouragement to proceed with the work discussed in this essay. Maxwell Taylor states [17]:

'. . . there is an alternative . . . based on the prompt use of battlefield nuclear weapons of minimum yield without awaiting a palpable breakdown of a conventional defense. [It is] worth reviewing . . . if it offers any chance of avoiding reversion to the sterile strategy of massive retaliation'.

A different sort of encouragement is expressed in a lengthy review of reference 6 in the 1974 SIPRI Yearbook [18]. It concludes,

> At first sight, this policy (essentially the use of the New Nuclear Force) may seem attractive, particularly from the point of view of military tactics. But there are fundamental weaknesses in any policy depending upon the use of tactical nuclear weapons. Few would be confident that this or any other feasible policy would, in practice, prevent armed conflict involving the major powers from escalating to an all out nuclear war so long as any nuclear weapons regardless of type, are used. And there is no guarantee, or even likelihood, that the opponent will adopt similar tactics. In particular, the introduction of very low yield (less than a kiloton) nuclear weapons would blur the present distinction between conventional and nuclear weapons. It is of *paramount importance* that an absolute "firebreak" should be maintained between nuclear and conventional war.
> It is important that these objections to such a policy for the use of tactical nuclear weapons should be widely understood because strong pressures may arise in the near future for such a policy to be officially adopted, particularly in Europe.

Eastern force development

The USSR entered the nuclear arena somewhat later than the USA; however this tardiness could well have been compensated for by its willingness to adjust its military posture to meet political ends through an intense and open debate which has been well exposed in the unclassified literature [19, 20]. It is therefore reasonable to ask why they have not adopted the approach outlined in this chapter. The answer is immediate if the USSR chooses to present NATO with a preponderant offensive force capable of taking and holding NATO territory for either political coercion or aggression; to this end an offensively oriented force is required. Further, a defensive force as proposed here will hardly serve to maintain order within the WTO. As a consequence, one can expect the USSR strongly to oppose the new nuclear force and make every effort to strengthen its WTO conventional force and eliminate tactical nuclear weapons from *both* NATO and WTO forces [21].

The Soviet–China problem, on the other hand, might well be resolved by either country willingly restricting its military forces to defensive nuclear forces along the common border. Territorial ambitions by either party would then be thwarted. This could well serve the Soviet plan.

The following sections address the premise for the New Nuclear Force. Attention is focused upon the defensive nuclear force and the kind of conventional and nuclear weapons it would require. Conventional precision guided munitions and their roles when combined with nuclear weapons are discussed. A defensive force structure is outlined and criticized. The new nuclear force is then considered in the context of a new NATO force and as a basis for a US tactical nuclear umbrella, a military force designed to meet worldwide commitments with an effective and rational level of force.

II. The premise for the new nuclear force

An enlightened country† is forced continuously to reconsider the political role it must play and what military forces are required to establish and maintain that role. To this end the country must decide whether or not to arm its force with nuclear weapons, or if so armed, whether its posture should be changed.

This section attempts to document a course of thought that such a country might take in reaching such a decision. First the general military and political aspects of the proposed force are considered. The subsequent step requires the imposition of real world constraints or boundary conditions characteristic of the country – constraints which may deny all or part of the proposed nuclear capability briefly described in part I.

A consideration of the general military and political aspects focuses first upon the two military responsibilities of the country:

To minimize the prospect of an attack against its national treasure, in particular the destruction of its cities, a strategic nuclear force is required.

To protect itself from invasion a defensive nuclear force is postulated.‡

In addition to these two military responsibilities, a political responsibility should be added. There is an evident trend to accomplish political ends not by military aggression but through social deterioration and subversion [15]. Though the capability to meet this political responsibility is not addressed directly here, it is contended that the existence of a proper military force will frequently serve as the best deterrent and cure to such political ills.

General force features

The general military and political features of the new nuclear force which must be considered by the country are outlined as follows:

Force element independence. The strategic and defensive nuclear forces must be independent of one another. They must operate under different strategies with their own personnel, material, crisis management, communications and control. An operation involving both forces is controlled only at the highest level of political command authority. Only through such a prescription can this command authority release its defensive nuclear forces in a

---

† In addressing the subject of tactical nuclear warfare one naturally chooses as a model the European theatre since it presents practically the only source of experience. In this chapter an attempt is made to treat the subject in general fashion; hence the generic noun country is employed.

‡ The use of the word 'defensive' is possibly premature at this early point. It implies that the country has no aggressive plans. The introduction of nuclear weapons (as well as precision-guided conventional munitions) into the tactical area makes an offensive tactical military force obsolete whether the country has offensive or defensive motives. The preponderant firepower of nuclear weapons requires one to modify the cliché 'the best defence is a good offence' into the 'best defence is a good defence'.

responsive fashion without jeopardizing the release of its own or an attacker's strategic force.

In order to achieve this decoupling, it is necessary to make and advertise the following steps.

1. Such concepts as continuity of force, flexible response and limited strategic options must be discarded.

2. The strategic force is developed only in its most destructive and punitive role.

3. The defensive nuclear force should engage an attack at the border only on the defender's soil with weapons of such limited range, low yield and intended use as to present no offensive threat to an attacker. Enemy forward manoeuvre companies are the targets which best suit this constraint. Throughout the country conventional forces are employed to support the forward defence, and to defeat subversive military attacks and airborne attacks.

Force objectives. Where the objective of the strategic force is maximum destruction, the objective of the defensive nuclear force is to meet an attack in a militarily effective manner, while creating a minimum and politically acceptable amount of unwanted (collateral) damage. There is an historic fear of nuclear weapons which usually overcomes rational discussion of their use. It forces severe constraints on the level of permissible civilian casualties and property destruction in the execution of a defensive nuclear war.

Force invulnerability. The forces must not present a vulnerable target complex to the enemy, particularly one which is lucrative and threatening. Fixed missile sites, air force installations, nuclear storage sites, and massed military forces are such attractive targets. They invite pre-emptive nuclear attack. Such attacks spell defeat of the defender by the easy destruction of essential elements of his military force. These attacks create considerable destruction of the defender's national treasure and force consideration of the use of the strategic force.

A careful selection of invulnerable strategic systems between types and operational modes of missiles, aircraft, and more unconventional means is required. The submarine would seem to hold a current advantage, though the others should not be ignored.

All elements of the defensive nuclear force should be constructed with survivability in mind. Particular emphasis should be given to the nuclear weapons which can be most easily protected by high mobility, large numbers, low profile, and deployment sufficiently far from the border.

Deterrence. Both the strategic and defensive nuclear forces must be capable of generating a deterrence to political or military aggression. This deterrence is recognized as a product of two factors; the manifest capability of the force, and the assurance, particularly in the mind of the aggressor, that the force will be used promptly if challenged. Historically, deterrence has resided with the strategic force. It is essential to structure a defensive force of such high quality and responsiveness that it serves as the overriding deterrent.

The manifest capability of a highly invulnerable strategic force dedicated to a maximum punitive mission is tremendous. As a consequence one can and should be noncommittal as to the conditions under which it should be employed and still maintain a high level of deterrence. The manifest capability of the defensive nuclear force is identified with its war fighting ability which, in turn, will depend upon its prompt commitment as the enemy crosses the border. This commitment is easily tested by the aggressor's probing attacks. Thus, it is essential that the force be used upon any occasion that the border is subjected to aggression, i.e. the intensity of the aggressor's attack should have little to do with the promptness and quality of the response by the defence. An intelligence organization to evaluate continually the intention of potential aggressors is essential. A fraction of both forces must be retained on alert depending upon the intelligence estimate. Large-scale alerts of the defensive nuclear force must be routinely practised. Whatever procedures are demanded of the civilian population to reduce collateral damage must be understood by all concerned, and practised.

Threat independence. To a maximum degree both the strategic and defensive nuclear forces should be structured to meet any possible threat. As such they are absolute† forces in that the number of personnel, equipment, and modus operandi are not subject to significant change because there is no *military* reason for the change.

Responsiveness. Both forces must be of high quality and capable of operating to meet an attack in a matter of a few minutes. It should be recognized that both strategic and defensive wars may last of the order of hours. Consequently, they must be fought with equipment, ammunition and supplies already in place. Nuclear weapons must be ready to fire. Both forces must be controlled with simple (and independent) political-military command structures.

Research and development. A well-supported research and development programme should be sustained with a prejudice to carry the programmes up to the manufacture of prototype systems. In particular, anti-tank and anti-aircraft missiles with conventional warheads should receive more attention. Anti-missile capabilities should not be ignored; this is particularly true for countries with a small area and limited threat. It is difficult to plan forces with essential equipment which has not yet been perfected.

Arms control requirements. The force should be easily adaptable to restrictions regulated under arms control provision [10].

1. As the potential of the defensive nuclear force becomes apparent, the importance of strategic deterrence diminishes. One can then consider stronger international sanctions on the strategic force starting with such constraints as no first use and hopefully, terminating with the reduction and final elimination of this force. As long as the major deterrent rests with strategic forces, as it does today, such prospects cannot be considered. On the contrary, every step must be taken to preserve the credibility of the strategic forces.

2. A strong defensive nuclear force will leave an aggressor no alternative but to forsake his offensive force and employ his own defensive nuclear force.

---

† In sharp contrast to current conventional forces.

Such a benign defensive nuclear force confrontation may represent the ultimate goal for the arms control community, and a welcome relief from the current goal held by many of an assured strategic second strike capability.

3. Proliferation of nuclear weapon technology and capability will always present an aggravating issue. However, in the proposed situation where strategic forces have lost their importance and where defensive forces benignly confront one another, the issue is far less serious.

4. To the degree that the forces are designed and installed to meet any threat, an arms race is no longer an issue; i.e. there is no reason for increasing the size or quality of the force.

It follows that the arms control goal of the international community should be the implementation of the force prescribed in this chapter, and the reduction, and final elimination of strategic forces along with the establishment of those controls which ensure this abolition.

Force constraints

The real world constraints on the design of the force can be of overriding proportions in determining the premise for the force. These pressures are composed of a combination of constraints both internal and external to a country's boundaries. They are often irrational and can be so severe as to offer little opportunity for the average country to define, construct or control its own military force. The major constraints are the following:

— Bureaucracies are, in general, strongly committed to the status quo, and are for this and various other reasons opposed to a nuclear defensive force.

— Economic constraints are frequently posed, though this objection is far more imagined than real [22]. The proposed defensive nuclear force offers far greater military effectiveness for the money than the present conventionally oriented force and it is an absolute or 'fixed price' force. The costs of nuclear weapons specifically have been exaggerated, particularly if one observes the pitfalls outlined in Appendix 2.

— The technological capability of the country is a factor to be considered, though this aspect has also been exaggerated [22].

— The availability of fissile material is an example of a valid resource constraint. However, proper weapon and force design will significantly reduce the amount of fissile material required. Nevertheless, the country anticipating this nuclear adventure should give careful consideration to its sources of fissile material. Considering the intensity of the concern of the world community over nuclear weapon proliferation and the relative ease with which one can control fuel elements [23], plutonium may be rather difficult to acquire. On the other hand, new methods of isotope separation, the mandatory requirement for large numbers of fission reactors, and the resistance of many countries to any control, leave the door ajar.

— There are many other internal features of the country such as its topography, climate, general political persuasion, social description etc., which bear heavily on the problem. One feature should be given more serious thought than the few words which follow; namely, the defensive nuclear force of an island country, or a country with an extended coastline. For example, the invasion from the sea using surface ships would be the height of folly if one assumes any kind of appropriate nuclear defence force. The character of this force would be different than the one proposed here which defends along a land border. For example, it might employ far fewer, longer range missiles armed with somewhat higher yield warheads. Having taken such precautions, the attention of the defence should be focused more on an invasion from the air or upon subversive action.

Many overriding constraints are generated beyond a country's borders. Its geographical position; the strength and attitude of its allies, enemies, and neutrals; the importance of its natural resources; and its treaty obligations are some of the more important elements.

Force definition

To develop a prescription for a New Nuclear Force for a specific country as outlined in this chapter requires detailed and continuing debate within the country. Those discussions should engage wide participation and should be directed by the civilian authority with proper participation by the military services. A principal goal of these deliberations will be the definition of the rules which control the firing of nuclear weapons by the military organization ensuring on the one hand that military effectiveness is realized, on the other hand limiting collateral damage to an acceptable level. This limitation will also require rules for civilians, particularly those in the battle area: should they remain where they are; should they collect in population centres, or should they retreat from the battle area at an early time? (With a defensive force largely in place, the road network would be relatively free for such an exodus.) Continuing and detailed consultation and negotiations should be pursued with all other interacting countries, both allies and adversaries.

Strategic force options. Strategic forces, as exemplified in the forces of the major powers, are well described, when compared to the tactical forces. The acquisition may present a major problem to a country which is forced to address the many complicating issues outlined in the previous parts of the chapter, e.g. the forces can be technically complicated; they can be expensive; the strategic threat can be large; etc. On the other hand, the US strategic force requires only somewhat more than 10 per cent of its defence budget; also the US force (and that of the USSR) is a number of times bigger than is required to execute a punitive mission. In addition, the strategic technology exists, and some strategic systems are or will probably be available for purchase. To repeat, the acquisition of a strategic force poses the country a challenging problem. After it has considered all aspects it may choose one of a number of courses; it may choose to:

— Follow the route of the major powers and procure bombers, submarines, land-based missiles, and possibly other systems.

— Procure a cheaper more simple force; however, one which may be quite effective.

— Procure a small force to complement the force of an ally.

— Develop no strategic force but seek the protection of some ally.

— Ignore the problem created by the acquisition of a strategic force and depend only on its defensive nuclear force.

No doubt there are other choices. However, since the choice depends primarily upon the particular situation and not upon hardware development or force structure, the strategic force will not be addressed until the last sections of this chapter.

Defensive nuclear force. The situation with respect to the defensive nuclear force, on the other hand, is totally different. Such a force has yet to be described. Thus, the next parts of this chapter will be devoted to this task.

III. Weapon requirements for the defensive nuclear force

The heart of the defensive nuclear force is a guided missile tipped with a simple subkiloton fission warhead. This section addresses how and why this selection was made and completes the description of the weapon.† Conventional precision-guided munitions (PGMs) also play a significant role. The distinction between the roles of nuclear warheads and PGMs is presented.

Target and warhead selection

A principal objective of the defensive nuclear force is to defeat aggressor manoeuvre companies at the border, while creating a minimum and acceptable amount of collateral damage. In order for this company to operate as an offensive element it will be contained within an area of about a square kilometre. It is proposed that a subkiloton fission warhead can be employed to defeat such a target.

---

† Many of the numbers chosen will appear arbitrary. In fact, they have been selected after considerable debate and deliberation. It is believed that more detailed consideration will make little difference in the arguments.

One could choose a smaller target for sizing the yield of a nuclear weapon – for example, one or two vehicles. Certainly this could be approved from the point of view of cost; however, it does not take into account the principal advantage of the nuclear weapon, namely its ability to instantly paralyse an attack by defeating a much larger force than one tank with a single explosion. Attempting to defeat a rapidly advancing manoeuvre company, a vehicle or so at a time, implies an expensive, highly dense multilateral defence with an associated offensive capability for recapturing lost territory, a force much like the one required to support conventional precision-guided munitions [1, 14].

On the other hand, we can choose a larger target to size the yield. There are a number of reasons why this is undesirable.

— As the yield increases, the potential for collateral damage increases. Unless the damage is simply accepted, the areas in which the weapons can be used are correspondingly reduced.

— Lower-yield weapons give the defence more reaction time against an aggressor manoeuvre company, because fire can be called closer to the defence.

— Targets larger than a manoeuvre company can always be divided into smaller units (e.g., company size) and then attacked by a number of lower-yield devices, with a corresponding reduction in collateral effects.

— Lower yield weapons should be more easily released to the control of lower levels of military command.

— Higher-yield weapons obscure the distinction between defensive and punitive action, increasing the risk of escalation. This defeats the purpose of the defensive nuclear force.

Constraints to maintain a subkiloton warhead yield

In order to keep the yield of the nuclear weapons as low as possible while attacking a manoeuvre company a number of restrictions are imposed:

— Radiation is accepted as a kill mechanism. Until recently this mechanism has been resisted in some quarters in preference to blast kill probably because of the uncertainty associated with radiation kill, and the necessity for assigning high assurance to the performance of a single round. Certainly the ratio between that level of radiation which will produce death after several weeks and levels which produce prompt incapacitation is of the order of 50. However, the corresponding effects radii differ by only a factor of 2. Accepting radiation kill results in a number of advantages to the defensive nuclear force. The principal target, armoured vehicles, offer little protection from radiation, particularly when compared to blast. They usually attenuate the radiation by only a factor of 2 or 3. The corresponding change in the kill radius is very small, and remains much larger than the blast kill radius for tanks. By adding large amounts of shielding, radiation may be attenuated by a factor of 4 or 5, possibly more; however, this buys little in the way of vehicle survivability as measured by the reduction in tank kill radius.

— At the proposed subkiloton yield, simple, cheap fission warheads can be used to advantage since at these low yields a larger fraction of their energy output is in the form of prompt nuclear radiation. This advantage disappears as the fission yield requirement increases. The weapons will always be airbursts to maximize the effects of prompt nuclear radiation, and to essentially eliminate concerns over fallout, rainout and plutonium dispersal.

— The aggressor manoeuvre unit is not fired upon until its location is known through the detection of one or more of its vehicles.

— To minimize unwanted damage, the delivery error should be as small as possible; however, extreme accuracy under all weather conditions is presently unattainable, and, in any case, tactically unnecessary. Delivery errors of about 100 m represent an acceptable compromise.

— One must be willing to fire more than one nuclear round at a target. The target defeat criteria per nuclear round must not be too stringent.

Weapons should have only one low-yield option. Weapons with yield options greater than 1 kt are anathema to the defensive nuclear force, for which the advantages of higher yield options are considered illusory.

There are other constraints suggested in part II which bear repeating.

— Survivability. Emphasis should be given to nuclear weapons which can be most easily protected by high mobility, large numbers, low profile, and deployment sufficiently far from the border. Deployment in a band sufficiently far from the border provides the additional important advantage of bringing to bear many weapons on the same local area of the border. This not only improves the quantity and rate of fire, but reduces the total number of weapons in the stockpile.

— Exercise. Both nuclear and conventional forces must be frequently and realistically exercised in the local area where they will be employed in wartime. For nuclear fire missions, the nuclear warheads will be replaced with appropriate conventional charges. The cost of these exercises must fall within an approved budget.

— Fissile material availability. The availability of fissile material is an example of a valid resource constraint. Proper weapon and force design will significantly reduce the amount of fissile material required.

Selection and description of nuclear weapon

As defined in the preceding sections, the task of the delivery vehicle is to deliver a cheap, subkiloton fission bomb to a target on the defender's soil, from a firing area covering a depth between 25–100 km from the border, and within a CEP of 100 m of a target moving at battlefield speed. The delivery device must be relatively invulnerable and inexpensive.

Weapon evaluation. One can analyse in detail the following vehicles: aircraft, cannon, inertially-guided missiles, atomic demolition munitions (i.e. no vehicle). Aircraft are highly vulnerable on the ground and in the air; their bases form an unacceptable target complex inviting pre-emptive nuclear attack; they represent a strategic threat; they have difficulty delivering nuclear weapons with sufficient accuracy and in close proximity to defensive troops. Artillery-fired nuclear projectiles are quite expensive and are limited in range and accuracy. Efforts to extend the range with improved accuracy lead to complications and further expense. Inertially-guided missiles may have some promise, though present systems are not accurate, and are very expensive to buy and maintain. Atomic demolition munitions may have some limited value by forming barriers to aggressor advance; however it is probably safe to say that experience to date has demonstrated that their military utility is small when compared to the political problems they generate.

The guided missile.† The only vehicle that meets all the requirements is the guided missile, and it meets these requirements very well. A description of the missile system as extracted from Appendix B of Ref. 11 follows:

'There are a number of ways to configure a satisfactory missile and launcher system. The missile should use solid propellant, require little routine maintenance, and be as close to a "wooden" round as practical with the shipping container acting as part of the launcher. The missile should require only a minimum of routine and prefire checks. The size of the missile will depend on its range and the desire to minimize the quantity of special nuclear materials in the warhead. A review of standard US Army ground-handling equipment reveals little advantage in a missile weighing less than about 1000 lb unless it can be made man-portable. Warhead and also dynamic considerations lead to a missile diameter of about 14 in. A number of guidance schemes can provide the required accuracy at reasonable cost. To nullify enemy counter-measures, several schemes could be adopted for different missiles. One possible guidance scheme uses a terminal-guidance beacon with a small computer. The missile flies to an offset from the beacon. The CEP for this system is a strong function of the offset from the beacon, but for offsets of about 3 km, one can expect a CEP of 30 m. The missile can be launched at a constant angle of elevation, with range determined by thrust termination at the appropriate velocity. An azimuth launch error of 2° is allowable. The launch procedure thus amounts to slewing the missile to the approximate azimuth and programming its thrust-termination velocity. Without midcourse guidance, the system would have a maximum range of 60 to 70 km and still be able to hit the window needed for terminal guidance. Minimum range would be about 20 km. The missile and its container-launcher would weigh about 1500 lb. This weight implies that either four or six missiles could be mounted on a standard Army 5-ton truck (the advantage of using a standard vehicle, besides its lower cost and maintenance, is that it complicates the enemy's target-acquisition problem even further). The time of flight for this missile would be 2 to $3\frac{1}{2}$ minutes.

A fire mission could progress via data links which are all digital, assuming that the weapons have already been released. First, the forward observer sends the target coordinates to the FDC (fire

---

† The cost of the missile and its annual operational and maintenance cost depend strongly upon the method of calculation; however, the procurement cost for the type of missile needed here should be in the neighbourhood of an order of magnitude less than that for such missiles as Lance and Pershing and the operation and maintenance cost is less by a factor of 3 to 5.

direction center) computer. The FDC computer finds an available missile unit within range of the target and sends the range and azimuth to the nuclear fire unit and a warning signal to friendly troops. The fire unit slews the missile to the correct azimuth, sets the desired terminal velocity, and fires the missile, after which the time of launch is sent to the FDC computer. The observer can update target co-ordinates during the mission. About one minute before detonation, the FDC computer gives the needed information to the homing beacon. About 30 to 40 seconds before detonation, a signal from the beacon arms the missile and the mission goes to completion . . .'

Surface-to-surface Nike Hercules. To make the proposal more specific and to facilitate timely deployment, Nike Hercules in the surface-to-surface mode would serve quite well if certain state-of-the-art modifications were made and somewhat higher costs were acceptable. Figure 10.1 shows a Hercules battery stripped down, modified and deployed to meet the requirements of the nuclear defence force. A brief description of system modifications and operation follows:

— The target-acquisition and target-tracking radars would be removed.

— These radars would be replaced by forward observers who would feed original and updated target coordinates to a fire-control centre that, in turn, would allocate targets to specific missiles and missile-tracking radars and would also override inappropriate targets.

— A standard Army transport vehicle would carry the launcher, one missile, and the firing apparatus. Another would carry all the paraphernalia supporting the missile-tracking radar (Fig. 10.2).

— The launcher vehicle could be located independently of the other elements of the system. The missile-tracking radar would be located so it could time share and guide up to four missiles from different launchers. These radars could also be netted to increase their flexibility greatly and reduce the vulnerability of the system. The tracking beam is very narrow and carries coded commands to the missile, making the system very hard to jam.

— Finally, the analogue computers would be replaced with a modern digitized processor, and solid state electronics would be employed. This essential and rather easy operation would improve response time to approximately 7 minutes from time of initial request to explosion and would result in greater system accuracy, reliability, and reduced maintenance. (Missile CEP would probably be around 50 m and would certainly be well within the desired 100 m required for discriminate nuclear fire.)

A major issue is the cost of fielding and maintaining the system. The estimated cost per year per battery (12 launchers) would be 2.6 million dollars. This low figure results primarily from reducing the Hercules manning level to about 135 people per battery, which, is a conservative estimate of the number required. The technical performance and costs have been verified and coordinated with the missile system contractor.

Figure 10.1. Nike–Hercules surface-to-surface application

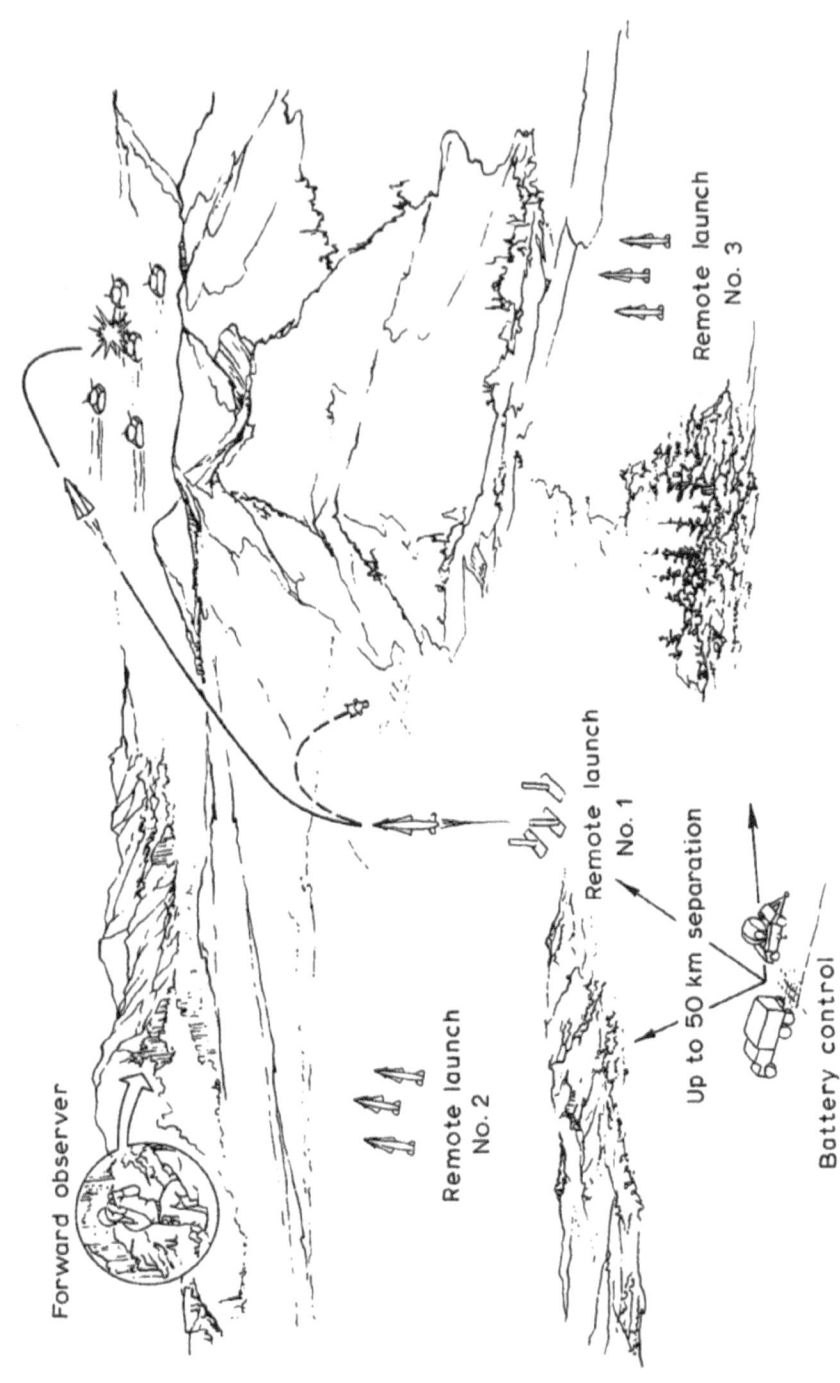

Forward observer

Remote launch No. 2

Remote launch No. 1

Remote launch No. 3

Up to 50 km separation

Battery control

Figure 10.2. Modified surface-to-surface mobile capability

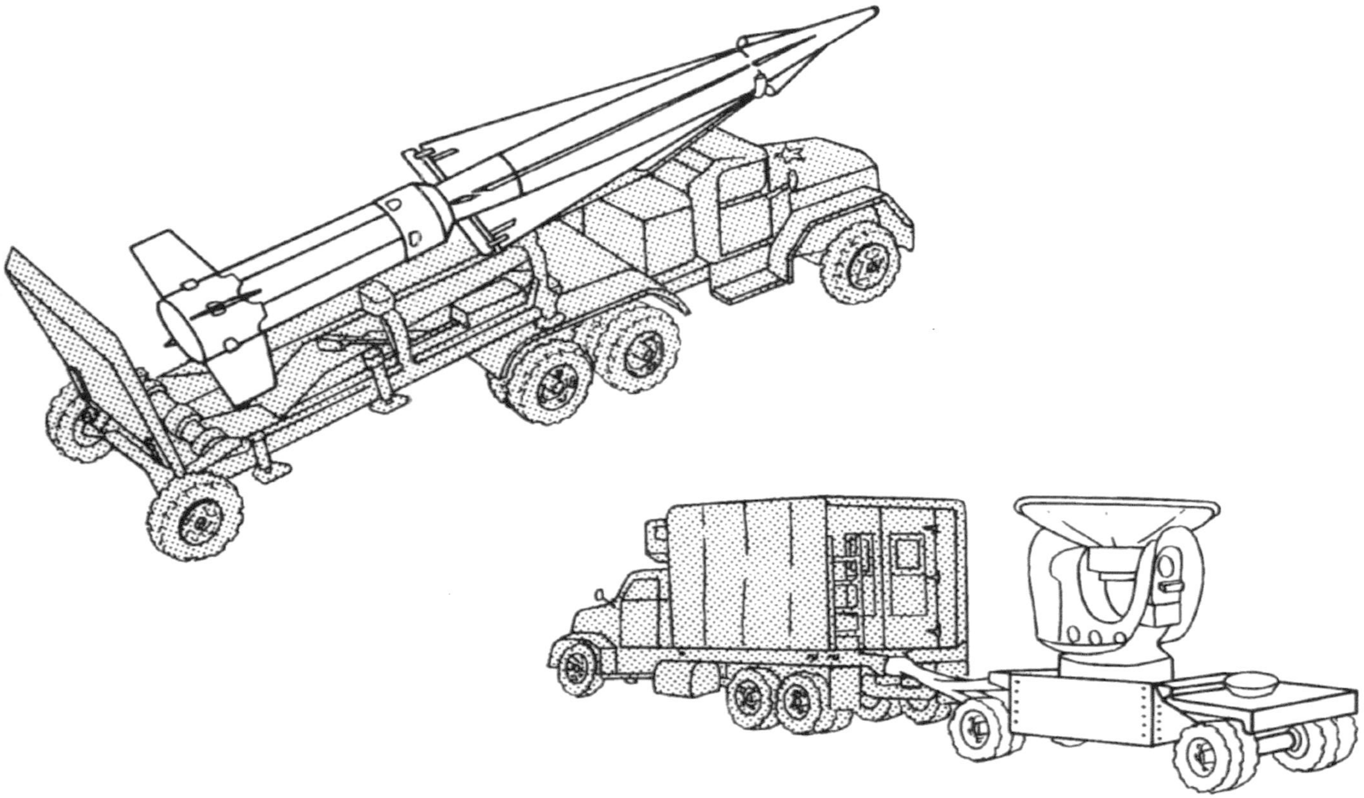

Nuclear warhead description

A consideration of the preceding sections of this chapter leads to a fission warhead of nominal yield of 0.5 kt and a diameter of 14 in. (Were Nike Hercules used, one could employ a 25–30 in diameter warhead which would give some advantage with respect to the quantity of fissile material.)

The nuclear warhead will be manufactured with removable fissile material which would quickly and easily be inserted just prior to missile launch. The intention is to build the explosive implosion and firing elements of the warhead as permanent parts of the missile. The nuclear material would require special handling. Since the nuclear material has been removed from the close proximity of the explosive the nuclear safety aspect of weapon handling has been essentially eliminated. Instead of protecting weapons in isolated but well identified bunkers, one now stores the fissile material in 'bank vault' enclosures in the Kasernes. Of course, some fraction of the missile vehicles with nuclear components will be continually dispersed in tactical positions which are frequently changed. Special precautions would be required to ensure the security of these vehicles particularly in peacetime when one might expect a third of them to be in the field. The following steps are suggested on the assumption that each such vehicle would operate within a 200 km² area:

1. Each vehicle would be protected by a platoon of about 6 troops (see part IV).

2. Each vehicle would be in communication with its Kaserne and with the surrounding (~ 8) units, ensuring prompt assistance.

3. The vehicle would be protected by *simple* devices for making the vehicle inoperable and difficult to enter. The fissile material would be enclosed in special vaults mounted on the vehicle. As a consequence quick removal of weapon components by unauthorized persons would be extremely difficult. Such precautions would place little strain on manpower and technology. They should hopefully meet the demands of the critic, since they are reasonable and since they are in some respects not far from current practice in NATO. The major advantage posed by this action would be the removal of the storage sites as prime targets for the aggressor, at the same time essentially eliminating the safety problems associated with unintended nuclear explosion and plutonium dispersal. As important, the prevention of the seizure of fissile material would be handled in a realistic and effective manner.

Precision-guided conventional munitions and their role vis-a-vis the nuclear missile

It is essential to incorporate high quality conventional PGMs into the defensive nuclear force to defeat tanks, aircraft, and personnel. There are review articles [1, 20, 24, 25] on the capabilities and status of PGMs, and they will not be discussed here except to recognize the following essential points.

— PGMs are presently in a developing state. Their capability already establishes their importance in any modern arsenal, however, it is not known how much better they can be made.

— PGMs are basically 'hitting weapons', i.e. to defeat a tank or aircraft with conventional explosives, the target must be hit with the munition. PGMs cannot now operate satisfactorily in inclement weather and do poorly at night. In time, they will improve; however, they will probably always fall short of desire.

— There is a sharp distinction to be drawn between the roles of low yield nuclear weapons and PGMs. To repeat, PGMs have limited capabilities and are fundamentally hitting weapons. Nuclear weapons, on the other hand, are area weapons. They do not require high precision delivery and they are designed to kill many targets at once promptly to paralyse an attack. Realizing this, the aggressor is dissuaded from attacking. PGMs may cause an aggressor high casualty rates which he will attempt to counter by increasing the size and configuration of his force and changing his tactics. They will add to the credibility of the defender's force but they alone will not change the fundamental political/ military position of two adversaries. The aggressor will still acquire the defender's real estate, which the defender will then be required to retake with a force similar to the one he has today [14, 15, 20]. Nuclear weapons permit the luxury to the defender of a static, simple, cheap force; PGMs do not.

— The fair position in this matter of the capability of conventional PGMs versus nuclear weapons for land warfare has yet to be determined in detail. In light of the previous arguments, it is hardly credible that the result will not be that the two types of munition will complement one another. There will be such continuing questions faced by the field commander as to how he will employ these weapons to address a particular military situation. This will be reflected in the quantitative construction of the defensive nuclear force, though it should have a modest qualitative effect.

— While the principal weapon for the border defence is the nuclear weapon with support from the conventional PGM, the role of air defence is totally restricted to the conventional munition. Conventional munitions are less expensive, are not encumbered with security and command and control problems, and they have about the same kill probability/missile as the nuclear munition. Arguments are made that nuclear warheads are required for defence against massed aircraft attack (an unlikely event) or missiles. This latter application requires careful investigation; however, it is not clear that the conventional munition approach would not be preferred.

Firing doctrine

Much of the substance of parts I and II can be summarized in a 'Firing Doctrine'. The National Command Authority might define its direction to its military force through such a document, an *example* of which follows.

FIRING DOCTRINE
for
DEFENSIVE NUCLEAR FORCE

This firing doctrine is intended to facilitate the execution of the defence of the country by defining a set of rules which are to be rigidly followed by the military force.

1. The military force should function under the assumption that the total professional military force is in position with its nuclear armament released for use by the National Command Authority prior to the crossing of the borders by the aggressor.

2. Warhead detonation will be confined to the territorial domain of the country. The defence should be structured to prevent the penetration of the country more than 10 km. The aggressor should be held under fire as long as he remains within the border of the country.

3. The nuclear warhead will depend upon radiation kill and will be restricted in yield to that quantity of radiation produced by less than a 1 kt fission device.

4. Nuclear fire should be employed against any aggressor penetrations of the border which strains the capability of the conventional elements of the force to meet the requirements spelled out in this document. Tactics, capabilities, and nuclear weapon stockpiles should be developed to defeat any conceivable enemy penetration.

5. Conventional munitions should be employed:

— In coordination with nuclear fire to defeat any aggressor forces at the border;

— To provide for defence against enemy aircraft at the border and throughout the country;

— To provide defence and protection of communities, facilities, and other areas within the country which are restricted from nuclear fire according to the collateral damage regulations defined in paragraph 6.

6. The following regulations are intended to restrict collateral damage to a minimum. Firing doctrine will conform to these regulations:

— The intention of the National Command Authority is to minimize civilian jeopardy to nuclear fire. Civilian population will be advised to move from the border region and from the principal invasion avenues at time of alert. Regulations for civilians will be formulated for other portions of the country in order to protect the population from harm and to support the military defence against air invasion and subversive attack. As the situation permits, military police will assume the responsibility for the protection of civilians and the evacuated territory;

— Maps will be prepared defining areas restricted from nuclear fire. These restricted areas will include communities and eligible facilities. Certain of these areas will be restricted from nuclear fire under any condition. Other areas may receive fire at the discretion of the military commander;

— Detailed regulations for civilian action, for the functioning of the military police, and for the definition of areas to be restricted from nuclear fire will be approved by the National Command Authority as will the maps defining the restricted areas.

7. This document and all supporting documents will be distributed throughout the country and to allies and potential aggressors.

Additional comment. A few additional comments are made in support of the firing doctrine. In part II it was stated: 'Thus, it is essential that the force be used upon any occasion that the border is subjected to aggression, i.e., the intensity of the aggressor's attack should have little to do with the promptness and quality of the response by the defence'. In support of this statement the field commander should be aware that

— one need expend only a few prompt nuclear rounds to establish the intention of the defence,

— where a forward observer sees a small force there may be a much larger one concealed but within the lethal range of the nuclear weapon,

— it is essential to the concept of the defensive nuclear force that an attack be stopped at the border. One can afford to be particularly extravagant with nuclear ammunition to ensure that an attack is stopped there, prior to heavier commitment and losses by either side, and

— the aggressor should be given no reason to believe that he can structure an attack in such a fashion as to seriously draw down the nuclear stockpile of the defence to his advantage.

IV. *The structure of a defensive nuclear force*

The organization of a defensive nuclear force will depend upon many factors. For almost any approach however, one must recognize the need for forces to carry out three relatively independent tasks: the defence of the country's border, the defence of the country at large, and the defence of the air space over the country. In this section attention is focused upon the force responsible for the defence of the border. There is no intent to play down the importance of the other two forces; however, their development would seem to be straightforward when compared to that of the border defence force. As a consequence these two forces are dismissed quite briefly.

Area defence force

The area defence force would be responsible for the following:

— Administration of the militia. All three forces would draw heavily on militia;

— Control of subversive elements;

— Civil defence. This could be a major undertaking involving, among other things, the management of civilians in the border areas;
— Defeating airborne incursions inside the border defence.

Air defence force

The air defence force would be responsible for defeating any aggressor object which penetrates the air space of the country. Close coordination will be required with the border defence force and that portion of the area defence responsible for handling airborne incursions, both of which will certainly choose to have some responsibility for their own air defence, particularly against lower flying planes. The size and composition of the air defence will depend upon the perception of the threat, and the country's proposed reaction. In time, a proper air defence could discourage the maintenance of an air force by aggressor countries. Missile defence should not be ignored. As technology improves this defeat of missiles and other unmanned systems could prove to be the major task of the air defence force.

Border defence force

The design of a border defence force is a task for the country's military organization based upon the direction given by the National Command Authority. This direction as outlined in parts II and III, can vary considerably. Certainly the task of forming the force is a formidable one with many ramifications strongly reflecting the attitude of the planners. This is evident in reference 10 where an initial attempt is made to lay out a detailed border defence plan for NATO Central Front. The general case considered here was taken in large part from reference 9.

The fundamental – indeed the essential – aspect of the border defence force is that it is defensive. It is structured as a true forward flexible defence with the objective of defeating enemy manoeuvre units as they cross the border. It is based upon preponderant nuclear and conventional firepower, and the promise of presenting no threat of invasion to an enemy.

This defensive force explicitly acknowledges a worst case threat posed by the presence of nuclear weapons in the adversary's forces. Accordingly, the border defence force must be designed to make it as difficult as possible to find lucrative targets for aggressor nuclear weapons used either in a preemptive disarming attack or against the defence in conducting an invasion attempt. The defenders would rely on their ability to deliver low-yield nuclear weapons on attacking forces in sufficient numbers to stall an attack and destroy the attackers before the attack succeeded in significantly penetrating their territory. Low-yield nuclear weapons delivered quickly with requisite accuracy could defeat attacks by massed ground-gaining elements without the self-defeating unwanted destruction attending the use of high-yield weapons.

The object in designing the force has been to present to aggressor planners a problem whose inherent uncertainties offer little prospect of successful resolution through the use of military force. Due primarily to the profound effects of nuclear weapons on offensive military operations, these uncertainties would be accompanied by the realization of the aggressor that the required defensive effort would not be so self-destructive as to keep the defenders from making that effort.

Before they could hope to degrade seriously the defensive capability of the proposed force, the aggressor would have to improve vastly, in some unforeseeable manner, his ability to find and destroy large numbers of small mobile elements dispersed over a wide area. The proposed force includes large numbers of conventional surface-to-air weapons to complicate further the target acquisition problem during those times when weather failed to do so. The difficulty of penetrating a deep defence armed with nuclear weapons could be made even more severe by the provision of protection for the defensive elements against the effects of nuclear weapons. How much protection to provide obviously depends on many factors, but even that provided by the field fortifications that can be constructed by the units themselves, augmented by a few combat engineers, would be substantial.

To provide a high degree of survivability, the area within which the force would be deployed to defend would extend about 100 km inside the border. Defensive elements would be dispersed throughout this area, with the density of nuclear delivery units greater to the rear, and that of close combat and target acquisition elements greater near the border. There would be no concentrations of men or material to draw nuclear attention.

Border defence force elements. The force would include the following:

— Composite combat units of about 30 men per unit would provide short-range target acquisition, local defence against infiltrators, a contribution to local air defence in the form of shoulder-launched ground-to-air seeker missiles, and PGMs to counter dispersed attacks. A limited number of units would have the primary responsibility of protecting nuclear fire units and fire direction centres. Elements of these units would acquire attacking manoeuvre units, call for the launching of low-yield nuclear warheads, and terminally guide the warheads onto the attackers. These units, a majority of which would be militia, would be organized in defensive zones of about 100 km width and depth along the border. As to the density of units for this purpose, an average of one unit per 4 km^2 of defensive zone with greater density toward the border and less to the rear, would pose a very difficult problem for hypothetical aggressor planners;

— nuclear fire units, each containing a launcher, four terminally guided missiles with nuclear warheads of less than a kiloton yield, communications gear, and a small crew, would launch missiles on request from composite combat units. The fire units would be dispersed in a wide band extending from about 25 km to 100 km from the border. The depth of this band, and the width and depth of the defensive zones, is established by the maximum and minimum ranges of the missiles. The number of delivery units required for the nuclear

weapons of the defence cannot be precisely calculated, but a density corresponding to one launcher per kilometre of front would set aggressor planners' problems of target acquisition and anticipated volume of fire that could not be
solved with high confidence. They could not be addressed at all without considering the use of aggressor nuclear weapons;

— command and control within the defensive zones would be greatly
decentralized to make the defence highly responsive to the moves of the
attackers. Such management would be carried out through the highly automated fire direction centres (FDCs), which coordinate the activities of the
composite combat units and the nuclear fire units, all according to some approved firing doctrine (see part III);

— logistic support for the force would also be greatly decentralized.
Since offensive operations against aggressor forces armed with nuclear weapons
would be for practical purposes impossible, no contingency for these operations
would be allowed for in the logistic structure. A large number of small, widely
dispersed, concealed supply points would constitute the essence of the logistic
structure. Combat elements would carry several days of supplies.

Defensive nuclear force cost

The costs of any force are dominated by the number of active military men
maintained in the force. In general, 10 members of the militia can be supported
for the cost of supporting one active soldier. The major opportunity for saving
thus lies in depending on militia for all that does not clearly require a full time
military man.

The acquisition costs of the arms for the proposed force, which would
have a much larger number of ground combat and surface-to-air PGMs than
are apparently being planned for the present forces, would be kept low by relying
on cheap, simple systems. The missile for nuclear delivery would have no stringent accuracy requirement to drive its cost up, and the simple subkiloton
fission warhead could be made very cheaply (Appendix 2).

The principal conclusion to be drawn from comparing costs of the present
and proposed forces is that the former contains many costly elements that
cannot easily be related to defence against ground invasion. The cost of the
proposed force would be considerably less than that of even the defensive
elements of the present forces [11].

V. Criticism of the new nuclear force

From what has been stated in the previous chapters it is obvious that a major
premise for the new nuclear force is that the avoidance of war is better served by
rational capabilities to fight a war for finite objectives than by forces that are
intended to intimidate or punish. Such a thesis which departs so far from the

norm is bound to generate a great deal of adverse criticism. Much of this comment is justified if for no other reason than that the entire proposal is inadequately developed and is outside anyone's experience, including that of those military experts upon whom one would depend to flesh out the concept. Unfortunately, some questions can not be fully addressed until the military planner carries out this task, something he will not be inclined to do until requested by political authority. A dilemma arises because the political authority does not know what questions he should ask. Hopefully this essay will offer assistance.

The basic questions usually posed are: (1) can the defensive philosophy of a defensive nuclear force be made sufficiently clear in the way the force is presented and configured? and (2) will a new nuclear force suffice to protect the country? Answers to the following specific questions are believed to constitute strong support for the new nuclear force.

Question 1. A missile of 100 km range tipped with a nuclear weapon (albeit subkiloton), coupled with a very effective fire control system, represents the core of a credible offensive capability, particularly were it supported by some form of mobile reserve force. Is not such a capability contrary to the defensive philosophy of the defensive nuclear force?

Answer. To a degree this is so and one must take every reasonable step to be sure that no intrusion of aggressor territory is intended or indicated. To this end the following steps are taken:

The force doctrine as endorsed by the country's National Command Authority and as it is reflected in all military training manuals is unequivocal on this point, i.e. there is no intention to invade aggressor territory. Contrary to the popular statement – intentions are important (as are capabilities).

The capability of the force is restricted. The infantry has only enough mobility to ensure that it does not present a fixed target. A large number of small, widely dispersed, concealed supply points would constitute the essence of the logistic structure. There would be no logistic capability to support an offensive drive.

It may be that some military planner would find reason to augment the proposed force with a limited reserve force to retain territory which has undergone an intense nuclear-biological-chemical attack. If so, this force would be trained to operate quickly to reestablish nuclear fire support before the territory can be occupied by the aggressor. There are large quantitative differences in the equipment and personnel required for this role and those required to mount a sustained attempt to seize territory actually held by other forces.

Question 2. A defensive force that engages the aggressor only on the defender's soil gives the aggressor a great advantage to mass his forces and execute preparatory fire on the defence at his leisure. Is this a tolerable situation?

Answer. This point is well taken. The aggressor certainly acquires a significant advantage if he can devote all the energy that he had expended on defence to his proposed aggression. Nevertheless, the answer to the question is an emphatic 'yes'! Massed forces give the aggressor no advantage over the defence when the defence has nuclear weapons; the aggressor only composes better nuclear targets. Such is not the case when the aggressor faces a conventional PGM defence. In fact, that is how he defeats such a conventional defence – by overwhelming it with his augmented offensive forces.

Preparatory fire by conventional or nuclear means may require the forward defence elements already concealed and dispersed, to build a number of protective bunkers [11]. This can be done cheaply and effectively. One cannot deny the temptation to the defence of counter battery fire with the proposed nuclear missiles against an aggressor's barrages. The results would be devastating. The manifest capability should be enough to discourage massed conventional artillery fire.

Question 3. How does the defence defeat a dispersed attack by the aggressor?

Answer. Two kinds of dispersed attack have been considered [13]. One type would send numbers of units with ground-taking capability against the defence in the expectation that the individual units would not be large enough to attract nuclear fire. Aggregation of the attacking units would not take place until defensive elements were so closely engaged that nuclear weapons used then against the attackers would also kill defenders, presumably precluding their use.

There are several points to be made with respect to this type of attack. In the first place, both the conventional and nuclear elements have the capability to defeat such small units, and the field commander will have made plans as to how the defence would be made. However, it is essential that nuclear weapon stockpiles be adequate and tactics sufficiently flexible to exclude the prediction by the aggressor of what would constitute a target suitable for nuclear fire on the part of the defence. The missile with its 100 km range optimizes this situation for the defence. Secondly, if the defence chooses to retreat to his bunkers, he is far better protected than the attackers. He can call nuclear fire to within 100 m of his own location. These tactics would not only take care of most of the uncertainty in the mind of the defence but it would be extremely discouraging to the aggressor and, at the least, disrupt any highly orchestrated plan required for such an attack.

In the second kind of dispersed attack, the aggressor would try to introduce into the fabric of the defence a large enough number of observers and other target acquisition means to discover and attack with supporting firepower as many defensive elements as necessary to nullify the defence. If successful, this tactic would enable the aggressor to employ his ground-gaining elements with impunity.

It is hard to imagine that the defensive infantry with its sensors, minefields, and modern equipment could not manage to defeat this kind of attack.

Question 4. Is not the nuclear defence force, like the Maginot line, susceptible to defeat?

Answer. There is no question that the defensive nuclear force is designed as a true border (or forward) defence. There is no intention to permit the aggressor to acquire the defender's territory. This attitude is only made possible with the firepower of nuclear weapons. Their introduction as a principal defence element should be treated as a mutation† in military force. Previous conventional experience has limited value in evaluation of the defensive nuclear force.

The significance of this question rests on its timing. During and after World War II it became popular to criticize the Maginot line. The level of criticism was unwarranted. The more so today when the effectiveness of offensive elements (tanks, armoured infantry, and aircraft) are being questioned by the advent of conventional PGMs. There is little question that these offensive elements will be overwhelmed by the effective development and use of nuclear weapons and PGMs.

Still there is reason to heed the question. Military planning and technical potentials should be exercised to the limit to maximize the capability of and the general confidence in the defensive nuclear force.

Question 5. In the past there has been an often expressed fear that countries which develop large effective military forces are inclined to use them. If the defensive nuclear force is as good as claimed in this essay would it not make the possessor 'trigger happy'?

Answer. The defensive nuclear force is a *defensive* force. It will not – cannot – be used except in defence on one's own territory; i.e. there is no potential for being trigger happy in the offensive sense. There is no denying that the defence would not be at all reluctant to use its subkiloton missiles against any intrusion of its territory. In that sense it is a 'trigger happy' force. However, detonating nuclear weapons in such a fashion is not a dangerous, escalating act that has the potential for precipitating strategic exchange. Thus the question in the context of the defensive nuclear force is a bit naive. Nevertheless, the question is most appropriate since it is usually posed as a consequence of experience with the present NATO/WTO forces. In that context, any NATO military action may be considered as irresponsible and the thought of detonating a nuclear weapon following any amount of consideration is hardly comprehensible. The latter point is recognized by the endorsed firebreak between conventional and tactical nuclear war. What is not appreciated is that the fear of escalation induces a second firebreak which strongly inhibits any NATO response. Thus the present force is not troubled by being trigger happy. Quite the contrary, it is paralysed. As such it ceases to be a sensible instrument of national or Alliance policy. Admittedly, and unfortunately, the situation is not this simple. The US position is based upon a continuity of force which evidently extends from abilities to counter minor conventional aggression to the maximum capabilities of the Single Integrated Operational Plan. It is contended that this force can be managed so that one can address any level of aggression over this range of force in a highly controlled manner. Of course the great fear is that under great pressures a US President might do just that. The consequences could be disastrous beyond imagination. Such prospects form the principal reason for the development of a new nuclear force for NATO.

---

† An expression attributed to General Beaufre.

Question 6. The New Nuclear Force is a very simple force. One would expect that in time an effective counterforce could be developed. Is this not so?

Answer. This begs the question as to what is really required of a military force structure to make it invulnerable, dependable, and durable. It is proposed that to meet these ends:

1. any action by the defence must be relatively invulnerable to any counter-action by the aggressor;

2. there must be duplication of any defensive capability not only in numbers of identical equipment, but in numbers of ways of doing the same thing;

3. there must be constant vigilance and improvement of the force with respect to tactics, equipment, and morale.

Whether a military force is simple or complicated probably bears on this matter, but it is not clear in what direction. No doubt there is a consensus that a force to be effective must be complex. However, in the extreme, a complex force might be capable of doing a great many things, few of them well, and many of which are nonessential. Admittedly a hallmark of the new nuclear force is simplicity. This is accomplished by reducing the mission to essentials – for example, the border defence force attacks forward manoeuvre elements – with the only weapon which can do this task well, the guided missile. At the same time each element of the system has the potential for meeting the criteria of invulnerability, repetition and vigilance. In short it should do the essential job extremely well. Its principal element of high deterrence rests on this simplicity through the advertisement of its obvious effectiveness.

Finally, however, one must concede that no force structure in history has survived though some have stood the test for long periods. It is impossible to say how the New Nuclear Force will evolve in time, and what change in the force would mark its demise. The most one might conclude is that it looks quite healthy under current investigation and there are no obvious prospects for its defeat. Most important is the fact that there appears no other choice but to develop and preserve this military option.

VI. NATO and the new nuclear force

The history of the NATO Alliance has been marked by considerable success. However, it is probably fair to say that as it has aged its problems have become more severe. Indeed, unless certain changes and accommodations are made which involve the basic objectives of the Alliance, its military doctrine, and the alteration of the responsibility of its members in response to these changes, NATO may cease to be an effective organ. Conversely, if such accommodations are made, such as the introduction of the New Nuclear Force, NATO can emerge as an effective and dynamic element of Western security.

The deterrence† of the existing NATO force rests almost completely upon its strategic elements. The conventional and tactical nuclear forces are relatively weak, their capabilities are not well understood, and they are not appropriately releasable or responsive to the threat [19]. Increasing and refurbishing these elements, improving air defences, or developing preventive measures against chemical attack only increase the assurance that in the event of aggression, the entire force will be struck by a pre-emptive nuclear attack. Its exposed nuclear storage sites, air, command, and logistic installations along with its massed forces present a highly vulnerable, essential, and lucrative target complex which ensure the destruction of the force in the initial minutes of a nuclear counter-force strike.

There are good reasons to criticize adversely the force other than its obsolescence and inadequacy. It is very expensive. It is a dangerous force; in the event of some unforeseen aggression the prospect of muddling to a strategic exchange is unacceptably high. The increasing inferiority of the NATO force in both its tactical and strategic elements nurtures an unhealthy political climate which may permit the Soviet Union to accomplish its political objectives without recourse to military aggression. Finally, the existing force can serve to promote arms control arrangements which may prove most ill-advised in the context of rational political-military postures.‡

NATO new nuclear force

In this section the NATO Alliance is discussed in the context of the only solution to the NATO problem, the new nuclear force. As a basis for discussion the following suggestions are proposed:

Force responsibility. The NATO defensive nuclear force would be the direct responsibility of those countries who border on the Iron Curtain. Each of these countries could supply the force prescription, manpower, hardware, and control for its own protection. While other Alliance members may be concerned about the defence of the border countries, none approach the concern of the countries themselves. Essential to any military force is a high *esprit de corps* which is a principal benefit associated with the protection of one's own real estate [26]. In particular, the present shared responsibilities for the defence of the Central Front may make political sense; however, realistic considerations require West Germans to compose the entire composite combat units or the bulk of the force. A major problem concerns the acquisition of low yield defensive nuclear weapons by these border countries. One can hardly exclude the possibility of one or more supplying its own weapons. At the moment this

---

† Here deterrence is defined as in part II as some product of the potential of the force multiplied by the assurance that the force will be used if necessary.

‡ For example, the NATO tactical nuclear weapon stockpile could well be drawn from 7000 to 1000 or removed altogether in the context of the present force [6]. It may be difficult to reestablish the numbers of weapons once they have been removed. For the new nuclear force the large numbers are essential. Agreements or understandings on no first use, firebreaks, and restricted nuclear fire zones also are generally highly ill-advised.

prospect would seem to be remote; hence it would be necessary to depend on the USA, France and the United Kingdom. One solution would involve these countries in the manning of the nuclear fire units. The entire nuclear fire force to man the Central Front would be composed of less than 10 000 professional troops.

Western Europe has an extensive shoreline to protect against invasion, which should be undertaken by the host countries with whatever support is required from the nuclear powers.

NATO council responsibilities. NATO council responsibilities would consist of:

— stipulating the immediate and long-range objectives and general guidance appreciating that the goal is to develop political and economic stability throughout the Alliance. For example, the present NATO strategy (MC 14/3) must be cancelled and new direction to NATO military forces must be formulated consistent with the concepts outlined above (see suggested Firing Doctrine in part III). The stated goal of flexibility to defeat any level of attack at the border must be most explicit. It must also recognize the existence of a discontinuity with respect to both military force and crisis management between this defensive nuclear force and the NATO strategic force. With the new nuclear force this discontinuity is natural, and escalation control within the defensive force is no longer at issue;

— redefining the roles of the members of the Alliance and ensuring that mutually acceptable outlays of responsibilities and assets are made;

— refining and controlling a common NATO strategic force. These will be difficult and continuing tasks. The purpose of this force would be to counter a punitive threat to western Europe by the Soviet Union. It might operate under a 'no-first-use' constraint. It would be targeted against Soviet cities. One sensible approach would be to employ only submarines, thus removing the target complexes posed by missile fields and aircraft bases;

— managing the procurement of military hardware, though the level of procurement, the degree of complication, and the importance of commonality are considerably reduced over those aspects of the existing force;

— ensuring the detailed peacetime consultation on the use of nuclear weapons in NATO. This planning must include the necessity for the availability of subkiloton defensive nuclear weapons in the hands of the field commander prior to an enemy attack;

— defining and coordinating a common air defence of western Europe.

Current force improvements. There are a number of things that could be done within the framework of *any* force structure to improve the posture of the Alliance:

1. A guided missile such as the one proposed in part III should be deployed. For various reasons the revised Nike Hercules would serve as a timely substitute. Of course, it would be essential to deploy the associated fire direction centres, and target acquisition teams. Most careful attention should be given to the proper use of sensors which are capable of locating in time and space manoeuvre units on NATO soil; particularly in an intense nuclear environment.

2. Nuclear sites should be eliminated, probably using the techniques outlined in part III.

3. NATO air defence system should be deployed based primarily upon surface-to-air missiles. A careful and objective philosophy should be developed which would take seriously into account the requirements of the system including its vulnerability, and the necessity to prevent real time reconnaissance.

4. Measures to protect against chemical agents should be carefully studied and implemented.

5. The cost-effectiveness of aircraft in a PGM environment should be objectively evaluated. This study should include the importance of specific roles, the use of conventional and/or nuclear munitions, and the contribution in various parts of the theatre.

6. The cost-effectiveness of the battle tank in both the nuclear and PGM environments [20, 26] also needs to be examined.

If these efforts and studies are objectively made and carried out to their rational limits, the advantage of a new nuclear force would become obvious; indeed its implementation would become a necessity. Tactics should be introduced in keeping with the border defence force. Initially, such a force might be superimposed upon existing conventional forces. As described herein, it would occupy relatively fixed defences, carefully selected and structured in peace-time at various distances from the border agreed to by the involved allies. As a first step, the ACE Mobile Force could be converted to the role of designing and evaluating such defences. The Standing Naval Force Atlantic might be converted to a small fleet of fast, nuclear missile armed boats in support of coastal defences.

Difficulties

The introduction of the proposed changes in NATO is fraught with difficulties.

It would be extremely difficult for the Alliance members to consider either the suggested change in political/military philosophy, the change in their NATO roles, or the associated adjustment in structure.

There would be strong resistance to placing essentially total dependence upon the use of nuclear weapons. It can be argued that the WTO, in its frustration over the success of the NATO new nuclear force may opt to use its strategic force to conquer western Europe: thus the NATO option of early surrender would have been denied. Though this argument must be recognized, it poses a risk which is minor when compared to the incentive for devastating preemptive attacks on the existing threatening, vulnerable and lucrative NATO targets.

Advantages

It is contended that this high price that NATO must pay has an impressive return: it buys an effective military force within current budgets, which minimizes the prospect of war; there is an associated political strengthening of the members of the Alliance and the Alliance itself; its introduction would present a tremendous step towards arms reduction and control as outlined in part II. The elimination of battle tanks and aircraft from the military inventory would constitute a significant step in this regard. In addition, it is probable that the removal of NATO fighter bombers and Pershings would be matched by a corresponding reduction in the Soviet strategic force which is targeted against western Europe; there could be a significant political–military penalty to the USSR. One could expect strong Soviet reaction to NATO's adoption of the posture [21]. They would find their present force obsolete and would have no good alternative. Against NATO, they could consider themselves forced to adopt a similar defensive force. Although such a move would establish a very well-defined equivalence with NATO forces, it might not serve the Soviet ideology; if one insists upon some strategy of '*détente*', a new nuclear force would serve as a far more comfortable reference frame for NATO than would the existing force, in fact it is probably the only rational posture.

Conclusion

Perhaps the most likely prospects lie with nations not involved in the NATO–Warsaw Pact relationship. The list of such countries contains many names and many of these countries have not been continuously exposed to the conditioning that prevents thinking about matters nuclear as other than un-mitigated evils. A few fulfil other requirements, not the least of which is an urgent necessity to survive in a hostile environment. It may be that such a country will lead the way in demonstrating the appropriate role of nuclear weapons. In the meantime, taxpayers of other countries will be left to wonder why it is considered necessary to spend such astronomical sums on forces so unsuited for the contingencies they are supposedly meant to meet.

VII. The US and a tactical nuclear umbrella

The USA could separate its worldwide political–military responsibilities into three categories:

1. It must provide for its own defence. It would be unnecessary to proceed through the 'drill' outline in part II, to conclude that the force designed for this purpose would focus on its strategic force. (The invasion of its borders hardly merits mention.)

— The strategic force should be maintained at a high level of quality and sufficiency to deter any conceivable attack on the US treasure. To ensure this objective the US Joint Chiefs should be charged in general fashion with the imperative of strategic second strike capability and an active research programme which explores all conceivable avenues open to the strategic force and which carries valid proposals through prototype development.

— The strategic force should be the subject of intense negotiation to multilaterally reduce the size of all nuclear strategic forces with their elimination as the ultimate objective.

— The strategic force should be constrained to a maximum degree within the limits of deterrent credibility. It is imperative that the pressures of a moment do not force the management of a crisis to frightening extremes as it did in the Cuban Missile Crisis and the 1973 Middle East Crisis. There is a well recognized deterrent dividend for these actions; but they forebode similar actions in the future which, in time, define the destruction of the existing culture. Overall political/military posture should not make such drastic action necessary.

— The debates over limited strategic use options should be quieted. This debate not only wastes time, particularly the valuable time of the US Congress, but it creates a dangerous familiarity which could well lead to the more easy release of the force.

2. The United States has an obligation to its NATO allies which was discussed in the last section.

3. Finally, there is the question of US commitment to the rest of the world and how much of this commitment can still be justified under present circumstances. In the light of its traumatic experiences since World War II, particularly in Korea and Vietnam, it would certainly be prudent to consider possible new courses of political/military action, including disengagement [28].

In addressing this third task it is to the point to outline a number of constraints:

— The forward positioning of high yield nuclear weapons on naval surface ships and on island bases is of questionable value. These weapons have continued to present almost an anathema. Enquiry into their use has never been well received and it has certainly proven unproductive. For those who feel strongly that we should depend only on the continuing capability of a deterrent strategic force, these weapons may have some limited value. On the other hand, those who hold with the potential of a defensive warfighting capability, may well conclude that the strategic role should better be restricted to submarines and a strategic force based in the continental USA.

— It will be difficult, and often impossible, to commit US troops outside the US and Europe. The commitment in Korea could be short-lived.

— It will be increasingly difficult to use clandestine or subversive means to accomplish political/military ends.

— Military budgets may diminish, which they should if this can be done without sacrificing significant political advantage.

In light of the present situation and these probable constraints, the US must choose between the following options: (a) It can continue as it has in the past. Unquestionably, this is the most probable course; (b) It can retreat behind its own borders. This may be the most sensible option, though it is the most unlikely choice today since many long standing political commitments and the role as champion of the Free World call forth an automatic and negative reaction. Possibly more important is the fact that such an option ignores a *raison d'être* for the traditional US military force; (c) a third option would involve the incorporation of the surface Navy, the Marines and the Tactical Air Force into what will be termed a US Tactical Nuclear Umbrella. This option represents a departure from the proposed new nuclear force; however, it is similar in many respects. It is this option that will be discussed.

Tactical nuclear umbrella guidelines

The detailed description of the tactical nuclear umbrella must be left to the military experts. This description would be based on guidelines which must be hammered out under the auspices of the highest civilian authority. The following guidelines are intended as an example:

1. Aircraft should be employed to deliver low-yield nuclear weapons and conventional precision-guided munitions. There would be no nuclear weapons in the arsenal with effects greater than that of a 1 kt fission device.

2. All targets should be within the borders of an ally.

3. Long-range aircraft under the control of the tactical air command should be based on US-owned island fortresses such as Guam. These bases would be strongly protected against missile, aircraft, and surface attack.

4. Fast Carrier Task Forces should be formed to present a highly effective political/military force. Other forces may depend upon missile boats delivering low yield nuclear weapons and PGMs.†

5. Intensive peacetime consultation alone and with allies should establish the rationale and, to a sensible limit, the details of operational behaviour in time of crisis. Vietnam-like decision making in Washington must be reduced to an absolute minimum.

6. There should be no question that this umbrella has high military potential and that it will be used. It represents a real war fighting capability.

7. The budget should be closely monitored; e.g. it may be that one needs only seven carriers instead of fourteen.

8. The US Army should be given *no* role. Its attention should be restricted to western Europe and the continental USA.

9. The role of the shore party should be assigned to the US Marines who would restrict their action to liaison and target acquisition.

10. There will be no requirement for paramilitary activities.

11. Use should be made of modern technology to maximize the effectiveness of the force, e.g. aircraft-delivered laydown bombs may be the most

---

† See e.g. ref. 26, p. 228, *Naval Lessons of the Yom Kippur War*, Rear Adm. B. Telem.

important (and possibly the only) legitimate use of tailored effect nuclear weapons.

Tactical nuclear umbrella advantages

The tactical nuclear umbrella would present a number of advantages:

— The capability of the force to discourage military aggression could be impressive. The tactical nuclear umbrella would represent a strong deterrent. It would have both a substantial war fighting capability and it would be releasable, i.e. the use of such a force would not commit the employment of the US strategic force. Crises would be managed under the tactical nuclear umbrella. The importance of the US strategic force would diminish.

— The ally could be content with a much reduced military force, assured that the US would afford protection under this tactical nuclear umbrella. One solution for the ally would be to reduce its military operation to an internal police force; the development of a nuclear capability of any kind would be unnecessary. As such the umbrella would serve to inhibit nuclear proliferation.

— It might give both the US Air Force and the surface Navy a *raison d'être* by the formulation of and contribution to rational solutions of real world problems. The mission would require most careful consideration.

— The US cost would be low and it would require a minimum investment. The equipment and people are essentially available.

— There would be no expensive US ground forces with their high potential for Vietnam-like entanglements.

Tactical nuclear umbrella disadvantages

There are also disadvantages:

— Both aircraft and carriers are vulnerable systems and they are becoming more so. Sooner or later, a carrier will be sunk with the potential loss of thousands of US sailors. It must be understood that such a loss must be accepted with stoicism, i.e. such an act must not suggest the release of the US strategic force, and it may call into question the credibility of the force. However, it would be no surprise were its usefulness to continue for many years, particularly if it were susceptible to improvement through doctrinal and technological advances.

— Certainly there would be objections to such a nuclear force which was so available for use and potentially was so effective. Many would say that the force will tend to make the US 'trigger happy'.

Scenarios

The reaction within the USA and the impact on allies and adversaries of the tactical nuclear umbrella is open to speculation; certainly it would be substantial. In order to cultivate some impression of these reactions, it may be valuable to formulate such scenarios as the following:

1. The presentation of the US tactical nuclear umbrella concept to the Japanese. Consider such a scenario under the condition where the US strategic force was ruled out of the dialogue and those in the case where the use of the strategic force was recognized as restricted only to the defence of the US.

2. The exchange between the US and South Korea following the US suggestion that the US tactical nuclear umbrella substitute for the United Nations force deployed in South Korea.

3. Discussions with allies on sharing the cost of the tactical nuclear umbrella. The USA no longer has the fiscal vitality to carry this burden by itself.

4. Discussions between the US and NATO flank countries over the use of their ports for the use of US fast carrier task forces.

5. Similar discussions with Sweden.

6. Consider how the events might have been changed as a consequence of the deployment of a tactical nuclear umbrella within striking distance of Cuba and Angola during the Angola incident.

7. Discussions with NATO over possible coordination of the NATO force and the tactical nuclear umbrella.

8. Discussions with Iran over the augmentation of their strong and increasing conventional force.

Conclusions

This section has tried to define the role of those US General Purpose Forces other than those employed in the NATO theatre. Attention to such articles as the one recently published by Admiral Stansfield Turner [29] indicates that it might fall short of the mark. Nevertheless, just as in the case of NATO, the US ought no longer to essentially ignore the use of tactical nuclear weapons in the design of its military forces.

Appendix 1. Mini-nukes

The key element of the new nuclear force is the tactical nuclear missile tipped with a subkiloton nuclear warhead. Such warheads have been included in the US stockpile for decades, but only recently have they been widely advertised under the name 'mini-nukes'. It is not clear where or when this term arose, though it was probably assigned to the efforts made by NATO Secretary General Manlio Brosio, during the early activity of the NATO Nuclear Planning Group (1967–70). The Secretary General was concerned over the many difficulties associated with a NATO nuclear weapon stockpile which was composed predominantly of higher yield weapons. The consequences were high collateral damage, a reluctance on the part of the United States to enter into peacetime consultation on their use, and a general lack of tactical military utility. The net consequence could be a removal of nuclear weapons from NATO with serious political impact on the NATO Alliance. His interest in the advantages of low-yield weapons was instrumental in focusing attention onto the broader aspects of this subject.

About the same time US Congressman Craig Hosmer published in the US Congressional Record two letters to then Secretary of Defense Clark Clifford suggesting changes to the US tactical nuclear stockpile. These letters stressed his opinions on the advantages of what has been called the 'neutron bomb'.

Wherever the name 'mini-nuke' originated, the subject of low-yield nuclear weapons did eventually arouse considerable concern within the defence and arms control community. Unfortunately, much of the available information upon which argument was based was distorted or simply incorrect.

Comments on Mrs Myrdal's CCD presentation

In this Appendix a portion of the widely read statement by the Swedish Representative to the CCD, H.E. Mrs Alva Myrdal, is presented and commented upon by the author with the new nuclear force in mind.†

Before making detailed comments on Mrs Myrdal's statement, a general point should be made. No doubt a major objective of both Mrs Myrdal and the author would be to see established a political/military order which would reduce the prospect of war, particularly a strategic nuclear war, to a minimum. The difference arises in the approach. Mrs Myrdal wishes to eliminate nuclear weapons; failing that, she would eliminate proliferation and restrain the present nuclear stockpiles. One can only suppose that she accepts conventional war as the alternative. The contention of the author is that there is no way to turn back time; nuclear weapons are a reality. One should incorporate them into proper military forces with the above goal in mind. In addition, the author contends that nuclear weapons properly postured will make the prospect for any war, nuclear or conventional, far less probable, and he offers what he feels is a far greater prospect for strategic nuclear disarmament. The price, of course, is, if war comes – and no one can guarantee that it will come, the certain use of low-yield nuclear weapons as an alternative to surrender or strategic nuclear war.

† CCD/PV.620 pp. 14–15, 9.8.73.

Mrs Myrdal: ... The ominous military technological trends which I have illustrated today threaten the Non-Proliferation Treaty, as there are signs of developments which might affect the willingness to subscribe to the Treaty often regarded as the main bulwark in the priority field of nuclear disarmament.

Since the very entering into force of NPT there has been widespread recognition that it was not likely to survive because of its inherently discriminatory nature, if the Super Powers did not take 'effective measures relating to cessation of the nuclear arms race at an early date'. The SALT agreements reached in Moscow last year and the intentions expressed in Washington this year have rightly been hailed as promising steps in the right direction. Simultaneously, however, developments seem to be under way, threatening to render the NPT even more discriminatory to the disadvantage of non-nuclear-weapon states. I am referring to news items that the major nuclear-weapon states are about to launch a new generation of tactical nuclear weapons systems, the so-called mini-nukes. Such a development would drastically aggravate the nuclear threat against non-nuclear-weapon States everywhere.

There has for a number of years been a common understanding that the effects of a strategic nuclear war would be such that no one could contemplate triggering such a war. The SALT I agreements in a sense codified this understanding. A similar understanding has developed as regards the tactical nuclear weapons. One reason for this is that the consequences of a nuclear campaign using tactical weapons for the civilian population living among the military targets would be similar to the effects of a strategic nuclear attack.

Author: This is true for the present NATO force. It is not true for the new nuclear force for which procedures controlling the level of permissible collateral damage would be established in peacetime (see part III). The collateral damage with the new nuclear force should be far less than the damage resulting from the use of the present forces in Europe. With the new nuclear force, the duration in time and space of an engagement would be greatly reduced, and the conflict would be much less likely in the first place. More important to the civilian population of the leading nuclear-weapon Powers would be the formidable risk with the present force of an escalation to a higher level of nuclear exchange. The new nuclear force is designed to prohibit escalation. Admittedly, the success of the new nuclear force rests largely in the ability to demonstrate convincingly that decoupling of the two forces has been achieved.

Mrs Myrdal: Therefore, it is so disturbing when we learn that ongoing research and development might lead to a new generation of tactical weapons with yields in the subkiloton range, overlapping the yields of the most powerful conventional charges, with extreme delivery precision, and with extra-accurate intelligence support.

Author: The guided missile will employ a fission warhead which uses elementary warhead technology. The missile will use the latest technology which will serve to increase somewhat the response time and accuracy of the system. However, a missile like the original Nike Hercules in its surface-to-surface mode would have served the purpose reasonably well; extreme accuracy is not required. The term extra-accurate intelligence is not understood.

Mrs Myrdal: Most important is, however, than an introduction of such mini-nuclear weapons would blur the present distinction between conventional and nuclear weapons. We are strongly of the view that an absolute 'firebreak' must be kept up between nuclear and conventional war.

Author: The new nuclear force requires a strong firebreak between the defensive nuclear force and the strategic force. There is no firebreak between conventional and nuclear weapons. It should be recognized that insistence upon a firebreak between conventional and nuclear forces induces a second firebreak between the use of no force at all and the conventional force, i.e. crisis generates paralysis of any military action. Stated differently, the conventional force loses what little effectiveness it had.

Mrs Myrdal: Obviously, the introduction of mini-nukes and a decline of the nuclear threshold would create enormous proliferation risks. The main purpose of the NPT – to reduce risks for nuclear war – would be countered.

Author: The point is that proliferation of nuclear technology in the context of the new nuclear force becomes far less important, and possibly less likely. (The issue of the threshold is treated later.) This would occur at a time when in many countries a growing nuclear industry will produce considerable stockpiles of plutonium.

Mrs Myrdal: Military arguments for acquiring nuclear weapons may again come to make themselves heard, albeit with different strength from nation to nation.

Author: This is true as in the case of Yugoslavia [27]. The resulting danger is that a nuclear war carried out in the present unenlightened reference frame with current forces would almost certainly constitute a catastrophe. The new nuclear force presents an intelligent and properly constrained approach to the problem. It may be that a country should develop a nuclear force. The argument should not centre on proliferation, but how one should deter war and ultimately eliminate war as a method of political mediation.

Mrs Myrdal: Without any doubt, such a new development in regard to tactical nuclear weapons would affect the very premises on which adherence to the NPT is based. In the view of the non-nuclear-weapon powers, it is the tactical nuclear threat, rather than the one pertaining to strategic nuclear weapons, that today causes anxiety on their part. It is the option to produce ordinary and tactical nuclear weapons, if only of the 'old-fashioned' Hiroshima-size, which is primarily 'sacrificed' by adhering to the NPT.

Author: This is a very perceptive statement. The option also includes the production of simple subkiloton weapons.

> *Mrs Myrdal:* There are enormous risks involved in unsettling the status quo between nuclear- and non-nuclear-weapon powers by introducing changes in the tactical nuclear weapons capabilities. If the nuclear-weapon powers were to enter upon a new race to 'improve the usefulness' of tactical nuclear weapons a fundamentally new situation would be developing above our heads. It would affect possible war theatres around the globe, and could certainly not be restricted to scenarios for possible confrontations between the Super Powers themselves.

Author: This is well stated; however, the risks exist no matter what direction is taken. The dangers are much greater if one proceeds in the current fashion.

> *Mrs Myrdal:* Fortunately, responsible statesmen in various countries have expressed themselves in favour of a continued high nuclear threshold and of regarding all nuclear weapons as qualitatively different from conventional ones.

Author: This is true; however, these 'responsible statesmen' are falling into a trap by mouthing a platitude. In the first place, we must have a clear definition of what one means by the height of a nuclear threshold. Let us define the term as the probability that nuclear weapons would not be employed in a given political/military confrontation. This probability is a product of the probability that one would not have a conflict multiplied by the probability that in the event of a conflict, nuclear weapons would not be used. Increasing the NATO conventional force would probably make a war less likely; however it would certainly increase the probability that in the event of a conflict, the WTO would use nuclear weapons. Thus, it is not clear what the increase in conventional NATO forces would do to the nuclear threshold. If it were possible to consider the same political situations with the new nuclear force, it is argued that the nuclear threshold would be much higher, simply because the probability of any conflict would be greatly reduced, i.e. the war would be strongly deterred.

If the nuclear threshold were assigned to the probability of not having *strategic* war, then the new nuclear force would indeed be superior since its tactical and strategic elements are strongly decoupled.

> *Mrs Myrdal:* As it would take a much shorter time to change the doctrines for the use of these new systems than to develop and procure them, my delegation considers it essential that the development of mini-nukes be stopped now.

Author: The opening clause is not true when one considers the necessary changes in force structure. The second part of the sentence is also incorrect; subkiloton nuclear weapons (mini-nukes) have been in stockpile for many years.

Mrs Myrdal: This is such a serious matter that I must implore the delegations of nuclear weapons Powers, and most directly our two Co-Chairmen, to reply rapidly and fully to the following questions:

(1) Is it true that a new generation of tactical nuclear precision weapons with subkiloton yields are being developed and tested?

(2) Are plans for the development of such systems a reason for further testing and thus an obstacle to the conclusion of a comprehensive test ban?

(3) Is it true that preparations are being made for the early deployment of such systems?

(4) Does the recent agreement between the US and the USSR on the prevention of nuclear war refer also to wars where no nuclear weapons are used except such mini-nukes?

(5) Does Security Council resolution 255 (1968) in the interpretation of the Co-Chairman refer also to nuclear aggression in which no nuclear weapons are used except such mini-nukes?

Complete clarity on these points of major concern must be given. I sincerely hope that the answers can be so reassuring that no State party to the NPT will interpret the situation as an extraordinary event which has 'jeopardized the supreme interests of its country', and act accordingly. This is, of course, quoting the withdrawal clause of the NPT. On the contrary, I hope that the answers will remove all fears. Efforts must be encouraged to stop a threatening tactical nuclear weapons arms race by means of appropriate agreements here in this Committee, at SALT II, and at the 1975 NPT review conference, when amendments could take care of dim or disputed points.

Author: Mrs Myrdal's statement along with the responses of the United Kingdom, United States of America, and the Soviet Union are reproduced in this book on pages 258–264. All of the responses were consistent and clearly in sharp disagreement with the author's position in support of the new nuclear force. A major concern of the author is that international agreements may be reached which may not overly constrain nations within the existing policy of nuclear nations but which could seriously endanger the prospects of other solutions which may be far superior.

Appendix 2. Nuclear-device cost

The small amount of readily available information on the cost of nuclear devices indicates very high costs. In 1965, the AEC quoted the cost of a 100 kiloton Plowshare device to be $460 000; but in 1975, the following was stated:†

---

† An analysis of the economic feasibility, technical significance, and time scale for application of peaceful nuclear explosions in the US, with special reference to the Gulf Universities Research Consortium report therein. Purchase Order No. AD5AD 23, Final Report to ACDA from the Program on Science, Technology, and Society, Cornell Univ., Ithaca, NY April 1975.

'More recent analyses suggests that the 1974 price would be somewhat larger and an estimate of $700 000 for a comparable device has recently been made'.

The following quotation regarding nuclear-weapon cost was taken from a recent publication:†

'... Virtually no reliable information is available on the cost of nuclear warheads, including those for Lance and Pershing. Enthoven, however, has suggested an average of $500 000 per warhead for tactical nuclear delivery systems, a figure that is borne out by the author's calculations. This estimate is also in line with the AEC's unsuccessful fiscal 1974 budget request for $904 million to develop and produce some 2000 new nuclear shells for the 8 inch Howitzer, with a cost per warhead of about $452 000 ...'

It is not surprising that many people have associated the design and manufacture of nuclear explosives with necessarily large expenditures of money. To the extent that this impression is correct, it is attributable to US production and pricing practices that emphasized quality, not cost reduction. This approach is understandable, since warhead costs for most important weapon systems, like the Polaris submarine, had represented a very small fraction of the total weapon-system cost – out of all proportion to their importance.

Of recent time there has been good reason for the US to reconsider device costs:

1. With the advent of MIRV, many re-entry vehicles can be carried on one missile with a corresponding increase of total warhead cost per missile.

2. Somewhat more to the point, increasing importance is being placed on tactical nuclear weapons and Plowshare devices. In both arenas, the devices are facing competition from more conventional approaches to the problem; nuclear weapons must compete with conventional-explosive PGMs;† Plowshare must compete with various other cheap explosive and rock-fracturing techniques.‡

3. The Plowshare and military uses of nuclear devices will fall increasingly under the scrutiny of powers other than the US. Of particular interest will be the opinions of the non-nuclear powers. It is important that they have honest appraisals of device cost. Yugoslavia should not continue under the impression that a nuclear device can be manufactured for 200 dollars [27]. However, one

---

† There is also the frequent comment (e.g. ref. [6], p. 66) that replacing an obsolete stockpile of tactical nuclear weapons is an expensive process involving 'major net budgetary outlays'. At best, this would seem to be a relative point; 10 000 tactical warheads even at an assumed price of $500 thousand comes to $5 billion, a fraction of the annual outlay for the NATO force. Properly designed, fielded, and amortized, the cost of the devices would be grossly reduced.

‡ The US is more carefully analysing its Plowshare activities, with particular emphasis on economics. Device cost is a principal factor in these economics.

should not expect them to be so naïve as to believe they need cost over $500 000, once the problem of acquiring fissile material is solved.

In order to reduce costs, the following are among the interrelated aspects of nuclear-device design that must be considered:

1. *Design restrictions:* Device design must be kept as simple as possible. Simplicity requires that we give special attention to minimizing restrictions of size, shape, weight, shelf life, temperature and acceleration environments, and overall design procedures so we can minimize the number and complexity of device parts, the use of expensive materials, and stringent specifications.

2. *Security, safety, and reliability:* Sensible security, safety, and reliability requirements should be met with technologies and procedures that minimize cost.

3. *Weapon output:* Advantages gained by special nuclear-device output should be weighed carefully against the complexity and increased cost incurred.

4. *Production numbers:* The advantages of a stockpile with many kinds of weapons should be balanced carefully against the simplicity and lower cost of an equivalent stockpile based on a few simple designs. The difference between making 10 or 10 000 devices of the same kind has a significant effect on device cost because of the production learning process and the per-unit reduction in other production and research and development costs.

5. *Special nuclear materials:* Because the procurement of special nuclear materials can present a serious and costly problem, it is important to minimize the amount used in the design of a device.

6. *System impact:* The cost of a nuclear device must be balanced against the cost of the rest of the system and of fielding and maintaining the system. Certainly, the incentive to develop a cheap nuclear device is lessened if there is no matching incentive to keep other costs in line.

7. *Fiscal practice:* Tight and intelligent fiscal control must be maintained over all aspects of device design, development, manufacture, and retirement if costs are to be significantly reduced and realistically assigned.

Depending upon how these parameters are treated, the cost of a nuclear device of a particular yield can vary over orders of magnitude. Certainly, failure to establish these design goals can result in high unit device cost. Examples are the small diameter devices used for gas stimulation, and nuclear cannon projectiles. The costs of these devices are grossly higher than those for say, a simple fission device of relatively large volume and weight, produced in large numbers (e.g. 10 000), adequate to meet the needs of a battlefield missile. Also, devices for other Plowshare applications, such as PACER, may be significantly less constrained and produced in very large numbers, hence, much less expensive.

References and notes to Chapter 10

1. Burt, R., *New weapons technologies, debate and directions.* Adelphi Paper 126, Summer 1976.

This article evaluates the technical features of new weapons and proceeds to discuss their impact on military and political behaviour. The article is well written; it is objective; it is complete, though it does not reach to the limits described in the New Nuclear Force.

2. Annual Defense Department Report FY 1976 and FY 1977, pp. 1–11.
3. *Foreign Affairs*, Vol. 53, No. 4, (July 1975), pp. 771–6.
4. Heisenberg, W., *The Alliance and Europe: Part I; Crisis Stability in Europe and Theater Nuclear Weapons*, Adelphi Paper Number 96, (1973).
5. Martin, L., Theater nuclear weapons in Europe, *Survival*, (November–December 1974).
6. Record, J., *US Nuclear Weapons in Europe, Issues and Alternatives*. Brookings Institution, (December 1974). Reprinted in part in *Survival*, (March–April 1975).
7. Bennett, W. S., Sandoval, R. R. and Shreffler, R. G., A credible nuclear-emphasis defense for NATO, *Orbis*, Vol. XVII, No. 2, (Summer 1973), p. 473.
8. Bennett, W. S., Sandoval, R. R. and Shreffler, R. G., *United States National Security Policy and Nuclear Weapons*, Los Alamos Scientific Laboratory Report LA-5785-MS, (November 1974).
9. Sandoval, R. R. and Shreffler, R. G., *Nuclear Weapons and Their Role in NATO Political and Military Posture*, to be published, (August 1976).
10. Sandoval, R. R., Consider the porcupine: another view of nuclear proliferation, *Bulletin of the Atomic Scientists*, Vol. 32, No. 5, (May 1976), pp. 17–18.
    Sandoval suggests that 'It remains to be seen whether some hitherto nuclear-naked country will opt for a nuclear defense, forego posing the risk of destruction to its potential enemies, and accept the risk that its enemies may find a reason to destroy it, though they could not capture it intact'. He goes on to say, 'With the defense of its borders entrusted to forces structured around the firepower of nuclear weapons, any nation not now a nuclear power, and not harboring ambitions for territorial aggrandizement, could walk like a porcupine through the forests of International affairs: no threat to its neighbors, too prickly for predators to swallow'.
11. Buden, D. *et al.*, *A Defense Force for NATO's Central Region*, Los Alamos Scientific Laboratory Report, LA 5991-MS, (December 1975).
    This report defines the elements of a NATO defence force similar in structure to the one posed in this essay. It summarizes a considerable effort by the authors which leads to one conclusion, that the proposed force would cost about half that currently being spent in NATO. This results primarily from the reduction of the professional force by 50 per cent and their replacement by militia.
12. Wellnitz, B. A., Panel Secretary, LASL Panel on Tactical Nuclear Warfare, Report on Third Meeting, 14–15 May 1974. Los Alamos Scientific Laboratory Report, LA-6059-MS, (September 1975).
13. Hayes, J. K., *Computer Simulation of Tactical Nuclear Warfare*, Los Alamos Scientific Laboratory Report, LA-5806-MS, (December 1974).
    This document describes a computer code to simulate the kind of nuclear engagement described in this essay. The sensitivity of the important parameters is studied through visual situation displays of the battlefield throughout the course of the battle. This work has been extended at Los Alamos and elsewhere.
14. Stratmann, P. and Herrmann, R., Limited Responses, Escalation Potential, and the Central Region, *circa* 1976.
    The status of this document is unknown. It may well present a candid view of the NATO as viewed by some Germans. Although it represents a very positive step away from the present force by describing the kind of force required to support conventional

PGMs, it is most noteworthy for exposing the unknowns, uncertainties, and deficiencies of such a force when viewed in the nuclear context.

15. Geneste, M., Les trois piliers de la paix, *Armées d'aujourd'hui* no. 10 (Mai 1976), pp. 58–66.

Marc Geneste is one of the most articulate and prolific French writers on the subject of nuclear warfare. In many articles, Geneste has championed the overwhelming defensive capabilities of nuclear weapons in tactical warfare. In most respects his ideas conform with those expressed in this chapter. He has been the early champion of the essential advantage offered by nuclear weapons over conventional PGMs as described in part III. The differences may reflect a French attitude. For example, Geneste does not feel the necessity for a well defined firebreak between the tactical nuclear force and the strategic force; he places greater emphasis on the need for a strategic force, and the importance of using the capabilities of the existing tactical conventional force.

16. Brodie, B., What price conventional capabilities in Europe? *The Reporter*, (23 May 1963).

17. Taylor, M., *Foreign Affairs*, (April 1974).

18. *World Armaments and Disarmament*, SIPRI Yearbook 1974, (Almqvist & Wiksell, Stockholm, 1974), p. 69.

19. Schlesinger, J. R., *The Theater Nuclear Force Posture in Europe*, A report to the US Congress in Compliance with Public Law 93-365, March 1975 (unclassified version). Section C1, Warsaw Pact Strategy, Doctrine, and Force Posture, pp. 9–11.

This reference presents an excellent and official US position on the great threat posed by the WTO.

20. Karber, P. A., The Soviet anti-tank debate, *Survival*, (May–June 1976), pp. 105–111.

This is an impressive review of the serious and penetrating debate of the role of the conventional anti-tank guided missile (ATGM), as viewed by the Soviet Union. The Soviet inquiry is 'how to overcome the challenge of anti-tank weapons and retain a high rate of advance against a strengthened NATO defensive capability'. The role of nuclear weapons is discussed briefly and summarized in a final paragraph:

> 'Evidently the Soviet Union only sees anti-tank weapons as a threat "when nuclear weapons are not used". Yet Soviet ground force commanders are clearly unwilling to discount the anti-tank problem by relying solely on nuclear weapons. For the present, the nuclear solution seems to have been ruled out by both political and operational considerations'.

There would seem to be nothing in this article that is contrary to what is presented in this chapter. The conclusion is that where the ATGM presents a very serious problem to a WTO offensive, a nuclear threat as posed by the defensive nuclear force would define a disaster. Clearly there is no advantage for the Soviet Union to stress low yield nuclear weapons. Their hope for success rests in not aggravating NATO out of their present anachronistic posture.

21. Milshtein, H., Tactical nuclear weapons: Problems of definition and applications. This volume, pp. 169–174.

The author describes the confusion that has prevailed over the distinction between tactical and strategic nuclear weapons. He deplores the fact that 'the difference between conventional and nuclear weapons is being obliterated and certain military commanders might find the possibility of using highly accurate nuclear weapons more attractive than using conventional armaments'. He concludes that '. . . TNWs must not be improved, but be reduced and completely removed from all probable theatres'. Further: 'The USSR is consistently working for the reduction of nuclear weapons in

Europe, which is in complete accord with its fundamental position on the limitation and reduction of nuclear weapon and on achieving nuclear disarmament'.

22. Gylden, N. and Holm, L. W., *Risks of Nuclear Explosives Production in Secret*, Swedish Report FOA 4, C 4567-T3; English translation ERDA-TR-45.
The authors outline what must be done to acquire fissile material, technical support and production capacity to manufacture nuclear warheads.

23. Agnew, H. M., Atoms for lease, *Bulletin of the Atomic Scientists*, Vol. 32, No. 5, (May 1976), p. 23.
The author proposes that fuel for nuclear reactors be leased, not sold.

> 'However, if they (nuclear suppliers) would simply all agree never to sell the fuel but simply to lease it based upon delivery of a certain amount of thermal energy, a great deal of concern by the world could be alleviated. It would also make much simpler the concept of regional recycling and waste disposal centers'.

24. Digby, J., *Precision-Guided Weapons*, Adelphi Paper 118, Summer 1975.
This paper is a good review of conventional precision guided munitions. The first part of the paper deals with the mechanics and effectiveness of the weapons. The second part deals with the likely effects of PGMs on force posture and the conduct of war. The author leaves little question that PGMs will have an increasing impact on modern warfare; however, his many reservations and his admittedly limited analysis leave a great deal of apprehension. Digby probably sees PGMs as a method for forestalling the use of nuclear weapons.

25. Morse, J. H., New weapon technologies: implications for NATO, *Orbis*, (Summer 1975), Vol. XIX, No. 2, pp. 497–513.
This is an interesting and thought-provoking article which is summarized in its last paragraph:

> 'A fundamental reassessment of NATO's mission to defend Europe, and the strategies and capabilities needed to carry it out, is long overdue. While the flexible response called for by present NATO strategy is sound in principle, its implementation needs new thinking. It is dangerous to retain the current shaky defense structure indefinitely. The means for revamping it are present, thanks to new and emergent weapons technologies'.

This article discusses many of the points made in this chapter. Unfortunately, the article essentially ignores nuclear weapons and the role they would play with PGMs.

26. *Military Aspects of the Israeli–Arab Conflict*, University Publishing Projects, Tel Aviv, 1975.
This book is in the category of 'required reading'. It reports on 25 presentations and question periods which took place at an international symposium in Jerusalem on 12–17 October 1975, under the patronage of the Minister of Israeli Defense, Mr Shemen Peres. In brief, the Yom Kippur War is reviewed two years after it was fought by those who played the essential roles in the Israeli side of the conflict. One can argue that some parts are not altogether objective, but all parts were certainly important, pertinent, and interesting.

27. Cedza, D. S., Yugoslavia and nuclear weapons, *Borba*, (7 December 1975), Reproduced in *Survival*, (May–June 1976), pp. 116–7.
The *Survival* introduction to this article is quoted:

> 'While Yugoslav defense strategy remains based on the concept of territorial defense undertaken by large sections of the population, officials have recently expressed a keen interest in new weapons technologies, such as precision-guided missiles, to defend

against attack. In the article, which appeared in the Communist party newspaper Borba, Dimitrije Seserinac Gedza sees important advantages in both civilian resistance and new conventional weapons, but argues that the acquisition of nuclear weapons may become necessary in the light of Eastern and Western nuclear capabilities'.

The final paragraphs of the article are quoted:

> '... However, should the use of mass terror be contemplated, or should nuclear or other weapons for mass destruction be used, our country may, in the framework of the general defense concept, re-consider its attitude towards the question of non-proliferation of nuclear weapons. Because today the possibility exists – both in the East and in the West – of manufacturing nuclear weapons costing a few hundred dollars, instead of a few hundred million dollars as in the past. Cheap and easy manufacture of "mini-nuclear" weapons, capable of destroying entire units or headquarters of the aggressor, would have a sobering effect on anyone contemplating invasion of our country, and it is no exaggeration to state that mass terror comparable to that practised during the last war cannot ever again be carried out on the territory of Yugoslavia.
>   If the unity of the people, its readiness to resist any aggressor and to defend the achievements of the thirty years of development are added to all this, it is possible to look into the future of Yugoslavia with confidence and equanimity'.

An obvious concern is that Yugoslavia has no example to follow in the development of its nuclear defence. This essay may inject some rationality into their approach.

28. Cohen, S. and VanCleave, W., The Asian Pacific Region: implications for US policy, 1975–80, *Orbis*, Vol. 19, No. 3 (fall 1975).

Although this article addresses the Far East, it has wider geographic application. In many respects its contents conform with the opinion expressed in this chapter. Its introduction well covers the post World War II history.

29. Admiral S. Turner, The naval balance: not just a number game, *Foreign Affairs*, Vol. 55, No. 2, (January 1977).

This excellent article discusses the roles of a navy to perform in strategic deterrence, naval presence, sea control (assertion and denial), and the projection of power ashore. The capabilities of the USA and USSR navies in these four areas are discussed.

30. Collins, A. S., Jr., Tactical nuclear warfare and NATO: viable strategy or dead end? *NATO's Fifteen Nations*, (June–July 1976), pp. 73–87.

General Collins adopts the attitude of many senior officers in the United States Army, that tactical nuclear weapons are an anathema to the battlefield. He contends that nuclear weapons should be relegated to deterring nuclear war on the battlefield by threatening to strike the Soviet Union, albeit at their border. He supposes that this would free the Army to fight its conventional war without fear of nuclear interruption.

Bibliography on tactical nuclear weapons (in Europe)

M. Leitenberg

The bibliography requires several essential notes as to the limitations of its coverage. For the most part it contains references to tactical nuclear weapons in the European context. The literature surveyed is nearly entirely English language, but even within that limitation, relatively few references originate from the United Kingdom, French, USSR, West German of Indian publications, which might have appeared in English. Thus, the literature surveyed for the most part appeared in the United States.

The bibliography omits any coverage of references to the Multilateral Force (MLF) or to a European Nuclear Force (ENF). By and large, it also omits the very extensive literature on 'limited war', in which references to the possible use of tactical nuclear weapons often appear though without much substantive discussion in such cases. References to military manuals are also omitted.

Finally, only a very small section of references on specific kinds of tactical nuclear weapons appears at the end of the bibliography; that is on individual weapon types such as the Pershing, Sergeant, Lance, Pluton, Frog, Scud, Asroc, Falcon, Genie, depth charges, etc. The references pertaining to these which were supplied represent only a very small fraction of those that would be available in the open literature on such individual weapons.

Sections:

(A) Bibliographies
(B) US Congressional Hearings and Reports, and Reports by US Federal Agencies
(C) Books, Monographs, and Book Chapters (listed by year)
(D) Journal Papers (listed by year)
(E) Individual Weapon Systems (very short list)

(A) Bibliographies

1. *Nuclear Weapons and the Atlantic Alliance, A Bibliographic Survey*, DA Pam 20-66, Headquarters, Department of the Army, Washington, DC, (December 1965), 193 pp.

2. *Nuclear Weapons and NATO, Analytical Survey of Literature*, DA Pam 50-1, Headquarters, Department of the Army, Washington, DC, (January 1970), 450 pp.
3. *Nuclear Weapons and NATO, Analytical Survey of Literature*, Department of Defense, Washington, DC, (January 1975), 546 pp.

(It should be pointed out that despite their titles, these three bibliographic volumes contain only a very small percentage of references concerning tactical nuclear weapons.)

(B) US Congressional hearings and reports, and reports by US Federal Agencies

1. *Military Applications of Nuclear Technology, Part I*, Hearings, Subcommittee on Military Applications, Joint Committee on Atomic Energy, 93rd Congress, First Session, (April 1973), 54 pp.
2. *Military Applications of Nuclear Technology, Part II*, Hearings, Subcommittee on Military Applications, Joint Committee on Atomic Energy, 93rd Congress, First Session, (May, June 1973), 135 pp.
3. *US Security Issues in Europe: Burden Sharing and Offset, MBFR and Nuclear Weapons.* A Staff Report, Subcommittee on US Security Agreements and Commitments Abroad, Committee on Foreign Relations, US Congress, (2 December 1973), 27 pp.
4. *To Consider NATO Matters*, Hearing, Joint Committee on Atomic Energy, 93rd Congress, First Session, (19 February 1974), 28 pp.
5. *The Theater Nuclear Force Posture in Europe*. A Report to the United States Congress, Secretary of Defense, Schlesinger, J. R., (1974?), 30 pp.
6. *The Atlantic Alliance*, Hearings, Subcommittee on National Security and International Operations, Committee on Government Operations, US Senate (in seven parts and appendices). See particularly, Part 2, testimony of Gen. L. Norstad, 1966. Also appeared as *The Atlantic Alliance*, Jackson Subcommittee Hearings, and Findings, Senator H. M. Jackson, (ed.), (F. A. Praeger, New York, 1967), 308 pp.
7. Hearings, Special Subcommittee on North Atlantic Treaty Organization Commitments, Committee on Armed Services, House of Representatives. 92nd Congress, First and Second Sessions, (1972), 1095 pp.
8. *The American Commitment to NATO*, Report of the Special Subcommittee on North Atlantic Treaty Organization Commitments, Joint Committee on Atomic Energy, 93rd Congress, First Session, (1972), 54 pp.
9. *US Forces in NATO*, Hearings, Committee on Foreign Affairs, House of Representatives, 93rd Congress, First Session, (June–July 1973), 440 pp.
10. *US Forces in Europe*, Hearings, Subcommittee on Arms Control, International Law and Organization, Committee on Foreign Relations, US Senate, 93rd Congress, First Session, (July 1973), 385 pp.
11. *Nuclear Weapons and Foreign Policy*, Hearings, Committee on Foreign Relations, US Senate, 93rd Congress, (March–April 1974), 316 pp.

Items containing only a small proportion of material on tactical nuclear weapons.

12. *Security Agreements and Commitments Abroad*, Report to the Committee on Foreign Relations, US Senate, (21 December 1970), 28 pp.
13. *Policy, Troops, and the NATO Alliance*, Report to the Committee on Armed Services, US Senate, (2 April 1974), 14 pp.
14. *Development, Use, and Control of Nuclear Energy for the Common Defense and Security and for Peaceful Purposes*, First Annual Report to the US Congress, Joint Committee

on Atomic Energy, (30 June 1975), 104 pp.; Second Annual Report, (30 June 1976), 197 pp.

15. *Need to Reexamine Some Support Costs Which the US Provides to NATO*, Report to the Congress, Comptroller General, US General Accounting Office, ID-75-72, (25 August 1975), pp. 23–24.

(C) Books, monographs and book chapters

1954:

1. Reinhardt, Col. G. C. and Kintner, Lt. Col. W. R., *Atomic Weapons in Land Combat*, (Military Service Publishing Co., Harrisburg, Pennsylvania, 1953–54), 250 pp.

1955:

2. Reinhardt, Col. G. C., *American Strategy in the Nuclear Age*, (University of Oklahoma Press, Norman, 1955), 235 pp.

1957:

3. Kissinger, H., *Nuclear Weapons and Foreign Policy*, (Harper, New York, 1957), 463 pp.
4. Spier, Hans, The atomic exercise 'Carte Blanche': 1955, Chapter 10 in *German Rearmament and Atomic War*, (Row, Peterson & Co., Evanston, Illinois, 1957), pp. 182–93.

1958:

5. Hoag, M. W., *The Place of Limited War in NATO Strategy*, (RAND P-1566, December 1958), 45 pp.
6. Miksche, Lt. Col. F. O., *Atomic Weapons and Armies*, (Praeger, New York, 1958), 222 pp.

1959:

7. Burns, A. L., NATO and nuclear sharing, Chapter 7 in *NATO and American Security*, Knorr, K. (ed.), (Princeton University Press, Princeton, 1959), pp. 151–75.
8. Mataxis, Col. T. C. and Goldberg, Col. S. L., *Nuclear Tactics, Weapons and Firepower in the Pentomic Division, Battle Group, and Company*, (Military Service Publishing Co., Harrisburg, Pennsylvania, 1959), 254 pp.
9. Rathjens, George, NATO strategy: total war, Chapter 4 in *NATO and American Security*, Knorr, K. (ed.), (Princeton University Press, Princeton, 1959), pp. 65–97.
10. Snyder, G. H., *Deterrence by Denial and Punishment*, Center of International Studies, Princeton University, Research Monograph No. 1, (2 January 1959), 39 pp.

1960:

11. Lowenstein, H. and Zuhlsdorff, V., The nuclear defense of the West, Chapter 9 in *NATO and the Defense of the West*, (Praeger, New York, 1960), pp. 97–111.

1962:

12. Hastings, W. H., *Limited War Patterns: I. Southeast Asia* (1963) (U) (RAND RM-2961-ISA, July 1962), 66 pp.
13. Knorr, K. and Read, T. (eds.), *Limited Strategic War*, (Praeger, New York, 1962),

14. Osgood, R. E., *NATO, The Entangling Alliance*, (University of Chicago Press, 1962), 420 pp.
15. Strachey, J., A reversal of NATO's nuclear strategy, Chapter 7 in *On The Prevention of War*, (Macmillan, London, 1962), pp. 106–16.

1964:

16. Hoag, M. W., Nuclear strategic options and European force participation, Chapter 10 in *The Dispersion of Nuclear Weapons, Strategy and Politics*, Rosecrance, R. (ed.), (Columbia University Press, New York, 1964), pp. 222–58.
17. Wessel, A. E., *Some Implications of Strategic Concepts for Western European Nuclear Weapons*, (RAND P-2904, April 1964), 22 pp.

1965:

18. Armbruster, F. and Singer, M., *Shielding Emphasis Deployment for Tactical Nuclear War*, HI-586, (Hudson Institute, August 1965).
19. Schelling, T. C., Nuclears, NATO and the 'New Strategy', Chapter 9 in *Problems of Nuclear Strategy*, Kissinger, H. (ed.), (Praeger, New York, 1965), pp. 169–85.
20. Stanley, T. W., NATO in transition, *The Future of the Atlantic Alliance*, (Praeger, New York, 1965), 412 pp.

1966:

21. Marshall, A. W., *Determinants of NATO Force Posture*, (RAND P-3280, January 1966), 22 pp.
22. Vandevanter, Gen. E. B. Jr., *Studies on NATO, An Analysis of Integration*, (RAND RM-5006 PR, August 1966), 65 pp.

1967:

23. Fox, W. T. R. and Fox, A. B., *NATO and the Range of American Choice*, (Columbia University Press, 1967), 345 pp.
24. Schlesinger, J. R., *European Security and the Nuclear Threat Since 1945*, (RAND P-3574, April 1974), 25 pp.
25. Schlesinger, J. R., *ibid*.

1968:

26. *State of European Security, The Tactical Use of Nuclear Weapons and the Defense of Western Europe*, Report, Assembly of the Western European Union, Document 440, (2 May 1968), 16 pp.

1969:

27. Marshall, A. W., NATO defense planning: the political and bureaucratic constratins, Chapter 9 in *Organization and Management, A Systems Approach*, Rosensweig, J. E. and Kast, F. (eds.), (McGraw-Hill, New York, 1969–70), pp. 353–68.

1970:

28. Lee, W. T., *Influence of NATO/Europe on Soviet Military Policy*, (Stanford Research Institute, Arlington, Virginia, SSC-ISR-TN-2, 30 April 1970), 46 pp.
29. Shreffler, R. E. and Bennett, W. S., *Tactical Nuclear Warfare*, (Los Alamos Scientific Laboratory, LA-4467-MS, June 1970).

345

30. Wolfe, T. W., Soviet thinking on theater warfare in Europe, Chapter 10 in *Soviet Power and Europe, 1945–1970*, (Johns Hopkins University Press, 1970), pp. 195–216.

1971:

31. Beaufre, Gen. A., *Problems of Strategy and European Security: The Problem of Tactical Nuclear Weapons*, Stanford Research Institute, Franco–American Symposium on Problems of Strategy and European Security, (25 August 1971), 23 pp.

32. Gilinsky, V., *Arms Control Aspects of the Deployment of Tactical Nuclear Weapons in Europe*, Southern California Arms Control and Foreign Policy Seminar, (October 1971), 43 pp.

33. Nelson, Senator G., *Report on Tactical Nuclear Weapons*, MCPL, Members of Congress for Peace Through Law, Washington, DC, (Congressional Record, Senate, S11626-S11628, 20 July 1971).

34. Newhouse, J. *et al.*, *US Troops in Europe, Issues, Costs and Choices*, (The Brookings Institution, Washington, DC, 1971), 177 pp.

35. Schultze, C. L., Tactical nuclear weapons in Europe, in *Setting National Priorities, the 1972 Budget*, (The Brookings Institution, Washington, DC, 1971), pp. 94–102.

36. *Tactical Arms Limitation in Europe*, (Verbal Presentations and Discussions), Fifteenth Pugwash Symposium, Lahti, Finland, (22–24 August 1971), 204 pp., privately distributed.

37. *Taktik des Allgemeinen Gefechts in Kernwaffenkrieg*, (Deutscher Militärverlag, Berlin, 1971), 428 pp.

38. Weizsäcker, C. F. von (ed.), *Kriegsfolgen und Kriegsverhütung*, (Carl Hanser Verlag, Munchen, 1971), 699 pp.

1972:

39. Biddle, W. F., Tactical nuclear weapons, Chapter 17 in *Weapons Technology and Arms Control*, (Praeger, New York, 1972), pp. 255–62.

40. Boylan, E. S., *Some Thoughts on Tactical Nuclear Weapons*, (Hudson Institute, HI-1613-P, 22 March 1972), 20 pp.

41. Burrows, B. and Irwin, C., The threat and the strategy, Chapter 3 in *The Security of Western Europe: Towards A Common Defense Policy*, (Charles Knight, London, 1972), pp. 57–83.

42. Cliffe, T., *Military Technology and the European Balance*, Adelphi Paper No. 89, (August 1972), 58 pp. (only certain sections).

43. Miettinen, J. K., *European Security Balanced by Tactical Nuclear Weapons?*, 22nd Pugwash Conference on Science and World Affairs, Oxford, England, (September 1972), 31 pp., mimeographed.

44. Nerlich, U., *Some Comments on Modernization of Nuclear Stockpiles in Europe: A German View*, (Hudson Institute, HI-1626-D, 24 April 1972), 11 pp.

45. Owen, D., The evolution of NATO's nuclear strategy, Chapter 11 in *The Politics of Defense*, 1972, pp. 151–67.

46. Tactical (Nuclear) Arms Limitation in Europe, Conference Report, Fifteenth Pugwash Symposium, Lahti, Finland, 22–24 August 1971, *Pugwash Newsletter*, (January 1972), Vol. 9, No. 3, pp. 79–87.

47. Yefremov, A. Ye., *Europe and Nuclear Weapons*, (Moscow, 1972), 391 pp.; (translation, Joint Publication Research Service, JPRS 58481, 14 March 1973).

1973:

48. Canby, S. L., *Policy Implications of Restructuring NATO's Military Forces*, California Arms Control and Foreign Policy Seminar, Helsinki Seminar, (June 1973), 29 pp.

49. Heisenberg, W., *The Alliance and Europe. Part I: Crisis Stability in Europe and Theatre Nuclear Weapons*, IISS, Adelphi Paper No. 96, (Summer 1973), 35 pp.

50. Joshua, W., *Nuclear Weapons and the Atlantic Alliance*, National Strategy Information Center, New York, 1973, 60 pp.

51. Kaplan, M. A., *The Rationale for NATO*, American Enterprise Institute – Hoover Institution, Washington, DC and Stanford, California, 1973, 94 pp.

52. Miettinen, J. K., *Recent Developments in Tactical Nuclear Weapons, Non-Proliferation and European Security*, Symposium on Tactical Nuclear Weapons in Europe, Dutch Pugwash Group, (3 November 1973), 9 pp., mimeographed.

1974:

53. Canby, S., *The Alliance and Europe. Part IV: Military Doctrine and Technology*, IISS, Adelphi Paper No. 109, (Winter 1974–5), 42 pp.

54. Lawrence, R. D. and Record, J., *US Force Structure in NATO, An Alternative*, (The Brookings Institution, Washington, DC, 1974), 136 pp.

55. Nerlich, U., *Die Nuklearen Strike-Verbande Der NATO, Als Gegenstand Von Ost-West-Verhanglungen*, (Forschungsinstitut für Internationale Politik und Sicherheit, 1974), 155 pp.

56. Record, J., *US Nuclear Weapons in Europe, Issues and Alternatives*, (The Brookings Institution, Washington, DC, 1974), 70 pp.

1975:

57. Maziere, Gen. U. de, *Rational Deployment of Forces on the Central Front*, Assembly of Western European Union, 21st Ordinary Session, Document 663, (2 April 1975), 59 pp.

58. Nerlich, U., *The Alliance and Europe. Part V: Nuclear Weapons and East–West Negotiations*, IISS, Adelphi Paper No. 120, (Winter 1975–6), 35 pp.

59. Record, J., *Sizing Up The Soviet Army*, (The Brookings Institution, Washington, DC, 1975), 51 pp.

1976:

60. Douglass, J. D. Jr., *The Soviet Theater Nuclear Offensive*, Studies in Communist Affairs, Vol. I, (US Air Force, US GPO, Washington, DC, 1976), 127 pp.

(D) Journal papers

1946:

1. Nickerson, H., Atomic military theory. Some reflections on pre- and post-atomic military theory, *Journal of the Royal Artillery*, Vol. 73, No. 3, (July 1946), pp. 218–24.

1950:

2. Gavin, Maj. Gen. J. M., The tactical use of the atomic bomb, *US Army Combined Forces Journal*, Vol. 1, No. 4, (November 1950), pp. 9–11.

1952:

3. Atomic warfare, *The Army Quarterly* (UK), Vol. 63, No. 2, (January 1952), pp. 131–33.

4. Barclay, C. N., Atomic warfare. Some ideas on possible future developments, *Brassey's Annual* (Chapter V), 1952, pp. 59–70.

1953:

5. Reinhardt, Col. G. C., Sea power's role in atomic warfare, *United States Naval Institute Proceedings*, Vol. 79, No. 12, (December 1953), pp. 1279–87.

1954:

6. Aillert, Col., Des problèmes de tactique et de stratégie atomiques, *Revue Militaire d'Information*, No. 243, (10–25 December 1954), pp. 18–22.
7. Brodie, B., Nuclear weapons: strategic or tactical?, *Foreign Affairs*, Vol. 32, (January 1954), pp. 217–29.
8. Fantastic weapons: what some of the talk is all about, *Life Magazine*, date unknown, pp. 121–24.
9. Pamart, Col., Les transmissions et les expériences atomiques, *La Revue des Transmissions*, No. 60, (November–December 1954), pp. 39–40.
10. Reinhardt, Col. G. C., Tactics for atomic war, *Ordnance*, Vol. 38, No. 204, (May–June 1954), pp. 936–38.
11. Rowny, Col. E. L., Ground tactics in an atomic war, *The Army Combat Forces Journal*, (August 1954), pp. 18–22.

1955:

12. *Army Quarterly* (UK), *Special Issue*, Vol. 69, No. 2, January 1955.
    — Some aspects of defence in atomic warfare, pp. 58–67.
    — 'Goose egg', pp. 68–72.
    — The organization of the army for nuclear war, pp. 73–74.
    — A logistical concept for an atomic war, p. 75.
    — Administration in nuclear war, pp. 75–6.
    — The case for armour in nuclear war, p. 77.
    — The nuclear legion, p. 78.
    — Nuclear defence, p. 79.
13. Current Comment, The tactical side of atomic warfare, Letter to Col. Reinhardt on article in BAS 11, No. 2 from Eugene Rabinowitch and the reply from Colonel Reinhardt, *Bulletin of the Atomic Scientists*, Vol. 11, No. 5, (May 1955), pp. 191–2.
14. Phillips, B. Gen. T. R., Our point of no return, *The Reporter*, Vol. 12, No. 4, (24 February 1955), pp. 14–18.
15. Reinhardt, Col. G. C. and Kintner, Lt. Col. W. R., The tactical side of atomic warfare, *Bulletin of the Atomic Scientists*, Vol. 11, No. 2, (February 1955), pp. 53–58.
16. White, T. H., The atomic battlefield: conversation with a soldier, *The Reporter*, Vol. 12, No. 3, (10 February 1955), pp. 29–32.

1956:

17. *Army, Special Issue*, 6, No. 8, March 1956.
    — Requirement: guided missile for the army, p. 10.
    — The IRBM: the army has the know-how, pp. 11–12.
    — Mr Brucker 'Carries the ball', pp. 12–13.
    — The IRBM: An artillery support weapon, pp. 14, 16, 53.
    — The case for tactical atomic weapons, pp. 24–5.
18. Eddleman, Lt. Gen. C. D., Men, missiles and atomics on the future army battlefield, *Army*, Vol. 7, No. 5, (December 1956), pp. 24–30.
19. Hadley, A. T., Low-yield atomic weapons: a new military dimension, *The Reporter*, Vol. 14, No. 8, (19 April 1956), pp. 23–5.

20. Harvard University, Defense Policy Seminar, 1955–6. *Planning for a Peripheral War*, (Reading Assignment No. 15, compiled by Robert W. Berry, 23 February 1956).
21. Taylor, Gen. M. D., 'Safety Lies Forward' – technologically and tactically, *Army*, Vol. 7, No. 5, (December 1956), pp. 20–24.

1957:

22. King, J. E. Jr., Nuclear plenty and limited war, *Foreign Affairs*, Vol. 35, No. 2, (January 1957), pp. 238–56.
23. Le Hagre, Col., Ground forces in amphibious operations, exerpted from *La Revue Maritime* (France), October 1956 and reprinted in *Military Review*, Vol. 36, No. 2, (February 1957), pp. 104–7.
24. Mattimoe, Comdt. C. M., Atomic war tactics, exerpted from *An Cosantóir* (Ireland), January 1956 and reprinted in *Military Review*, Vol. 36, No. 2, (February 1957), pp. 88–94.

1958:

25. Beaumont, Lt. Col. H. C., Missile division, *Army*, Vol. 8, No. 11, (June 1958), pp. 49–52.
26. Berry, B. Gen. J. A. and Maxatis, Col. T. C., Test evaluation – the King Cole (CPX-FTX) experience, *Army*, Vol. 9, No. 10, (October 1958), pp. 47–9.
27. Christy, Maj. J. V., What's a battle group?, *Army*, Vol. 9, No. 10, (October 1958), pp. 57–61.
28. Hoag, M. W., NATO: deterrent or shield?, *Foreign Affairs*, Vol. 36, No. 2, (January 1958), pp. 1–15.
29. Kohanson, Lt. S. C., Pentomic counterfire methods, *Army*, Vol. 8, No. 8, (March 1958), pp. 40–1.
30. Kissinger, H. A., Missiles and the Western Alliance, *Foreign Affairs*, (April 1958), pp. 383–400.
31. Murphy, C. J. V., The NATO alliance goes nuclear, *Fortune*, (February 1958), pp. 98–103, 234, 236, 239–40, 242.
32. Poncet, Lt. Col., Guerre atomique et transmissions, *La Revue des Transmissions*, No. 77, (March–April 1958), pp. 10–30.
33. Rathjens, G. W. Jr., Notes on the military problems of Europe, *World Politics*, Vol. 10, No. 2, (January 1958), pp. 182–201.
34. Vissering, Maj. Gen. N. H., Sea transport in atomic war, *Army*, Vol. 8, No. 8, (March 1958), pp. 25–7, 60.
35. Watkins, Maj. H. B. C., Atomic weapons in the land battle, Chapter XIX, *Brassey's Annual*, (1958), pp. 177–84.
36. Wynne, Capt. G. C., Pattern for limited (nuclear) war: the riddle of the Schlieffen plan – III, *Journal of the Royal United Service Institute*, Vol. 103, No. 610, (May 1958), pp. 215–22.

1959:

37. Baldwin, H. W., Limited war, *The Atlantic*, (1959), pp. 35–43.
38. Gordon, L., NATO in the nuclear age, *The Yale Review*, Vol. 48, No. 3, (March 1959), pp. 321–35.
39. Klinkrade, A., The Schlieffen plan – nuclear war and the Soviet and German armies, *Journal of the Royal United Services Institute*, Vol. 104, No. 613, (February 1959), pp. 77–80.

40. Majumdar, Lt. Col. B. N., The atomic army, exerpted from *The Infantry Journal* (India), (October 1958) and reprinted in *Military Review*, Vol. 39, No. 6, (September 1959), pp. 107–9.
41. Peal, Maj. C. D. M., Nuclear fire planning at divisional level, *The Journal of the Royal Artillery*, Vol. 86, No. 3, (Winter 1959), pp. 173–75.
42. Poncet, Col., Guerre atomique et transmissions (3rd article), *La Revue des Transmissions*, No. 82, (January–February 1959), pp. 8–26.
43. Silvasy, Lt. Col. S., Give it guidance, *Military review*, Vol. 39, No. 6, (September 1959), pp. 63–73.
44. White, Maj. S. N., Fusion produces greater mass and terrific energy, *The Royal Engineers Journal*, Vol. 73, No. 3, (September 1959), pp. 285–303.
45. Younger, Maj. A. E., The effects of atomic warfare on engineer operations, *The Royal Engineers Journal*, Vol. 73, No. 1, (March 1959), pp. 11–7.
46. Younger, Maj. A. E., Field engineers in atomic warfare, *The Royal Engineers Journal*, Vol. 73, No. 2, (June 1959), pp. 163–68.

1960:

47. Baar, J., Army fears tactical missile gap, *Missiles and Rockets*, Vol. 6, No. 9, (May 1960), pp. 11–2.
48. DePuy, Col. W. E., The case for a dual capability, *Army*, Vol. 10, No. 6, (January 1960), pp. 32–40.
49. Duncan, Maj. Gen. N. W., The employment of nuclear weapons on the battlefield, *The Royal Armoured Corps Journal*, (UK), Vol. 14, No. 4, (October 1960), pp. 189–90.
50. Hayes, Col. J. H., Mushrooms are poison, *Army*, Vol. 11, No. 1, (August 1960), pp. 67–73.
51. Kirk, Capt. R. L., If he doesn't get the word, *Army*, (November 1960), pp. 64–7.
52. Kissinger, H. A., Limited war: conventional or nuclear? A reappraisal, *Daedalus*, (Fall 1960), pp. 800–17.
53. Lemnitzer, Gen. L. L., Forward strategy reappraised, *Army*, Vol. 11, No. 2, (September 1960), pp. 41–3.
54. Nuclear sharing, *Orbis*, Vol. 4, No. 1, (Spring 1960), pp. 3–5.
55. Richardson, R. C. III, Atomic bombs and war damage, *Orbis*, Vol. 4, No. 1, (Spring 1960), pp. 39–51.
56. Spiedel, Gen. Hans, Mission and needs of NATO's shield, *Army*, Vol. 11, No. 2, (September 1960), pp. 33–8.

1961:

57. Asprey, Capt. R. B., Wintershield: war in a nuclear climate, *Army*, Vol. 11, No. 10, (May 1961), pp. 40–6.
58. le Cheminant, Gp. Capt. P., Tactical deterrence or limited war?, *Brassey's Annual*, (1961), pp. 112–21.
59. Dupuy, T. N., Can America fight a limited nuclear war?, *Orbis*, Vol. 5, No. 1, (Spring 1961), pp. 31–42.
60. Howze, Maj. Gen. H. H., The land battle in an atomic war, *Army*, Vol. 11, No. 12, (July 1961), pp. 29–36.
61. New leadership on the free world, *Orbis*, Vol. 5, No. 1, (Spring 1961), pp. 3–14.
62. Schaffer, Lt. Col. R. W., Decentralize the Honest John battalion, *Army*, Vol. 11, No. 6, (January 1961), pp. 71, 74.
63. Whitten, H. P., Radiation exposure and the tactical mission, *Army*, Vol. 11, No. 12, (July 1961), pp. 53–62.

1962:

64. Air defence in the nuclear battle zone, *Interavia*, Vol. 17, No. 3, (March 1962), pp. 286–8.

65. Aron, R., The US nuclear monopoly and Europe: must we have blind faith, *The New Republic*, Vol. 147, No. 25, (22 December 1962), pp. 9–10.

66. Birrenbach, K., The reorganization of NATO, *Orbis*, Vol. 6, No. 2, (Summer 1962), pp. 244–57.

67. Blue Water – Britain's air-transportable nuclear missile, *Interavia*, Vol. 17, No. 3, (March 1962), pp. 284–5.

68. Buchan, A., NATO divided: nuclear weapons, Europe and the US, *The New Republic*, Vol. 147, No. 26, (29 December 1962), pp. 13–6.

69. Dougherty, J. E., European deterrence and Atlantic unity, *Orbis*, Vol. 6, No. 3, (Fall 1962), pp. 371–421.

70. Eryx, Tactical missiles in nuclear defence, *Interavia*, Vol. 17, No. 3, (March 1962), pp. 280–5.

71. Healey, D., The US nuclear monopoly and Europe: what could Britain do?, *The New Republic*, Vol. 147, No. 25, (22 December 1972), p. 10.

72. Kintner, W. R. and Possony, S. T., NATOs nuclear crisis, *Orbis*, Vol. 6, No. 2, (Summer 1962), pp. 217–43.

73. Kissinger, H. A., The unsolved problems of European defense, *Foreign Affairs*, (July 1962), pp. 515–41.

74. Lippman, W., The US nuclear monopoly and Europe: only one driver can sit at the wheel, *The New Republic*, Vol. 147, No. 25, (22 December 1962), pp. 7–8.

75. MacDonald, J. A., The Marine Corps today, *Naval Review 1962–1963*, pp. 127–29, 131.

76. Miksche, Col. F. O., The European shield: are atomic weapons the most ideal?, *NATO's Fifteen Nations*, Vol. 7, No. 4, (August–September 1962), pp. 15–22.

77. Murphy, C. J. V., NATO at a nuclear crossroads, *Fortune*, Vol. 66, No. 6, (December 1962), pp. 85–7, 214, 219–20, 222, 225.

78. Stanley, T. W., NATOs nuclear debate: Washington's view, *The Reporter*, (5 July 1962), pp. 19–21.

79. Time to think, *Interavia*, Vol. 17, No. 3, (March 1962), p. 279.

1963:

80. Acheson, D., The practice of partnership, *Foreign Affairs*, Vol. 41, No. 2, (January 1963), pp. 247–60.

81. Ball, G. W., The nuclear deterrent and the Atlantic alliance, *The Atlantic Community Quarterly*, (Summer 1963), pp. 199–204.

82. Basler, Hptm. E., Die Beanspruchung von Bauwerken durch Nuklearwaffen (Schluss), *Technische Mitteilungen für Sappeure, Pontoniere und Mineure* (Zurich), Vol. 27, No. 4, (January 1963), pp. 156–65.

83. Hoag, M. W., Nuclear policy and French intransigence, *Foreign Affairs*, Vol. 41, No. 2, (January 1963), pp. 286–98.

84. Kissinger, H. A., Strains on the alliance, *Foreign Affairs*, Vol. 41, No. 2, (January 1963), pp. 261–85.

85. Kissinger, H. A., NATO's nuclear dilemma, *The Reporter*, (28 March 1963), pp. 22–38.

86. McHaney, Col. G. M., What is tactical use of nuclear weapons?, *Army*, Vol. 13, No. 9, (April 1963), pp. 40, 45–7.

87. Messmer, P., French military problems, *The Atlantic Community Quarterly*, (Summer 1963), pp. 185–6.

88. Mulley, F. W., Nuclear weapons: challenge to national sovereignty, *Orbis*, Vol. 7, No. 1, (Spring 1963), pp. 32–40.

89. Nixon, R. M., From a military alliance to a political confederation, *The Atlantic Community Quarterly*, (Summer 1963), pp. 205–6.

90. Possony, S. T., Toward nuclear isolationism?, *Orbis*, Vol. 6, No. 4, (Winter 1963), pp. 623–44.

91. Read, T., Nuclear tactics for defending a border, *World Politics*, Vol. 15, No. 3, (April 1963), pp. 390–402.

92. Rose, F. de, Atlantic relationships and nuclear problems, *Foreign Affairs*, Vol. 41, No. 3, (April 1963), pp. 479–90.

93. Rose, F. de, Atlantic relationships and nuclear problems, *The Atlantic Community Quarterly*, (Summer 1963), pp. 187–98.

94. Rostow, E. V., A new start for the alliance, *The Atlantic Community Quarterly*, (Summer 1963), pp. 207–10.

95. Shtemenko, Col. Gen. S. M., Combat training of ground troops for modern war, *Army*, Vol. 13, No. 8, (March 1963) (reprinted from *Red Star*, 3 January 1963), pp. 47–52.

96. Sixsmith, Maj. Gen. E. K. G., Review of the military situation in Europe, Chapter II in *Brassey's Annual*, (1963), pp. 6–10.

97. Slessor, Air Marshall Sir John, Control of nuclear strategy, *Foreign Affairs*, Vol. 42, No. 1, (October 1963), pp. 96–106.

98. Stanley, T. W., Decentralizating nuclear control in NATO, *Orbis*, Vol. 7, No. 1, (Spring 1963), pp. 41–8.

99. Strachey, J., Reversing NATO strategy, *Encounter*, No. 103, (April 1963), pp. 8–19.

100. Yool, W. M., West European defence, Chapter V in *Brassey's Annual*, (1963), pp. 25–33.

1964:

101. Carrison, Capt. D. J., Defense against nuclear attack at sea, *United States Naval Institute Proceedings*, Vol. 90, No. 5, (May 1965), pp. 35–43.

102. Hoag, M. W., Rationalizing NATO strategy, *World Politics*, Vol. 17, No. 1, (October 1964), pp. 121–42. Includes review of:
    — Buchan, A. and Windsor, P., *Arms and Stability in Europe*, (Praeger, New York, 1963).
    — Steel, R., *The End of Alliance*, (The Viking Press, New York, 1964).
    — Strauz-Hupé, R., Dougherty, J. E. and Kintner, W. R., *Building an Atlantic World*, (Harper & Row, New York, 1963).

103. Kissinger, H. A., Coalition diplomacy in a nuclear age, *Foreign Affairs*, Vol. 42, No. 4, (July 1964), pp. 525–45.

104. Loosbrock, J. F., NATO's nuclear dilemma, *Air Force Magazine*, Vol. 47, No. 4, (April 1964), pp. 38, 41–42.

105. Moore, Gen. J. E., NATO today: an analysis of its nuclear and conventional power, *Army*, Vol. 15, No. 1, (August 1964), pp. 27–33.

106. Mulley, F. W., NATO's nuclear problems: control or consultation, *Orbis*, Vol. 8, No. 1, (Spring 1964), pp. 21–35.

107. Pergent, J., Les unités supérieures des forces terrestres, *Revue Militaire d' Information*, No. 358, (March 1964), pp. 26–31.

108. Pierpont, C., Effets lumino-thermiques de l'arme nucléaire, *Revue Militaire d' Information*, No. 362, (July–August 1964), pp. 23–31.

1965:

109. Dabros, Capt. W. J., The credibility of the deterrent and its implications for NATO, *United States Naval Institute Proceedings*, Vol. 91, No. 7, (July 1965), pp. 28–35.

110. Hassel, K.-U. von, Organizing western defence, *Foreign Affairs*, Vol. 43, No. 2, (January 1965), pp. 209–16.

111. Owen, H., NATO strategy: what is past is prologue, *Foreign Affairs*, Vol. 43, No. 4, (June 1965), pp. 682–90.

1966:

112. Bader, W. B., Nuclear weapons sharing and 'the German problem', *Foreign Affairs*, Vol. 44, No. 4, (July 1966), pp. 693–700.

113. Evelegh, Maj. J. R. G. N., The conventional demolition belt in nuclear confrontation, *The Journal of the Royal United Services Institute*, Vol. 111, No. 644, (November 1966), pp. 328–9.

114. Heymont, I., The NATO nuclear bilateral forces, *Orbis*, Vol. 9, No. 4, (Winter 1966), pp. 1025–41.

115. O'Ballance, Maj. E., Nuclear land mines, *Ordnance*, Vol. 51, No. 278, (September–October 1966), pp. 165–6.

116. Thomas, J. R., Limited nuclear war in Soviet strategic thinking, *Orbis*, (Spring 1966), pp. 184–212.

117. Whitaker, Maj. D. R., The use of nuclear explosives in defence planning, *Journal of the Royal United Services Institute*, Vol. 111, No. 642, (May 1966), pp. 156–8.

1967:

118. Amme, C. H., NATO strategy and flexible response, *United States Naval Institute Proceedings*, (May 1967), pp. 59–69.

119. Amme, C. H., The soldier and THE BOMB, *Army*, Vol. 19, No. 9, (September 1967), pp. 29–37.

120. Hockaday, A., Nuclear management in NATO, *NATO Letter*, (May 1967), pp. 3–7.

121. Walters, R. E., The role of nuclear weapons for the West, *Journal of the Royal United Services Institute*, Vol. 112, (August 1967), pp. 249–52.

1968:

122. Barker, Lt. Col. A. J., The command and control of nuclear weapons, *Brassey's Annual*, (1968).

123. Bolef, D. I. and Antell, M., Tactical nuclear weapons, *Environment*, (May 1968), pp. 91–6.

124. Camerson, R. A., NATO and the nuclear reality, *Air Force Magazine*, (August 1968), pp. 52–6.

125. Dupuy, Col. T. N., Tactical nuclear combat, *Ordnance*, No. 291, (November–December 1968), pp. 292–6.

126. Orphan, R. C., Nuclear artillery, *Ordnance*, (May–June 1968), pp. 564–6.

127. Pasti, Gen. N., Opinions on NATO nuclear strategy, *NATO's Fifteen Nations*, Vol. 13, No. 1, (February–March 1968), pp. 20–4.

1969:

128. Beaufre, Gen. A., Nuclear weapons and Asia, Chapter 2 in *Chanakya Defence Annual*, (1969), pp. 33–46.

129. Carey, R., British thinking on tactical nuclear deterrence in Europe, *The World Today*, Vol. 25, No. 4, (April 1969), pp. 172–7.

130. Hoag, M. W., What new look in defense, *World Politics*, Vol. 12, No. 1, (October 1969), pp. 1–28.

131. Possony, S. T., NATO's defense posture, *Ordnance*, Vol. 54, No. 295, (July–August 1969), pp. 41–5.

1970:

132. Eekelen, W. F. van, Development of NATO's nuclear consultation, *NATO Letter*, Vol. 18, Nos. 7–8, (July–August 1970), pp. 2–6.

133. Garnett, J., BAOR and NATO, *International Affairs*, Vol. 46, No. 4, (October 1970), pp. 670–81.

134. Hinterhoff, Maj. E., NATO's nuclear strategy and Eastern Europe, *NATO's Fifteen Nations*, Vol. 15, No. 5, (October–November 1970), pp. 42–8.

135. Karber, P. A., Nuclear weapons and 'flexible response', *Orbis*, Vol. 14, No. 2, (Summer 1970), pp. 284–97.

136. May, D., The hidden arsenal, TAC nukes: a more personal delivery, *Washington Monthly*, (December 1970).

137. Miller, M. J. Jr., Soviet nuclear tactics, *Ordnance*, Vol. 54, No. 300, (May–June 1970), pp. 624–7.

138. Schmitz, Hon. J. G., NATO and the neutron bomb, remarks in the *Congressional Record* (Extension of Remarks), (30 December 1970), E10880.

139. Whetten, L. L., The Warsaw Pact threat in the 1970s, *NATO's Fifteen Nations*, Vol. 15, No. 5, (October–November 1970), pp. 20–8.

140. Black, B.Gen. E. F., NATO's unmentionable option: tactical nuclear weapons, *Washington Report* (American Security Council), (21 December 1970). Also reprinted in *Congressional Record*, (30 December 1970), pp. E10880–E10882.

1971:

141. Bidwell, S., The use of small nuclear weapons in war on land, *Army Quarterly*, Vol. 101, No. 3, (1971), pp. 295–306.

142. Cohen, S. T., Tactical nuclear weapons and US military strategy, *Orbis*, Vol. 15, No. 1, (Spring 1971), pp. 178–93.

143. Geneste, Lt. Col. M. E., A common western nuclear doctrine?, *Military Review*, Vol. 49, No. 9, (September 1971), pp. 3–12.

144. Goodpaster, Gen. A. J., The military situation in Europe, *NATO Letter*, Vol. 19, Nos. 3–4, (March–April 1971), pp. 10–15, (extracts of a speech presented to the Royal United Services Institute, London).

145. Karber, P. A., Nuclear shield for NATO, *Ordnance*, Vol. 55, No. 306, (May–June 1971), pp. 529–31.

146. Lawrence, R. M., On tactical nuclear war – I, *Revue Militaire Générale*, No. 1, (January 1971), pp. 46–59.

147. Lawrence, R. M., On tactical nuclear war – II, *Revue Militaire Générale*, No. 2, (February 1971), pp. 237–61.

148. Mackintosh, M., Soviet aims and capabilities in Europe, Address to the Royal United Services Institute, 2 December 1970, reprinted in *Journal of the Royal United Services Institute for Defence Studies*, (March 1971).

149. Trettner, Gen. H., Tactical nuclear weapons for Europe, *Military Review*, Vol. 51, No. 7, (July 1971), pp. 43–9.

150. Vigor, P. H., Soviet military exercises, Lecture given at the Royal United Services Institute, 31 March 1971, reprinted in *Journal of the Royal United Services Institute for Defence Studies*, No. 663, (September 1971).

151. Whetten, L. L., A European view of NATO strategy, *Military Review*, (September 1971), pp. 25–37.
152. Réchin, P., De l'efficacité tactiques des feux nucléaires, *Revue Militaire Générale*, No. 6, (June 1971), pp. 92–100.

1972:

153. Beecher, W., Over the threshold: clean tactical nuclear weapons for Europe, *Army*, Vol. 22, No. 7, (July 1972), pp. 17–20.
154. Brenner, M. J., Decoupling, disengagement and European defense, *Bulletin of the Atomic Scientists*, Vol. 28, No. 2, (February 1972), pp. 38–42.
155. Brown, N., The tactical air balance in Europe, *The World Today*, Vol. 28, No. 9, (September 1972), pp. 385–92.
156. Canby, S. L., NATO muscle: more shadow than substance, *Foreign Policy*, No. 8, (Fall 1972), pp. 38–49.
157. Dorn, J., Europas nukleare schwierigkeiten, *Revue Militaire Générale*, No. 2, (February 1972), pp. 162–82.
158. Erickson, J., 'Shield-72': Warsaw Pact military exercises, *Journal of the Royal United Services Institute for Defence Studies*, Vol. 4, No. 117, (December 1972).
159. Hahn, W., Nuclear balance in Europe, *Foreign Affairs*, Vol. 50, No. 3, (April 1972), pp. 501–16.
160. Ryan, Gen. J. D., United States strategic and tactical air forces, *NATO's Fifteen Nations*, (August–September 1972), pp. 17–25.
161. Van Veen. Maj. E., Theatre air forces and tactical nuclear weapons, *NATO's Fifteen Nations*, (August–September 1972), pp. 35–41.
162. Van Veen, Major E., The Soviet tactical air force and tactical nuclear weapons, *NATO's Fifteen Nations*, Vol. 17, No. 4, (August–September 1972), pp. 42–8.
163. Wiegele, T. C., Nuclear consultation processes in NATO, *Orbis*, Vol. 16, No. 2, (Summer 1972), pp. 462–87.

1973:

164. Bennett, W. S., Sandoval, R. R. and Shreffler, R. G., A credible nuclear-emphasis defense for NATO, *Orbis*, Vol. 17, No. 2, (1973), pp. 463–79.
165. Joshua, W., A strategic concept for the defense of NATO, *Orbis*, Vol. 17, No. 2, (Summer 1973), pp. 448–62.
166. Komer, R. W., Treating NATO's self-inflicted wound, *Foreign Policy*, (Winter 1973–4), pp. 34–38.
167. NATO Nuclear Planning Group, *Final Communiqué*, (NATO Press Service, 16 May 1973), Press Release-M-NPG-1(73)17.
168. Polk, J. H., The realities of tactical nuclear warfare, *Orbis*, Vol. 17, No. 2, (1973), pp. 4
169. Selm, H. B., Nuclear policy-making in NATO, *NATO Review*, No. 6, (1973).
170. Richardson, R. C., Can NATO fashion a new strategy, *Orbis*, Vol. 17, No. 2, (Summer 1973), pp. 415–38.
171. Silvestri, S., The military aspects of European security, *Lo Spettatore Internatzionale*, Vol. 8, No. 3, (July–September 1973), pp. 199–222.
172. Yakushin, D., NATO nuclear plans: past and present, *International Affairs*, (November 1973), pp. 105–6. (Reviews V. G. Milayev, *The Nuclear Policy of the USA in NATO*, International Relations Publishers, Moscow, 1973.)

1974:

173. Faire, Col. S., A tactical nuclear strategy for NATO, *NATO's Fifteen Nations*, (April–May 1974), pp. 59–60.
174. Gladwyn, Lord, Nuclear weapons and Europe (Letter to the Editor), *Survival*, Vol. 16, No. 2, (March–April 1974), pp. 94–5.
175. Goodpaster, Gen. A. J., NATO and US forces: challenges and prospects, *Strategic Review*, Vol. 2, No. 1, (Winter 1974), pp. 6–17.
176. Gray, C. S., Mini-nukes and strategy, *International Journal*, Vol. 29, No. 2, (Spring 1974), pp. 216–41.
177. Gray, C. S., Deterrence and defence in Europe: revising NATO's theatre nuclear posture, *Journal of the Royal United Services Institute*, (December 1974).
178. Hagglund, Maj. G., United States NATO strategy, *Military Review*, (January 1974), pp. 39–49.
179. Hunt, Brig. K., Deterrence, *NATO Review*, No. 1, (1974), pp. 9–13.
180. Leitenberg, M., Nuclear nervousness, *New Scientist*, (1 August 1974), pp. 227–8.
181. Martin, L., Theatre nuclear weapons and Europe, *Survival*, Vol. 16, No. 6, (November–December 1974), pp. 268–76.
182. Nukes in NATO, *Aviation Week and Space Technology*, (8 April 1974), p. 7.
183. Pincus, W., Congress and tactical nukes, *New Republic*, (12 October 1974), pp. 19–20.
184. Pincus, W., Why more nukes, *New Republic*, (9 February 1974), pp. 15–7.
185. Record, J., US tactical nuclear weapons in Europe: 7000 warheads in search of a rationale, *Arms Control Today*, Vol. 4, No. 4, (April 1974), pp. 1–4.
186. Record, J., To nuke or not to nuke: a critique of rationales for a tactical nuclear defense of Europe, *Military Review*, No. 10, (October 1974), pp. 3–13.
187. Small yield nuclear weapons, Statement by Ambassador Martin to the Conference of the UN Committee on Disarmament, 23 May 1974.
188. York, H., Balance of terror in Europe, *Bulletin of the Atomic Scientists*, Vol. 32, No. 5, (May 1976), pp. 8–17.
189. Norman, L., The reluctant dragons, NATO's fears and the need for new nuclear weapons, *Army*, (February 1974).

1975:

190. Barber, Col. R. E., The myth of Soviet nuclear war strategy, *Army*, (June 1975), pp. 10–7.
191. Booth, K., Security makes strange bedfellows: NATO's problems for a minimalist perspective, *Journal of the Royal United Services Institute*, (December 1975).
192. Bowman, Brig. Gen. R. C., NATO in a time of crisis, *Air Force Magazine*, (April 1975), pp. 49–54.
193. Brenner, M. J., Tactical nuclear strategy and European defence: a critical reappraisal, *International Affairs*, (January 1975).
194. Cohen, S. T. and Lyons, W. C., A comparison of US–Allied and Soviet tactical nuclear force capabilities and policies, *Orbis*, Vol. 19, No. 1, (Spring 1975), pp. 72–92.
195. Erickson, J., Soviet military capabilities in Europe, *Journal of the Royal United Services Institute*, No. 3, (March 1975), pp. 22–30.
196. Goodpaster, Gen. A. J., NATO strategy and requirements, 1975–1985, *Survival*, Vol. 17, No. 5, (September–October 1975), pp. 211–6.
197. Holst, J. J., Implications of flexible options policy for alliance strategy and arms control, *NUPI-Notat* No. 89, (March 1975), pp. 1–16.
198. Menual, S., The use of nuclear weapons in the European theatre, *NATO's Fifteen Nations*, (April–May 1975), pp. 30–8.

199. Milton, Gen. T. R., The mystique of NATO's nukes, *Air Force Magazine*, Vol. 58, No. 1, (January 1975), pp. 26–7.

200. Record, J., with assistance of Anderson, T. I., Tactical nuclear weapons in Europe: alternative postures, *Survival*, Vol. 17, No. 2, (March–April 1975), pp. 73–80.

201. Schneider, B., Big bangs from little bombs, *Bulletin of the Atomic Scientists*, Vol. 31, No. 5, (May 1975), pp. 24–9.

202. Schwartz, D. N., The role of deterrence in NATO: defense strategy, *World Politics*, Vol. 28, No. 1, (October 1975).

203. Sinnreich, R. H., NATO's doctrinal dilemma, *Orbis*, Vol. 19, No. 2, (Summer 1975).

204. Vigor, P. H. and Donnelly, C. N., The Soviet threat to Europe, *Journal of the Royal United Services Institute*, (March 1975), pp. 5–9.

205. The purpose of Pluton, address by the French Prime Minister, M. Chirac, (10 February 1975) in *Defense Nationale*, (May 1975), reprinted in *Survival*, Vol. 17(5), (September–October 1976), pp. 226–8.

1976:

206. Bettit, Capt. E. D., Soviet tactical doctrine and capabilities and NATO's strategic defense, *Strategic Review*, Vol. 4, No. 3, (Fall 1976), pp. 95–105.

207. Collins, Lt. Gen. A. S. Jr., Tactical nuclear warfare and NATO: viable strategy or dead end?, *NATO's Fifteen Nations*, (June–July 1976), pp. 73–87.

208. Geneste, M., Pour une nouvelle stratégie de l'avant: Le projet de Los Alamos, *Défense Nationale*, Vol. 32, (October 1976), pp. 81–92.

209. Gray, C. S., Theater nuclear weapons: doctrines and postures, *World Politics*, Vol. 28, No. 2, (January 1976).

210. Miettinen, J. K., Time for Europeans to debate the presence of tactical-nukes, *Bulletin of the Atomic Scientists*, Vol. 32, No. 5, (May 1976), pp. 18–22.

211. Nichols, Col. D. L., Who needs nuclear tacair?, *Air University Review*, Vol. 27, No. 3, (March–April 1976).

212. TNFs: Critical US defense requirements, *Commanders Digest*, Vol. 19, No. 14, (1 July 1976), pp. 2–8.

213. Mery, Gen. G., Comments to the Institute des Hautes Etudes de Defense Nationale, *Defense Nationale*, June 1976, reprinted in *Survival*, Vol. 18(5), (September–October 1976), pp. 226–8.

214. France's Defense Policy, address by V. Giscard d'Estaing, President of the French Republic, at the Institute des Hautes Etudes de Defense Nationale, Paris (1 June 1976), mimeograph, 14 pp. Reprinted in *Survival*, Vol. 18(5), (September–October 1976), pp. 228–30.

1977:

215. Miettinen, J. K., Tactical nuclear weapons in Europe and trends in their development, *Bulletin of Peace Proposals*, Vol. 8, No. 1, (1977), pp. 32–46.

216. Comments by General Guy Mery in *Defense* Bulletin de l'Institute des Haute Etudes de Defense Nationale, (May 1977), pp. 19–23.

(E) Individual weapon systems

Sergeant

1. Improved Sergeant missile electronics ready to test, *Aerospace Technology*, Vol. 21, No. 2, (17 July 1967), pp. 51–2.

2. Erwin, J. B., Army fires Sergeant as field use nears, *Aviation Week and Space Technology*, Vol. 77, No. 20, (12 November 1962), pp. 82, 86, 88 and 89.
3. Leete, J. J., Pushbutton Missile: Sergeant increases flexibility and firepower of field commanders, *Ordnance*, Vol. 47, No. 256, (January–February 1963), pp. 471–74.

Pershing

1. Klass, P. J., Guidance device set for Pershing tests, *Aviation Week and Space Technology*, Vol. 102, No. 19, (12 May 1975), pp. 45 ff.
2. Powers, Lt. Col. P. W., Blackjack the giant killer, *Army*, Vol. 13, No. 10, (May 1963), pp. 17–24.
3. Project SWAP, *Armed Forces Management*, Vol. 15, No. 8, (May 1969), pp. 74 ff.
4. Pershing, launcher system field-tested, *Aviation Week and Space Technology*, Vol. 78, No. 16, (22 April 1963), pp. 90 ff.
5. Anderton, D. A., Tight control keeps Pershing program on schedule, *Aviation Week and Space Technology*, Vol. 75, No. 15, (9 April 1962), pp. 80, 81, 83 and 84 ff.
6. Staudt, H. R., Pershing field maintenance, *Ordnance*, Vol. 48, No. 261, (November–December 1963), pp. 329–31.
7. Andrews, W., New Pershing may get multiple warheads, *Aerospace Technology*, Vol. 21, No. 4, (14 August 1967), pp. 16–7.
8. Rudd, Lt. Col. E. A., The Pershing is 1-A, *Ordnance*, Vol. 53, No. 290, (September–October 1968), pp. 179 ff.

Lance

1. US, Senate, *Department of Defense Appropriations for Fiscal Year 1973*, Hearings before a Subcommittee of the Committee on Appropriations, US Senate, 92nd Congress, 2nd Session, Part 2, (1972).
2. Page, R. S., MGM-52C LANCE-A second generation battlefield missiles, *Aerospace International*, Vol. 9, No. 2, (March–April 1973), pp. 20–3.
3. Andrews, W., Propulsion problems may delay Lance missile production a year, *Aerospace Technology*, Vol. 21, No. 13, (18 December 1967).
4. Army's next artillery missile slowed in Vietnam squeeze, *Space/Aeronautics*, Vol. 48, No. 3, (August 1966), pp. 22 and 27 ff.
5. Alexander, G., Army emphasizes close support missiles, *Aviation Week and Space Technology*, Vol. 80, No. 11, (16 March 1964), pp. 157 ff.
6. The Lance tactical missiles – a new artillery weapon system for NATO, *International Defense Review*, Vol. 6, No. 2, (April 1973), pp. 199–203.
7. Oldfield, Col. B., USAF (Ret.), The Lance – 'Shoot and Scoot', *NATO's Fifteen Nations*, (April–May 1975), pp. 78–81.

Atomic Demolition Munitions

1. Orphan, R. C. and Gard, R. P., Nuclear demolitions, *Ordnance*, Vol. 53, No. 293, (March–April 1969), pp. 490–3.
2. Stowe, Maj. W. W., USA, Atomic demolition munitions, *National Defense*, (May–June 1975), pp. 467–70.

Index